T0073457

Enrique Herrera-Viedma, José Luis García-Lapresta, Janusz Kacprzyk,
Mario Fedrizzi, Hannu Nurmi, and Sławomir Zadrożny (Eds.)

Consensual Processes

Studies in Fuzziness and Soft Computing, Volume 267

Editor-in-Chief

Prof. Janusz Kacprzyk
Systems Research Institute
Polish Academy of Sciences
ul. Newelska 6
01-447 Warsaw
Poland
E-mail: kacprzyk@ibspan.waw.pl

Further volumes of this series can be found on our homepage: springer.com

Enrique Herrera-Viedma,
José Luis García-Lapresta, Janusz Kacprzyk,
Mario Fedrizzi, Hannu Nurmi,
and Sławomir Zadrożny (Eds.)

Consensual Processes

 Springer

Editors

Prof. Enrique Herrera-Viedma
Universidad Granada
Depto. Ciencias de la
Computación e
Inteligencia Artificial
C/ Periodista Daniel Saucedo
Aranda s/n
18071 Granada, Spain
E-mail: viedma@decsai.ugr.es

José Luis García-Lapresta
Universidad de Valladolid
Departamento de Economía
Aplicada
Avenida Valle Esgueva 6
47011 Valladolid, Spain
E-mail: lapresta@eco.uva.es

Janusz Kacprzyk
Polish Academy of Sciences
Systems Research Institute
ul. Newelska 6
01–447 Warsaw, Poland
E-mail: kacprzyk@ibspan.waw.pl

Prof. Dr. Mario Fedrizzi
Università di Trento
Fac. Economia
Dipto. Informatica e Studi
Aziendali
Via Inama 5 38100 Trento Trento, Italy
E-mail: fedrizzi@cs.unitn.it

Hannu Nurmi
University of Turku
Department of Political Science
and
Contemporary History
Turku, Finland
E-mail: hnurmi@utu.fi

Sławomir Zadrożny
Polish Academy of Sciences
Systems Research Institute
ul. Newelska 6
01-447 Warsaw, Poland
E-mail: zadrozny@ibspan.waw.pl

ISBN 978-3-642-20532-3   e-ISBN 978-3-642-20533-0

DOI 10.1007/978-3-642-20533-0

Studies in Fuzziness and Soft Computing   ISSN 1434-9922

Library of Congress Control Number: 2011927922

Typeset & Cover Design: Scientific Publishing Services Pvt. Ltd., Chennai, India.

Printed on acid-free paper

9 8 7 6 5 4 3 2 1

springer.com

Preface

Consensus has been a much talked about word for centuries, maybe millennia since people have always been aware of its importance for arriving at proper decisions which have had a long lasting impact on life of groups of people, countries or even civilizations. Needless to say that the growing complexity of the present world has made the word to be used so frequently nowadays.

The first question is, as always, the one that relates to the very meaning of the word. In general, one can say that by consensus is meant, on the one hand, a general agreement within the group of people, or agents, both human and software, in a more general setting. Clearly, in the strict meaning the general agreement has been viewed in a "yes-no" way, that is, in the sense of full and unanimous agreement. Since this may be an unreachable ideal, people has been for a long time trying to make this definition more realistic and have replaced this ideal concept by a more realistic one that encompasses all forms of partial, graded, etc. agreements within a group.

The second sense of this word is related to a process of reaching consensus which aims at reaching an agreement, possibly a high one, of all, or most, agents. It can involve the resolution and/or mitigation of some minor objections concerning options or aspects in question or individual agents involved.

It is easy to see that the second sense of consensus does involve the first sense because the process of reaching an agreement must be related to some assessment of what, and to which extent the current agreement within the group exists.

This volume is concerned with consensus reaching processes which may occur both within human groups and groups of intelligent agents. We adopt, first of all, a modern and realistic definition of consensus that considers consensus not necessarily as full and unanimous agreement but as agreement to some extent. This calls for some soft computational tools and techniques and we mainly use fuzzy and possibilistic ones to be able to account for imprecision in the very meaning of many concepts, issues and properties that play a role in both that more realistic definition of consensus and the very process of consensus reaching.

The volume is intended to provide a comprehensive coverage of various issues related to consensus and consensual processes.

Part I focuses on consensus from dynamical points of view. The chapters are concerned with fundamental issues of consensus dynamics:, notably aggregation, ranking, opinion changing propensity, preference modeling, etc.

D. Eckert and Ch. Klamler ("Distance-Based Aggregation Theory") consider the problem of aggregating several objects into an object that best represents them which is a central problem in diverse fields exemplified by economics, sociology, political science, statistics, biology, etc. This problem[s] has also been extensively dealt with in the theory of social choice which analyses the aggregation of individual preferences into a collective preference. In this context, the idea of a consensus is normatively particularly appealing. A natural way to operationalize the consensus among a group of individuals is by means of a distance function that measures the disagreement between them, and this approach is followed in the paper. Thus, in particular, the construction of aggregation rules based on the minimization of distance functions inherits the normative appeal of consensus. Clearly, the distance-based approaches to aggregation theory are not limited to the construction of aggregation rules, but can fruitfully be applied to the comparison of aggregation procedures as well as to the geometric representation and to generalizations of the aggregation problems as discussed in the paper. A discussion of related problems such as the complexity of distance-based aggregation rules is also included.

G. Beliakov, T. Calvo and S. James ("On Penalty-Based Aggregation Functions and Consensus") consider the problem of aggregating individual preferences in order to arrive at a group consensus in contexts like elections where a candidate must be chosen that best represents the individuals' differing opinions, sport competitions and the fusion of sensor readings. In these applications the aggregated result should be as close as possible to the individual inputs giving rise to the need of methods that minimize this difference and this is what penalty-based aggregation functions are about drawing upon various notions of a "difference". One of more difficult issues when aggregating preferences is how to best assign numerical scores if only ordinal or pairwise preferences are given. In terms of penalty-based aggregation, the problem is further complicated by determining how the penalties should be calculated. The rationale behind using certain functions and certain measures of distance or penalty is context dependent. To use arithmetic means in group decision making, for example, implies a faith in the individuals to make accurate and sincere judgments that will not skew the results. Medians are not as susceptible in this way. When using aggregation methods, how consensus is interpreted will affect how it is achieved. A measure of distance or deviation from this value, or the imposition of a penalty for not having consensus has been studied in various forms, The authors draw upon the results on the penalty-based aggregation functions and penalty functions in general. They present some well used definitions of penalty and show how some aggregation functions correspond to minimizing the overall penalty associated with a given input vector. They consider some alternative frameworks of penalty and also introduce the idea of aggregating penalties using the OWA operator. It is shown that the penalty-based aggregation functions provide a natural framework for mathematical interpretations of consensus.

I. Contreras, M.A. Hinojosa and A.M. Mármol ("Ranking Alternatives in Group Decision-Making with Partial Information. A Stable Approach") propose in their paper some procedures for constructing global rankings of alternatives in situations in which each member of a group is able to provide imprecise or partial information on his/her preferences about the relative importance of criteria that have to be taken into account. The authors first propose an approach based on the assumption that the final evaluation depends on the complete group since no possibility exists that the group might split into coalitions that look for more favorable solutions for the [the] coalition members. To this end, the partial information on criteria weights provided by each individual is transformed into ordinal information on the alternatives, and then the aggregation of individual preferences is addressed within a distance-based framework. In a second approach, the possibility of coalition formation is considered, and the goal is to obtain rankings in which disagreements of all the coalitions are taken into account. These rankings will exhibit an additional property of collective stability in the sense that no coalition will have an incentive to abandon the group. This last approach may be of interest in political decisions where different sectors have to be incorporated into a joint evaluation process aiming at a consensus across all possible subgroups.

M. Brunelli, R. Fullér and J. Mezei ("Opinion Changing Aversion Functions for Group Settlement Modeling") consider opinion changing aversion (OCA) functions which are used to quantify the decision makers' resistance to opinion changing. The authors obtain a collective representation of preferences by solving a non-linear optimization problem to minimize the total level of disagreement and seeking an optimal, consensual, solution, the least disagreed one. Whenever such a consensual solution has to be found, a single valued, nonnegative cost function, an opinion changing aversion (OCA) function, is assigned to each decision maker and then the overall cost is minimized. The authors focus on the quadratic OCA functions and show that the group decision (or settlement) boils down to the center of gravity of the opinions of the decision makers. It is shown that if each expert has a quadratic opinion changing aversion function, then the minimum-cost solution is the weighted average of the individual optimal solutions where the weights are the relative importances of the decision makers. The authors consider the minimum-cost solutions for group settlements under crisp and fuzzy budget constraints.

S. Montes, D. Martinetti, S. Díaz and S. Montes ("Statistical Preference as a Tool in Consensus Processes") deal with a so called statistical preference, a modern method of comparing probability distributions in the setting of consensus processes in which the intensities of preference can be expressed by means of probability distributions instead of single values. Since classical methods do not provide the possibility of comparing any pair of probability distributions, statistical preference is considered in the paper. One of its most remarkable advantages is that it allows to compare any pair of probability distributions. The authors study in depth some properties of this method and the relationship between the most commonly employed stochastic dominance and statistical preference. They also consider some of the most important families of distributions and analyze statistical preference among probability distributions in the same families.

Part II is concerned with issues underlying the meaning, composition and outcomes of individual and group decision making as well as social choice that are of relevance for consensus and consensual processes. It includes concepts and models dealing with social choice, group characterization and identification, veto power distribution, etc.

M. Regenwetter and A. Popova ("Consensus with Oneself: Within-person Choice Aggregation in the Laboratory") follow their former efforts to [to] cross-fertilize individual and social choice research[,] and apply behavioral social choice concepts to individual decision making. Though repeated individual choice among identical pairs of choice alternatives often fluctuates dramatically over even very short time periods, social choice theory usually ignores this because it identifies each individual with a single fixed weak order. Behavioral individual decision research may expose itself to Condorcet paradoxes because it often interprets a decision maker's modal choice (i.e., majority choice) over repeated trials as revealing their "true" preference. The authors investigate the variability in choice behavior within each individual in the research lab. Within that paradigm, they look for evidence of Condorcet cycles, as well as for the famed disagreement between the Condorcet and Borda aggregation methods. They also illustrate some methodological complexities involved with likelihood ratio tests for Condorcet cycles in paired comparison data.

D. Dimitrov ("The Social Choice Approach to Group Identification") gives an overview of selected topics from the theory of group identification intended to answer the question: given a group of individuals, how to define a subgroup in it?. The problem of group identification is then viewed as a process of group formation. As a starting point the author uses different axiomatic characterizations of the ``libera'' rule for group identification whereby the group consist of those and only those individuals who view themselves as members of the group. The focus of the paper is then on consent rules and recursive procedures for collective determination in which the opinions of other individuals in the society also count. Finally, the author addresses recent developments in the literature with respect to gradual opinions and group identity functions.

A. Laruelle and F. Valenciano ("Consensus versus Dichotomous Voting") consider consensus [meant] as a general, maybe unanimous, agreement among possibly different views. Reaching a consensus is often a complex and difficult process involving adjustments, concessions, threats and bluffing, with no general rules, and dependent on the particular context. In which social rules, customs, past experience and communication constraints play a role. By contrast, dichotomous voting rules are in principle simple mechanisms for making decisions by using a vote to settle differences of view: the winning side enforces the decision to accept or reject the proposal on the table. Thus, these rules may be viewed in their spirit completely opposed to the idea of consensus. Nevertheless, it is often the case that a committee whose only formal mechanism to make decisions is a specified dichotomous voting rule reaches a consensus about an issue. Moreover, in many such cases the final vote is a purely formal act ratifying the agreement resulting from a consensual process and dichotomous voting rules which are a means of making decisions by using votes to settle differences of view. A natural question is therefore: How then can it often be the case that a committee whose only formal mechanism for decision-making is a dichotomous voting rule reaches a consensus? In this paper, based on a game-theoretic model developed in the authors' previous papers, an answer to this question is provided.

J. Mercik ("On a priori Evaluation of Power of Veto") considers primarily the evaluation of power when some players have the right to veto, i.e. to stop the action of others permanently or temporarily. In certain cases, it is possible to calculate a value of power of veto attributed to the decision maker and to give the exact value of the power index as well. In other cases, it is only possible to compare the situation with and without veto attribute. In this paper the author analyzes the power of a player with a right to veto, expecting that the difference between the power of the player with veto and his or her power without veto makes it possible to evaluate directly or indirectly the power of veto itself.

Part III focuses on the environment and substantive content of consensus in various fields as well as provides an overview of approaches to measuring the degree of consensus, and some related topics.

H. Nurmi ("Settings of Consensual Processes: Candidates, Verdicts, Policies") considers the setting of social choice theory which basically deals with mutual compatibilities of various choice criteria or desiderata, and thus provides a natural angle to look at methods for finding consensus. The author distinguishes between three types of settings of consensus-reaching. Firstly, one may be looking for the correct decision. This is typically the setting where the participants have different degrees of expertise on an issue to be decided. Also jury decision making falls into this category. Secondly, the setting may involve the selection of one out of a set of candidates, for instance for a public office. Thirdly, one may be looking for a policy consensus. This setting is otherwise similar to the candidate choice setting, but usually involves more freedom in constructing new alternatives. The author first provides a review of these settings and relevant results in each one of them, and then discusses the implications of some choice paradoxes to consensus-reaching methods.

M. Martínez-Panero ("Consensus Perspectives: Glimpses into Theoretical Advances and Applications") gives a survey of polysemic meanings of consensus from several points of view, ranging from philosophical aspects and characterizations of several quantification measures within the social choice framework, paying also attention to aspects of judgment aggregation as well as fuzzy or linguistic approaches, to practical applications in decision making and biomathematics, to name a few. More specifically, the author first presents some philosophical aspects of consensus essentially focused on the doctrine that men are joined together within a society by a contract with explicit or hidden agreements, as Rousseau believed. Then, he outlines some further developments and connections, such as the link between Rousseau and Condorcet. The author also distinguishes between the concept of consent and the more technical and recent idea of consensus as appearing in modern political science and sociology. Next, he deals with several formal approaches to consensus mainly from the social choice framework, and some distance based, fuzzy or linguistic points of view. Moreover, he points out some aspects of an emergent research field focused on judgment aggregation, and concludes with a presentation of some applications as signs of the power of consensus-based methods in practice, a reference to the way of aggregating different estimates of each candidate through a median-based voting system.

J. Alcalde-Unzu and M. Vorsatz ("Measuring Consensus: Concepts, Comparisons, and Properties") study approaches of how to measure the similarity of preferences in a group of individuals which is what they mean by consensus. First, the consensus for two individuals is determined and then the average over all possible pairs of individuals in the society is calculated. In the dual approach, first, the consensus between two alternatives is determined and then the average over all possible pairs of alternatives is calculated. The authors show that the choice between the two measures used in the above processes reduces to the choice between different monotonicity and independence conditions. Finally, some recent approaches are surveyed that take into account the fact that alternatives which are on the average ranked higher by the members of the society are more important for the social choice and should therefore be assigned a higher weight while calculating the consensus.

J.L. García-Lapresta and D. Pérez-Román ("Measuring Consensus in Weak Orders") consider the problem of how to measure consensus in groups of agents when they show their preferences over a fixed set of alternatives or candidates by means of weak orders (complete preorders). Consensus is here related to the degree of agreement in a committee, and agents do not need to change their preferences. The authors introduce a new class of consensus measures on weak orders based on distances, and analyze some of their properties paying special attention to seven well-known distances. They extend Bosch's consensus measure to the context of weak orders when indifference among different alternatives is allowed, and consider some additional properties like a maximum dissension (in each subset of two agents, the minimum consensus is only reached whenever preferences of agents are linear orders and each one is the inverse of the other), reciprocity (if all individual weak orders are reversed, then the consensus does not change) and homogeneity (if we replicate a subset of agents, then the consensus in that group does not change). Then, the authors introduce a class of consensus measures based on the distances among individual weak orders paying special attention to seven specific metrics: discrete, Manhattan, Euclidean, Chebyshev, cosine, Hellinger, and Kemeny.

L. Roselló, F. Prats, N. Agell and M. Sánchez ("A Qualitative Reasoning Approach to Measure Consensus") introduce a mathematical framework, based on the absolute order-of-magnitude qualitative model, which makes it possible to develop a methodology to assess consensus among different evaluators who use ordinal scales in group decision-making and evaluation processes. The concept of entropy is introduced in this context and the algebraic structure induced in the set of qualitative descriptions given by evaluators is studied. The authors prove that it is a weak partial semilattice structure which under some conditions takes the form of a distributive lattice. The definition of the entropy of a qualitatively-described system enables us, on the one hand, to measure the amount of information provided by each evaluator and, on the other hand, to consider a degree of consensus among the evaluation committee. The methodology presented makes it possible to manage situations when the assessment given by experts involves different levels of precision. In addition, when there is no consensus within the group decision, an automatic process measures the effort needed to reach consensus.

M. Xia and Z. Xu ("On Consensus in Group Decision Making Based on Fuzzy Preference Relations") propose a method to derive the multiplicative consistent fuzzy preference relation from an inconsistent fuzzy preference relation. The fundamental characteristic of the method proposed is that it can get a consistent fuzzy preference relation taking into account all the original preference values without translation. Then, the authors develop an algorithm to transform a fuzzy preference relation into the one with the weak transitivity by using the original fuzzy preference relation and the constructed consistent one. After that, the authors propose an algorithm to help the decision makers reach an acceptable consensus in group decision making. It is worth pointing out that the group fuzzy preference relation derived by using the method proposed is also multiplicative consistent if all individual fuzzy preference relations are multiplicative consistent. The results obtained are illustrated by some examples.

S. Zadrożny, J. Kacprzyk and Z.W. Raś ("Supporting Consensus Reaching Processes under Fuzzy Preferences and a Fuzzy Majority via Linguistic Summaries and Action Rules") deal with the classic approach to the evaluation of the degree of consensus due to Kacprzyk and Fedrizzi (1986, 1988, 1989) in which a soft degree of consensus has been introduced as a degree to which, for instance, ``most of the important individuals agree as to almost all of the relevant options''. The fuzzy majority is equated with a fuzzy linguistic quantifiers (most, almost all, ...) and handled via Zadeh's classic calculus of linguistically quantified propositions and Yager's OWA (ordered weighted average) operators. The consensus reaching process is run by a moderator who may need a support which is provided by a novel combination of: first, the use of the a soft degree of consensus due, and then the linguistic data summaries, in particular in its protoform based version proposed by Kacprzyk and Zadrożny to indicate in a natural language some interesting relations between individuals and options to help the moderator to identify crucial (pairs of) individuals and/options which pose some threats to the reaching of consensus. Third, using results obtained in the authors' recent paper, additionally a novel data mining tool, a so-called action rule proposed by Raś and Wieczorkowska is employed. The action rules are used in the context considered to find the best concessions to be offered to the individuals for changing their preferences to increase the degree of consensus.

Part IV includes contributions which deal with the implementation of theoretical models within decision support systems for running consensus reaching sessions, notably in the Web environment, and some more important application areas, including broadly perceived multicriteria decision making.

I.J. Pérez, F.J. Cabrerizo, M.J. Cobo, S. Alonso and E. Herrera-Viedma ("Consensual Processes Based on Mobile Technologies and Dynamic Information") present a prototype of a group decision support system based on mobile technologies and dynamic information. It is assumed that the users can run the system on their own mobile devices in order to provide their preferences anytime and anywhere. The system provides consensual and selection support to deal with dynamic decision making situations. Furthermore, the system incorporates a mechanism that makes it possible to manage dynamic decision situations in which some information about the problem is not constant throughout the time. It provides a more realistic decision making setting through high dimensional or dynamic set of alternatives, focussing the discussion on a subset of them that changes in each stage of the process. The experts' preferences are represented by using a linguistic approach. Therefore, the authors provide a new linguistic framework that is mobile and dynamic, to deal with group decision making problems.

L. Iandoli ("Building Consensus in On-line Distributed Decision Making: Interaction, Aggregation and Construction of Shared Knowledge") discusses the possibility of exploiting large-scale knowledge sharing and mass interaction taking place on the Internet to build decision support systems based on distributed collective intelligence. Pros and cons of currently available collaborative technologies are reviewed with respect to their ability to favor knowledge accumulation, filtering, aggregation and consensus formation. In particular, the author focuses on a special kind of collaborative technologies, a so called online collaborative mapping, whose characteristics can overcome some limitations of more popular collaborative tools, in particular thanks to their capacity to support collective sense-making and the construction of shared knowledge objects. The author discusses some contributions in the field and argues that the combination of online mapping and computational techniques for belief aggregation can provide an interesting basis to support the construction of systems for distributed decision-making.

F. Mata, J,.C. Martínez and R. Rodríguez ("A Web-based Consensus Support System Dealing with Heterogeneous Information") show a novel Web application of a consensus support system to carry out consensus reaching processes with heterogeneous information, i.e. the decision makers may use different information domains (in particular: numeric, interval-valued and linguistic assessments) to express their opinions. The software application developed has the following main characteristic features: it automates virtual consensus reaching processes in which experts may be put in different places, experts may use information domains near their work areas to provide their preferences and it is possible to run the system on any computer and operating system. This application may be seen as a practical development of a theoretical research on consensus modeling. It could be used by any organization to carry out virtual consensus reaching processes.

J. Ma, G.-G. Zhang and J. Lu ("A Fuzzy Hierarchical Multiple Criteria Group Decision Support System – Decider – and its Application") discuss Decider, a Fuzzy Hierarchical Multiple Criteria Group Decision Support System (FHMC-GDSS) designed for dealing with subjective, in particular linguistic, information and objective information simultaneously to support group decision making particularly focused on evaluation. The authors introduce first the fuzzy aggregation decision model, functions and structure of the Decider. The ideas of how to resolve decision making and evaluation problems encountered in the development and implementation of Decider are presented, and two real applications of the Decider system are briefly illustrated. Finally, some further future research in the area are briefly outlined.

D. Ben-Arieh and T. Easton ("Product Design Compromise Using Consensus Models") discuss the costs associated with decision making using group consensus, and then describe three methods of reaching a minimum cost consensus assuming quadratic costs for a single criterion decision problem. The first method finds the group opinion (consensus) that yields the minimum cost of reaching throughout the group. The second method finds the opinion with the minimum cost of the consensus providing that all experts must be within a given threshold of the group opinion. The last method finds the maximum number of experts that can fit within the consensus, given a specified budget constraint. In all of them the consensus process is defined as a dynamic and interactive group decision process, which is coordinated by a moderator, who helps the experts to gradually move their opinions until a consensus is reached. The work focuses on product design compromise and discusses how group consensus can be used in this process, and demonstrates the importance of the consensus process to the product design compromise process, and presents there models as mentioned above that can be used to obtain such a compromise.

We wish to thank all the contributors for their excellent work. All the contributions were anonymously peer reviewed by at least two reviewers, and we also wish to express our thanks to them. We hope that the volume will be interesting and useful to the entire research community working in diverse fields related to group decision making, social choice, consensual processes, multiagent systems, etc. as well as other communities in which people may find the presented tools and techniques useful to formulate and solve their specific problems.

We also wish to thank Dr. Tom Ditzinger and Mr. Holger Schaepe from Springer for their multifaceted support and encouragement.

November 2010

E. Herrera-Viedma
J.L. García-Lapresta
J. Kacprzyk
H. Nurmi
M. Fedrizzi
S. Zadrożny

Contents

Part III: Various Aspects of Consensus, Its Measuring and Reaching

Part IV: Modern Trends in Consensus Reaching Support, and Applications

Part I: Basic Issues

Distance-Based Aggregation Theory

Daniel Eckert and Christian Klamler

1 Introduction

The problem of aggregating several objects into an object that represents them is a central problem in disciplines as diverse as economics, sociology, political science, statistics and biology (for a survey on aggregation theory in various fields see Day and McMorris [17]). It has been extensively dealt with in the theory of social choice (see Arrow et al. [5]), which analyses the aggregation of individual preferences into a collective preference. In this context, the idea of a consensus is normatively particularly appealing. A natural way to operationalize the consensus among a group of individuals is by means of a distance function that measures the disagreement between them. Thus, in particular, the construction of aggregation rules based on the minimization of distance functions inherits the normative appeal of consensus.

Distance-based approaches to aggregation theory are, however, not limited to the construction of aggregation rules, as the ones surveyed in section 3 (after a brief overview of the formal framework), but can fruitfully be applied to the comparison of aggregation procedures (section 4) as well as to the geometric representation (section 5) and to the generalizations of aggregation problems, which are dealt with in sections 6 and 7. A discussion of related problems such as the complexity of distance-based aggregation rules concludes the survey.

2 Formal Framework

2.1 Preferences

A natural and widely used way to represent individual preferences is by means of binary relations over a set of alternatives. For a set $X = \{a, b, c, ...\}$ of

Daniel Eckert · Christian Klamler
Institute of Public Economics
University of Graz

E. Herrera-Viedma et al. (Eds.): Consensual Processes, STUDFUZZ 267, pp. 3–22.
springerlink.com   © Springer-Verlag Berlin Heidelberg 2011

alternatives a preference is then a binary relation $R \subseteq X \times X$. Typically, preferences are assumed to be complete, reflexive and transitive[1] binary relations on a finite set of alternatives. In the following, the set of all complete, reflexive and transitive preferences is denoted by $\mathcal{R}$, while the set of all binary relations is denoted by $\mathcal{B}$.

2.2 Distance

A distance function on a set assigns a numerical value to any pair of objects from that set. As in this survey we are interested in distances over the set of preferences and binary relations in general, we begin this section by formally defining a distance function on the set $\mathcal{B}$ of all binary relations.

Definition 1. Let $R_1, R_2 \in \mathcal{B}$, the function $d : \mathcal{B} \times \mathcal{B} \to \Re$ is called *distance function* on $\mathcal{B}$.

Thus, for any two preferences R_1 and R_2, $d(R_1, R_2)$ represents the distance between R_1 and R_2. To give an idea in what way distance should be measured, we will define certain reasonable properties for such a distance function:
    The first axiom represents the basic properties of a distance.

Axiom 1. *1.* $d(R_1, R_2) \geq 0$ *(non-negativity)*
2. $d(R_1, R_2) = 0$ *if and only if* $R_1 = R_2$
3. $d(R_1, R_2) = d(R_2, R_1)$ *(symmetry)*
4. $d(R_1, R_2) + d(R_2, R_3) \geq d(R_1, R_3)$ *(triangle inequality)*

The second condition guarantees a certain neutrality among the alternatives under consideration.

Axiom 2. *If* R_1' *is derived from* R_1 *and* R_2' *from* R_2 *via a permutation of* X, *then* $d(R_1, R_2) = d(R_1', R_2')$.

The third condition ensures that distance can only depend on parts of the preferences which are different.

Axiom 3. *Let* $X = S \cup T$ *with* $S \cap T = \emptyset$ *and* aR_ib *for all* $a \in S$, $b \in T$ *and* $i \in \{1,2\}$. *If* R_1 *and* R_2 *fully agree upon* S, *written* $R_1|S = R_2|S$, *then* $d(R_1, R_2) = d(R_1|T, R_2|T)$. *The analogous needs to hold if* $R_1|T = R_2|T$.

Finally, a unit of measurement is chosen.

Axiom 4. *If* $R_1 \neq R_2$, *then* $d(R_1, R_2) \geq 1$, *i.e. the minimal positive distance is 1.*

---

[1] A preference $R \in \mathcal{B}$ is complete, if for every pair of distinct alternatives $a, b \in X$, aRb or bRa; it is reflexive if for every alternative $a \in X$, aRa; and it is transitive if for every triple of alternatives $a, b, c \in X$, aRb and bRc implies aRc.

Kemeny [29] considered the following distance function d_k between pairs of preferences.

Definition 2. A distance function $d : \mathcal{R} \times \mathcal{R} \to \Re_+$ is called *Kemeny distance* if for all R_1, R_2, $d(R_1, R_2) = |(R_1 \backslash R_2) \cup (R_2 \backslash R_1)|$.

This distance function d_K is hence the cardinality of the symmetric difference between R_1 and R_2. Differently speaking, it counts (twice) the number of inversions of pairs in the respective preferences R_1 and R_2.[2] This can be seen in the following example:

Example 1. Let $X = \{a, b, c\}$. The individual preferences are stated in the following table 1 where alternatives are ranked from more preferred (top) to less preferred (bottom).

Table 1 Rankings for 3 alternatives.

R_1	R_2	R_3
a	b	c
b	a	b
c	c	a

Preference R_1 differs from preference R_2 by a difference in the relative ranking between alternatives a and b, i.e. a full inversion between those two alternatives leads from one preference to the other. Hence, the symmetric difference between R_1 and R_2 is $(R_1 \backslash R_2) \cup (R_2 \backslash R_1) = \{(a, b), (b, a)\}$, leading to a Kemeny distance of $d_K(R_1, R_2) = 2$. Moving from R_1 to R_3, every single pair of alternatives needs to be fully inversed. As there are 3 such pairs, $d_K(R_1, R_3) = 6$. As R_3 is the exact opposite of R_1, the distance between those two preferences is also the maximal distance between two preferences over 3 alternatives.

The Kemeny distance is not only an intuitively plausible way to measure distances between preferences: Interestingly (see Kemeny and Snell [30] for a nice proof), the Kemeny distance d_K is also the only distance function to satisfy the above axioms.

Theorem 5. *(Kemeny [29]) Distance function d is the Kemeny distance d_K if and only if it satisfies axioms 1 to 4.*

---

[2] Interpreting the Kemeny distance via number of inversions is slightly problematic when considering weak orders, i.e. allowing for indifferences as in that case full inversions might not occur.

2.3 Aggregation

Many scientific disciplines are concerned with the aggregation of preferences into a social or group preference. Let N be a set of n individuals. Formally, $f : \mathcal{R}^n \to \mathcal{R}$ denotes an aggregation rule assigning to any profile $p = (R_1, ..., R_n) \in \mathcal{R}^n$ of individual preferences a group preference $f(p) \in \mathcal{R}$.

Probably the most common way to aggregate individual preferences is via simple majorities. For any profile $p \in \mathcal{R}^n$ and alternatives $a, b \in X$, let $m_{ab}(p) = |\{i \in N : aR_ib\}| - |\{i \in N : bR_ia\}|$, i.e. the majority margin[3] between a and b.

Definition 3. The aggregation rule $m : \mathcal{R}^n \to \mathcal{B}$ is the *simple majority rule* if for all $p \in \mathcal{R}^n$, $a\ m(p)\ b$ if and only if $m_{ab} \geq 0$.

However, simple majority rule does not fit the above definition of an aggregation rule f, as for some profiles the outcome will be cyclic, i.e. a majority prefers a over b, a majority prefers b over c but a majority also prefers c over a. This problem of majority inconsistency is known as the Condorcet paradox.

In general the major goal in this sort of aggregation theory is to find a certain *consensus* among the individuals in N. However, this requires a concise specification of what consensus means. Simple majority rule interprets consensus on a pairwise basis, i.e. looking for the largest agreement on any pair of alternatives. Kemeny [29] suggested to use the distance function d_K to arrive at a consensus ranking, i.e. the ranking whose Kemeny distance to all the preferences of the individuals has the lowest sum. Therefore we will also need to extend the Kemeny distance to measure distance between a profile $p = (R_1, ..., R_n)$ and a preference R by $\underline{d}(p, R) = \sum_{i=1}^{n} d(R_i, R)$. In the case of the simple majority relation being a linear order, this is identical to the previously specified consensus ranking. Kemeny's suggested approach to finding a consensus can now be formalized as follows:

Definition 4 (Kemeny Rule). An aggregation rule $f : \mathcal{R}^n \to \mathcal{R}$ is the *Kemeny rule* f_K if and only if for all $p \in \mathcal{R}^n$, $f(p) = \{R \in \mathcal{R} : \forall R' \in \mathcal{R}, \underline{d}(p, R) \leq \underline{d}(p, R')\}$.

Obviously a solution to this problem might not be unique as can also be seen in the following example.

Example 2. Let $X = \{a, b, c\}$, $|N| = 5$ and consider the following preference profile, where a number specifies how many individuals have the respective preference.

Given the preferences in table 2, we can obtain the consensus ranking via measuring the distance of each of the 6 possible rankings to the above profile. The respective distances can be found in table 3, where the distance of each ranking to the above profile is stated in the bottom line.

---

[3] Whenever clear from the context, we will not specifically refer to the profile p.

Table 2 Rankings for 3 alternatives.

2	2	1
a	c	b
b	a	c
c	b	a

Table 3 Distances from profile in Table 2.

a	a	b	b	c	c
b	c	a	c	a	b
c	b	c	a	b	a
12	14	18	16	12	18

From table 3 we observe that there are two consensus rankings for the above profile, i.e. rankings which are of smallest distance from the profile. Those are the ranking with a above b above c (which we will write abc from now) and the ranking cab which both are of Kemeny distance 12 from the profile. Although what is considered worst in the first Kemeny ranking is considered best in the second, at least with respect to possible chosen alternatives we could conclude that alternative b is rather not a good one.

The Kemeny rule is a so-called *Condorcet extension* (see Fishburn [24]), as its outcome is identical to majority rule as long as there are no preference cycles. Actually, the profile in the previous example leads to a preference cycle as a is socially preferred to b, b to c and c to a.

Another interesting connection to Condorcet's own considerations - more than 200 years ago - on how to resolve cycles has been provided by Young [60]. He gives an excellent explanation of why the Kemeny rule can be seen as the solution to this problem that Condorcet was aiming at. Condorcet argued on the basis of individuals being more likely to have the "correct" ranking between any pair of alternatives. Based on that assumption, Young [60] shows that "Condorcet's method can be interpreted as a statistical procedure for estimating the ranking of the candidates that is most likely to be correct." (p. 1232)

As previously for the Kemeny distance, also for the Kemeny rule there exists a strong axiomatic foundation established by Young and Levenglick [61]. Consider the following axioms for aggregation rules:

Axiom 6. *Let $V \subseteq N$ and (S, T) be a partition of V, i.e. S and T are disjoint "electorates". Denote a profile p restricted to the set $W \subseteq N$ of individuals by $p|W$. An aggregation rule f is consistent if and only if for all $p \in \mathcal{R}^n$, $f(p|S) \cap f(p|T) \neq \emptyset$ implies $f(p|V) = f(p|S) \cap f(p|T)$.*

Consistency requires that any (set of) preference(s) considered best in two disjoint electorates needs to be considered best also when those electorates decide jointly.

Axiom 7. *For any profile $p \in \mathcal{R}^n$, and aggregation rule f let $a \; \hat{f}(p) \; b$ denote the fact that $a \; f(p) \; b$ and for all $c \in X$, $c \; f(p) \; b \rightarrow c \; f(p) \; a$ and $a \; f(p) \; c \rightarrow b \; f(p) \; c$, i.e. a and b are neighbors in the preference ranking determined by f. An aggregation rule f is called Condorcet if $m_{ab} > 0$ implies not $b \; \hat{f}(p) \; a$ and $m_{ab} = 0$ implies $a \; \hat{f}(p) \; b$ if and only if $b \; \hat{f}(p) \; a$.*

The Condorcet property guarantees that whenever there is a majority preferring alternative a over b, then it cannot be the case that the aggregation rule ranks b immediately above a. This has some intuitive appeal in the sense that otherwise a simple switch of those two alternatives - without affecting any other alternative or pair of alternatives - would benefit a majority of the electorate.

Axiom 8. *Let π be a permutation of X and denote by $\pi(p)$ the permuted profile p. An aggregation rule f is said to be neutral if for all $p \in \mathcal{R}^n$, $a \; f(p) \; b$ if and only if $\pi(a) \; f(\pi(p)) \; \pi(b)$.*

Neutrality ensures that alternatives are treated equally, i.e. aggregation results do not depend on the names of the alternatives.

Theorem 9. *(Young and Levenglick [61]) The Kemeny rule is the only aggregation rule that satisfies the axioms neutrality, Condorcet and consistency.*

3   Other Distance-Based Aggregation Rules

The Kemeny rule can be seen as minimizing the number of pairwise inversions - hence the Kemeny distance - to a certain goal state. The defined goal state for this rule is a unanimity ranking, i.e. the same ranking for all voters. Distance aspects are therefore explicit in the definition of the Kemeny rule. This can also be seen in figure 1.

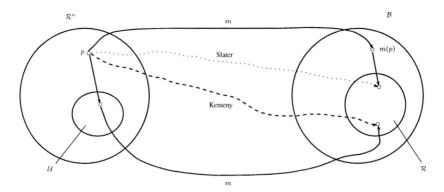

Fig. 1 Various distance based rules

The big circle on the left corresponds to the domain, which is the set of all preference profiles, i.e. $\mathcal{R}^n$. The big right circle corresponds to the codomain which is the set $\mathcal{B}$ of all binary relations on X (not necessarily complete, reflexive or transitive). The set of all complete, reflexive and transitive binary relations $\mathcal{R}$ is represented by the small circle in the codomain and can be seen - in general - as the set of acceptable social outcomes.[4] The set of all unanimous profiles is denoted by $\mathcal{U} \subset \mathcal{R}^n$ and represented by the small circle in the domain. Obviously, majority rule assigns to any profile $p = (R, R, ..., R) \in \mathcal{U}$ the social preference R which necessarily needs to be in $\mathcal{R}$.

However, there are also other aggregation rules that explicitly use distances in their definition. A widely known rule is the *Slater rule* (see Slater [58]). Instead of putting its focus on the domain, as in the case of the Kemeny rule, it draws attention to the codomain. Intuitively, the Slater ranking is the binary relation in $\mathcal{R}$ which is closest to the simple majority relation according to the Kemeny distance. Graphically this can also be seen in figure 1, and can be formally defined as follows:

Definition 5 (Slater rule). An aggregation rule $f : \mathcal{R}^n \to \mathcal{R}$ is the *Slater rule* f_S if and only if for all $p \in \mathcal{R}^n$, $f(p) = \{R \in \mathcal{R} : \forall R^{'} \in \mathcal{R}, d_K(m(p), R) \leq d_K(m(p), R^{'})\}$.

A third well known rule explicitly using distances, and maybe the historically first to do so (dating back to 1876), is the *Dodgson rule* (see Black [11]). The underlying criterion is the distance of any alternative from being a Condorcet winner, i.e. being preferred to any other alternative by a simple majority. The ranking of the alternatives according to those distances is the Dodgson ranking. Formally this can be stated as follows:

Definition 6 (Dodgson rule). Let $\mathcal{R}_x^n$ be the set of profiles in which $x \in X$ is a strict Condorcet winner and let $D : \mathcal{R}^n \times \mathcal{R}^n \to \Re_+$ be a distance function on the set of profiles such that for all $p, p' \in \mathcal{R}^n$, $D(p, p') = \sum_{i \in N} d_K(R_i, R_i^{'})$. An aggregation rule $f : \mathcal{R}^n \to \mathcal{R}$ is the *Dodgson rule* f_D if and only if for all $p \in \mathcal{R}^n$, and all $a, b \in X$, a $f(p)$ b if and only if $\min_{p' \in \mathcal{R}_a^n} D(p, p') \geq \min_{p' \in \mathcal{R}_b^n} D(p, p')$.

Hence, in contrast to the goal of unanimity - as in the Kemeny rule - the goal in the Dodgson rule is - in a certain sense - alternative wise simple majority.

Example 3. (Example 10 continued) Given the profile in table 2, we know that this leads to a Condorcet cycle with a preferred to b, b preferred to c and c preferred to a. The previous three distance-based aggregation rules give the following outcomes in table 4.

---

[4] This is the case whenever we consider - what Kenneth Arrow called - social welfare functions, i.e. aggregation rules that lead to a social preference and not only a social choice of certain (winning) alternatives. Reinterpretations of this picture for other sorts of aggregation rules are feasible.

Table 4 Outcomes for profile in Table 2.

Kemeny		Slater			Dodgson
a	c	a	b	c	ac
b	a	b	c	a	b
c	b	c	a	b	

The Kemeny rule and the Slater rule will provide multiple outcomes for certain profiles and therefore lack a certain decisiveness. Dodgson will be more decisive but lacks other essential properties (see Brandt [13]).

We saw for the above explicitly distance-based aggregation rules, that - in general - they differ in how they minimize the distance to a certain subset of the domain or the codomain. However, also other - at first sight not distance-based - rules have been characterized using distances and goal states. One of the first such results was obtained by Farkas and Nitzan [23] who proved that ranking the alternatives according to the profile's Kemeny distance to the closest profile that has the respective alternative top ranked by all individuals, is equivalent to using the Borda rule[5]. This links the Borda rule with the well known Pareto criterion. If, instead of looking at the profile, one measures the distance of an alternative from the simple majority relation to the closest binary relation in $\mathcal{B}$ that has this alternative top ranked and ranks the alternatives according to that distance, then the derived rule will be the Copeland rule[6] as shown by Klamler [34]. Meskanen and Nurmi [44] extend those results and show that a variation of the distance measure and/or the goal state provides a distance-based characterization of many well known aggregation rules. Most recently, Elkind et al. [21] extended the list of voting rules with distance characterizations by the maximin rule, Young's rule, Approval voting and most of the scoring rules. In Elkind et. al [22] they aimed at providing general results regarding distance characterizations of voting rules. As they are able to show that essentially any voting rule can be given a distance characterization, they conclude that this property as such cannot be considered as anything desirable. Hence they proceed by investigating the properties of the distance functions used for those characterizations to conclude certain qualities of voting rules.

---

[5] The Borda rule assigns points to the alternatives in the individual preferences, starting with $|X| - 1$ points for the top ranked alternative down to 0 points for the bottom ranked alternative. Summing the points over all individuals then gives the Borda ranking.

[6] The Copeland rule is defined as follows: for all $a, b \in X$, a is ranked at least as high as b in the Copeland ranking if $|\{c \in X : m_{ac} \geq 0\}| - |\{c \in X : m_{ac} \leq 0\}| \geq |\{c \in X : m_{bc} \geq 0\}| - |\{c \in X : m_{bc} \leq 0\}|$. Various versions of the Copeland rule are used as tournament solutions.

4 Distance-Based Comparisons of Voting Rules

In the social choice literature one finds many comparisons of aggregation rules based on (more or less) reasonable properties those rules satisfy or violate (see e.g. Fishburn [24] or Nurmi [50]). However, besides this axiomatic approach, there also exists a distance-based approach to compare outcomes of different aggregation rules. In particular the question is to find an upper bound by ' how much the outcomes of aggregation rules will differ on the set of profiles. Various - in a certain sense mostly discouraging - results have been proved in recent years, underlining the suspicion that the voting outcome highly depends on the voting rule chosen.

Example 4. Consider the preference profile for $n = 34$ and $X = \{a, b, c, d\}$ in table 5.

Table 5 Preference Profile for $n = 34$ and $|X| = 4$.

$$
\begin{array}{cccccc}
9 & 5 & 5 & 5 & 5 & 5 \\
\hline
a & a & b & c & d & d \\
b & c & d & d & b & c \\
c & b & c & b & a & a \\
d & d & a & a & c & b \\
\end{array}
$$

This profile leads to the majority margins in table 6.

Table 6 Majority margins for profile in table 5.

$$
\begin{array}{c|c}
m_{ab} = 4 & m_{bc} = 4 \\
m_{ac} = 4 & m_{bd} = 4 \\
m_{ad} = -6 & m_{cd} = 4 \\
\end{array}
$$

As can easily be seen from the majority margins, there is a Condorcet cycle *abcda*. Condorcet extensions such as the Dodgson rule or the Kemeny rule or scoring rules such as the Borda count will still provide a social ranking of the alternatives.[7] The Kemeny ranking can be determined by finding the linear order that minimizes the sum of switches according to table 5. For this, the majority margins are sufficient. In contrast to this, for the Dodgson ranking, we need the full profile information, as majority margins do not provide information about how close two alternatives are to each other in the individual rankings. E.g. in the above profile in table 5, whenever a is ranked below d by someone, there is always another alternative between them. Hence, to make a the Condorcet winner, 4 voters need to change their rankings by making two switches each, meaning that alternative a is of distance 16

---

[7] The Kemeny rule does - in general - not provide a unique ranking in the case of Condorcet cycles.

from being a Condorcet winner. As can easily be identified, b will already be a Condorcet winner whenever 3 voters make one switch each in their preferences. Actually, the rankings for some voting rules are summarized in table 7, where $\succ$ denotes strictly preferred and $\sim$ denotes indifference.

Table 7 Summary of voting outcomes for profile in table 5.

Kemeny	Dodgson	Borda
$a \succ b \succ c \succ d$	$b \succ c \sim d \succ a$	$b \succ a \succ d \succ c$
Copeland	Plurality	Plurality − Runoff
$a \sim b \succ c \sim d$	$a \succ d \succ b \sim c$	$d \succ a \succ b \sim c$

As can be seen, outcomes for the same profile differ considerably. Many results show the huge variety in outcomes that can occur for the same preference profile. Two different approaches are possible. One approach puts the focus on positional aspects. For any preference profile, it measures distance between voting outcomes on the basis of the position of a specific alternative with one method (usually the top or bottom alternative in the respective ranking) in the ranking derived from the other method. The largest distance between outcomes can therefore be seen if, for some profile, the winner with one voting rule is the loser in the other rule.

A second possibility is to look at the full rankings derived from the voting rules. There, the largest possible distance between voting outcomes seems to occur whenever there exists a profile for which the derived rankings by the voting rules are exactly contrary to each other. Obviously, the strength of any result depends on how indifferences are handled, i.e. the strongest form would be opposite linear orders.[8] However, not many results in that strong form do exist, in particular not for explicitly distance based aggregation rules. A weaker version would allow for indifferences in the rankings, e.g. the weak order $a \succ b \sim c \succ d$ could be seen weakly opposite the ranking $d \succ c \sim b \succ a$, in which the winner in one ranking is the looser in the other ranking but there are indifferences in the middle. In general the following upper bounds on distances between outcomes have been proved[9] (see also Lamboray [40]):

- The top ranked alternative according to the Kemeny rule can be bottom ranked according to the Dodgson rule and vice versa (Klamler [32] and Ratliff [52]).
- The top ranked alternative according to the Dodgson rule can be bottom ranked by any scoring rule (Ratliff [53]).

---

[8] This would also lead to the largest possible distance according to the Kemeny distance.

[9] Many other results focusing on the relationship between different voting rules w.r.t. voting outcomes have been analysed. Of particular interest to the distance-based voting rules are the results by Saari and Merlin [56] that show that the Borda count always ranks the Kemeny winner higher than the Kemeny loser and the Kemeny rule always ranks the Borda winner above the Borda loser.

- The Dodgson ranking and the Borda ranking can be weakly opposite (Klamler [31]).
- The maximin ranking[10] can be weakly opposite to the Borda ranking and to the Copeland ranking (Klamler [33]).
- The top ranked alternative by the Slater rule can be bottom ranked by the Dodgson rule (Klamler [32]).
- The unique prudent order[11] can be weakly opposite to the Borda ranking and the Copeland ranking (Lamboray [40]).
- The top ranked alternative in a unique prudent order can be bottom ranked according to the Kemeny rule and the Slater rule (Lamboray [40]).

5 Distance-Based Analysis and Geometric Representation of Aggregation Problems

For a distance-based approach to aggregation theory, geometric representations of aggregation problems, as the one most prominently developed by Don Saari [54], are of obvious relevance. In particular, Saari maps preference profiles into points in a hypercube. E.g. for $X = \{a, b, c\}$ we get 3 pairs $\{a, b\}$, $\{a, c\}$ and $\{b, c\}$. Each pair represents one dimension. On each dimension a "1" indicates the first alternative being preferred to the second and a "-1" indicates the opposite. This leads to a three-dimensional hypercube with 8 vertices as in figure 2.

The vertex $(1, 1, -1)$ represents the preference in which $a \succ b$, $b \succ c$ and $a \succ c$. As there are only 6 linear orders for 3 alternatives but there are 8 vertices, 2 of the vertices represent cyclic preferences, namely vertices $(-1, -1, -1)$ and $(1, 1, 1)$.

To map profiles into this hypercube, Saari [54] uses the normalized majority margins for any pair of alternatives, i.e. for any profile p and alternatives $a, b \in X$, $x_{ab}(p) = \frac{m_{ab}(p)}{n}$. This leads to a mapping $x : \mathcal{R}^n \to [-1, 1]^3$ from profiles to majority margins. As there are more vertices of the cube (namely 8) then there are feasible linear orders (namely 6), certain regions in the hypercube cannot be reached by x. Actually it is the convex hull of those 6 vertices representing linear orders which forms the range of x. This polytope represents the set of all pairwise voting situations.

Given $x(p)$, there is now a straightforward way to determine the simple majority relation. It is the closest vertex to $x(p)$ according to Euclidean distance. Obviously, for each vertex there is a whole set of points for which this vertex is the closest vertex. This set is called ranking region of that vertex. Now, considering the representation polytope, one observes that it

---

[10] The maximin rule ranks the alternatives relative to their largest majority loss against any other alternative. Obviously it is a Condorcet extension.

[11] A prudent order can be seen as a ranking in which the strongest opposition against this ranking is minimal. See Lamboray [40] for a detailed discussion.

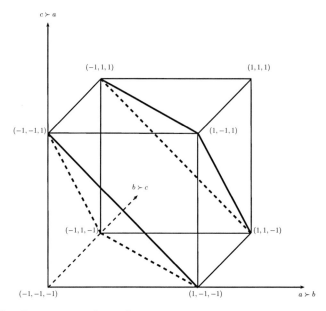

Fig. 2 Saari's representation cube

also passes through the ranking regions of vertices $(1,1,1)$ and $(-1,-1,-1)$ providing cyclic outcomes.

In this geometric representation, looking for consensus now implies avoiding situations of cyclic social outcomes in a reasonable fashion. One approach is to consider, for any profile p, the $l1$-distance[12] between $x(p)$ and the transitive ranking regions. The following can be proved:

Theorem 10. *(Saari and Merlin [56]) For any profile p, the Kemeny rule ranking is the ranking of the transitive ranking region which has the closest $l1$-distance to $x(p)$.*

An alternative geometric representation can be found in Zwicker [62]. For $|X| = 3$ he uses a hexagon (see figure 3) to represent the 6 feasible preferences.

He defines - similar to Saari's ranking regions - certain proximity regions, which are the sets of points in the hexagon closest to the same vertex. Now, if for any profile, we determine the mean point in the hexagon, the proximity region in which the mean falls, is the Borda outcome for this profile. In that respect, the Borda rule can be classified as a mean proximity rule (see Zwicker [62] for a detailed discussion).

Example 5. (Zwicker [62]) Consider the profile in table 8 represented also in figure 3.

---

[12] The $l1$-distance measures distance according to the restriction of only being allowed to walk parallel to the axisis. It is also often called Manhattan distance.

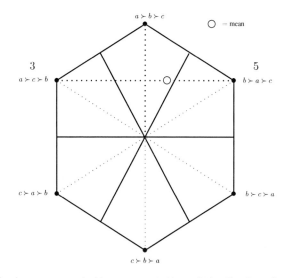

Fig. 3 Zwicker's mean proximity representation of the Borda rule.

Table 8 Preference Profile for $n = 8$ and $|X| = 3$.

$$
\begin{array}{cc}
3 & 5 \\
\hline
a & b \\
c & a \\
b & c
\end{array}
$$

The mean point, denoted by the circle, falls into the proximity region closest to $a \succ b \succ c$ which corresponds to the Borda counts of 11 for a, 10 for b and 3 for c. The closeness of the circle to the borderline between proximity region $a \succ b \succ c$ and $b \succ a \succ c$ indicates the similar Borda counts for those alternatives.

Zwicker [62] also shows that any abstract anonymous voting system is a generalized scoring rule if and only if it is a mean proximity rule. Other - earlier - results using the concept of median to characterize the Kemeny rule can be found in Barthélemy and Monjardet [9] (for a survey on median procedures see Hudry et al. [27]).

6   Alternative Choice Theoretical Structures

Preferences modelled by rankings of the alternatives are just one possibility. An attractive and alternative way to deal with aggregation problems whenever little structure is imposed on individual and/or social preferences are choice functions. The importance of choice functions in aggregation theory is discussed e.g. in Aizerman and Aleskerov [1], in Xu [59] and Aleskerov and Monjardet [4].

Definition 7. Let K denote the set of all non-empty subsets of X. A *choice function* is a function $C : K \to K$ such that for all $S \in K$, $C(S) \subseteq S$.

I.e. a choice function assigns to any set $S \in K$ a subset $C(S)$ of alternatives in S. An interesting aspect of choice functions is the possibility to use them to represent particular preferences which cannot be represented by a weak order. Consider e.g. the situation in which alternatives $X = \{a, b, c\}$ represent pieces of cakes where $a > b > c$, i.e. a is the largest piece and c the smallest. In conformity to certain social norms a possible preference might be to never choose the largest piece on the plate in case there is more than one (see Baigent and Gaertner [8]). This would lead to the following choices: $C(X) = \{b\}$, $C(\{a, b\}) = \{b\}$, $C(\{a, c\}) = \{c\}$, $C(\{b, c\}) = \{c\}$, and $C(\{i\}) = \{i\}$ for all $i \in X$. As can easily be verified, those preferences cannot be rationalized[13] by a weak order, i.e. there is no weak order according to which a rational individual can make all of the above choices.

Specific aggregation rules based on individual choice functions have been studied by Xu [59]. Despite the considered rules in those approaches, also distance based aggregation rules for choice functions can be defined. An analog of the Kemeny distance on binary relations for choice functions has been characterized by Klamler [36].[14] The general idea of measuring distance can be found already in Albayrak and Aleskerov [2] and Ilyunin et al. [28].

General choice function approaches have been used recently by Alcalde-Unzu and Vorsatz [3] and discussed in Martinez-Panero [42]. However, explicit use of distances on choice functions for the definition of choice aggregation rules is still open.

7 Distance-Based Approaches to Judgment Aggregation

As aggregation problems arise in many other areas than social choice distance-based approaches are not limited to the construction and analysis of voting rules. Obviously, whenever a distance can be meaningfully defined for individual inputs and be extended to profiles, distance-based aggregation rules can be constructed.

In particular, the recent literature on judgment aggregation (for a survey see List and Puppe [41]), extends the axiomatic analysis of aggregation

---

[13] On the topic of rationalizability of choice functions see e.g. Sen [57].

[14] Actually the characterization result is based on quasi-choice functions defined by $C : K \to K \cup \emptyset$. This distance function has an analog distance measure for weak orders whenever we restrict ourselves to rationalizable choice functions. See Klamler [35].

problems in social choice theory to the aggregation of individual judgments on logically interconnected propositions, - a problem that arises in expert panels, legal courts, boards etc. Given a formal language $\mathcal{L}$ a judgment aggregation problem is defined by an agenda $X \subseteq \mathcal{L}$ which is interpreted as the set of sentences on which judgments are to be made. An individual (or collective) judgment set is then the subset $J \subseteq X$ of accepted sentences in the agenda. Denoting by $\mathcal{J}$ the set of all logically consistent and complete judgment sets[15] an aggregation rule is a function $\mathcal{J}^n \to \mathcal{J}$ which assigns to every profile $\underline{J} = (J_1, ..., J_n)$ of individual judgment sets a collective judgment set $f(\underline{J})$.

In the same way as judgment aggregation provides a generalization of Arrovian social choice theory (see e.g. Dietrich [18]) and is prone to similar impossibility results, it is equally open for a distance-based approach. In fact, distance-based approaches have systematically been used in the related area of belief merging (for a survey see Konieczny and Pino Perez [38]), which deals with the aggregation of knowledge bases.

In a seminal paper, Pigozzi [51] has shown that distance-based rules imported from belief merging can help to deal with the analogue of Condorcet's voting paradox for judgment aggregation, namely the doctrinal paradox (Kornhauser and Sager [39]). This paradox owes its name to the logical connections between the issues established by legal doctrine: In particular, in a contract case a defendant is liable (L) if and only if there was a valid contract (V) and a material breach (B) of that contract. Thus the verdict L is equivalent to the conjunction of V and B.

The doctrinal paradox consists in the fact that propositionwise majority voting can lead to a logically inconsistent outcome, resp. that majority voting on the premises (V resp. B) leads to an outcome that is inconsistent with voting on the conclusion.

Table 9 Doctrinal Paradox

	V(alid contract)?	B(reach of contract)?	L(iability)?
Judge 1	yes	yes	yes
Judge 2	yes	no	no
Judge 3	no	yes	no
Court	yes	yes	?

Pigozzi [51] uses an aggregation rule based on the minimization of the sum of distances between judgment sets, or formally, for any profile of judgment sets $\underline{J} \in \mathcal{J}^n$ the socially accepted judgment set $f(\underline{J})$ is a (not necessarily unique) solution to $\min_{J \in \mathcal{J}} \sum_{i=1}^{n} d(J, J_i)$, where the distance between judgment sets is the arguably most natural one, namely the Hamming distance,

---

[15] A judgment set is complete if it contains for any proposition $p \in \mathcal{L}$ either p or its negation.

i.e. the number of sentences on which two judgment sets differ. The obvious similarity to Kemeny's archetypical rule is seen from the fact that these rules are equivalent in case that $\mathcal{J}$ is the set of all ranking judgments of the form "$a \succ b$" (Eckert and Mitloehner [19]).

Recently, Miller and Osherson [45] have introduced judgment aggregation analogues of the Dodgson rule, i.e. rules that select for any profile the outcome of the majority consistent profile that is closest to it, and investigated the properties of this and other types of rules for different classes of distances (for an extensive analysis including complexity issues of various distance-based rules in the framework of belief merging see Konieczny et al. [37]).

Another line of research investigates the reliability of different distance-based aggregation rules with the help of probabilistic models. In particular Hartmann et al. [25] additionally parameterize distance-based aggregation rules by the weight of the conclusion compared to the premises and establish the superiority of premise-based aggregation rules over a direct voting on the conclusion.

Further work on both lines of research could help establish distance-based approaches to judgment aggregation as an alternative to propositionwise judgment aggregation, the independence condition being not only responsible for the Arrovian type impossibility results but also less appealing in the context of aggregating logically interconnected sentences.

8   Discussion and Outlook

Distance-based approaches to aggregation can be seen as a promising way to capture the intuition of consensus. However, two problems should not be underestimated.

One problem related to voting rules, that has been studied intensively in recent years, is their computational complexity, in particular whether it is possible to calculate the outcome in polynomial time or not. Starting point was a seminal paper by Bartholdi, Tovey and Trick [10] analysing the computational complexity of the Kemeny rule and the Dodgson rule. They show that for both rules determining the social outcome is NP-hard, i.e. no efficient algorithm to do so in polynomial time is known. Two further approaches have been taken. The first focuses on making the results of Bartholdi et al. more accurate as those were only lower bounds (see Hemaspaandra et al. [26], Conitzer et al[16]). The other approach tries to approximate certain voting rules by determining the respective scores of alternatives only approximately. Such approximations for Dodgson's rule can be found in McCabe-Dansted et al. [43], Caragiannis et al. [14] or Caragiannis et al. [15].

Another problem challenges the very idea of distance-based aggregation. In a distance-based framework the condition of proximity preservation formulated by Baigent [6] seems particularly appealing, namely that smaller changes in profiles should not lead to larger changes in the outcome than larger changes in profiles. It however turns out that this condition is inconsistent with other very weak conditions on aggregation rules in the spirit of anonymity and unanimity respectively (see Baigent and Eckert [7]).

Acknowledgments

We are grateful to Andreas Darmann for providing very helpful input in the writing of this survey.

References

1. Aizerman, M., Aleskerov, F.: Theory of Choice. Elsevier, Amsterdam (1995)
2. Albayrak, S.R., Aleskerov, F.: Convexity of Choice Function Sets. Bogazici University Research Paper (2000), ISS / EC-2000-01
3. Alcalde-Unzu, J., Vorsatz, M.: The measurement of consensus: an axiomatic analysis. In: Documentos de trabajo (FEDEA 2008), vol. 28, pp. 1–32 (2008)
4. Aleskerov, F., Monjardet, B.: Utility Maximization, Choice and Preference. Springer, Berlin (2002)
5. Arrow, K.J., Sen, A.K., Suzumura, K. (eds.): Handbook of Social Choice and Welfare, vol. 1. Elsevier, Amsterdam (2002)
6. Baigent, N.: Preference proximity and anonymous social choice. Quarterly Journal of Economics 102, 161–169 (1987)
7. Baigent, N., Eckert, D.: Abstract Aggregations and Proximity Preservation: An Impossibility Result. Theory and Decision 56, 359–366 (2004)
8. Baigent, N., Gaertner, W.: Never choose the uniquely largest: a characterization. Economic Theory 8, 239–249 (1996)
9. Barthélemy, J.P., Monjardet, B.: The median procedure in cluster analysis and social choice theory. Mathematical Social Sciences 1, 235–267 (1981)
10. Bartholdi, J., Tovey, C.A., Trick, M.A.: Voting schemes for which it can be difficult to tell who won the election. Social Choice and Welfare 6, 157–165 (1989)
11. Black, D.: The Theory of Committees and Elections. Cambridge University Press, London (1958)
12. Bogart, K.P.: Preference Structures I: Distances between transitive perference relations. Journal of Mathematical Sociology 3, 49–67 (1973)
13. Brandt, F.: Some remarks on Dodgson's voting rule. Mathematical Logic Quarterly 55, 460–463 (2009)
14. Caragiannis, I., Covey, J.A., Feldman, M., Homan, C.M., Kaklamanis, C., Karanikolas, N., Procaccia, A.D.Rosenschein,J.S.: On the approximability of Dodgson and Young elections. In: Proceedings of the 20th ACM-SIAM Symposium on Discrete Algorithms, pp. 1058–1067 (2009)
15. Caragiannis, I., Kaklamanis, C., Karanikolas, N.,Procaccia, A.D.: Socially desirable approximations for Dodgson's voting rule. In: Proceedings of the 11th ACM Conference on Electronic Commerce (forthcoming)
16. Conitzer, V., Davenport, A.,Kalyanam, J.: Improved bounds for computing Kemeny rankings. In: Proceedings of the 21st National Conference on Artificial Intelligence, pp. 620–626 (2006)

17. Day, W.H.E., McMorris, F.R.: Axiomatic Consensus Theory in Group Choice and Biomathematics. SIAM, Philadelphia (2003)
18. Dietrich, F.: Arrow's Theorem in Judgment Aggregation. Social Choice and Welfare 29, 19–33 (2007)
19. Eckert, D., Mitlöhner, J.: Logical representation and merging of preference information. In: Multidisciplinary IJCAI-2005 Workshop on Advances in Preference Handling, pp. 85–87 (2005)
20. Eckert, D., Pigozzi, G.: Belief merging, judgment aggregation and some links with social choice theory. In: Delgrande, J., Lang, J., Rott, H., Tallon, J.-M. (eds.) Belief Change in Rational Agents: Perspectives from Artificial Intelligence, Philosophy, and Economics Dagstuhl, Germany, Dagstuhl Seminar Proceedings (2005)
21. Elkind, E., Faliszewski, P., Slinko, A.: On distance rationalizability of some voting rules. In: Proceedings of TARK, vol. 2009, pp. 613–622 (2009)
22. Elkind, E., Faliszewski, P., Slinko, A.: On the role of distances in defining voting rules. In: Proceedings of 9th Int. Conf. on Autonomous Agents and Multiagent Systems (2010)
23. Farkas, D., Nitzan, S.: The Borda rule and Pareto stability: a comment. Econometrica 47(5), 1305–1306 (1979)
24. Fishburn, P.C.: Condorcet social choice functions. SIAM Journal of Applied Mathematics 33, 469–489 (1977)
25. Hartmann, S., Pigozzi, G., Sprenger, J.: Reliable Methods of Judgment Aggregation. Journal of Logic and Computation 20, 603–617 (2010)
26. Hemaspaandra, E., Hemaspaandra, L.A., Rothe, J.: Exact analysis of Dodgson elections: Lewis Carroll's 1876 voting system is complete for paralles access to NP. Journal of the ACM 44, 806–825 (1997)
27. Hudry, O., Leclerc, B., Monjardet, B., Barthélémy, J.-P.: Metric and latticial medians. In: Bouyssou, D., et al. (eds.) Concepts and Methods of Decision Making. Wiley, Chichester (2009)
28. Ilyunin, O.K., Popov, B.V., El'kin, L.N.: Majority Functional Operators in Voting Theory. Automation and Remote Control 7, 137–145 (1988)
29. Kemeny, J.: Mathematics without numbers. Daedalus 88, 571–591 (1959)
30. Kemeny, J., Snell, L.: Mathematical Models in the Social Sciences. Ginn, Boston (1960)
31. Klamler, C.: The Dodgson ranking and the Borda count: a binary comparison. Mathematical Social Sciences 48, 103–108 (2004a)
32. Klamler, C.: The Dodgson ranking and its relation to Kemeny's method and Slater's rule. Social Choice and Welfare 23, 91–102 (2004b)
33. Klamler, C.: On the closeness aspect of three voting rules: Borda - Copeland - Maximin. Group Decision and Negotiation 14, 233–240 (2005a)
34. Klamler, C.: The Copeland rule and Condorcet's principle. Economic Theory 25, 745–749 (2005b)
35. Klamler, C.: On some distance aspects in social choice theory. In: Simeone, B., Pukelsheim, F. (eds.) Mathematics and Democracy: recent advances in voting systems and collective choice, pp. 97–104. Springer, Heidelberg (2006)
36. Klamler, C.: A distance measure for choice functions. Social Choice and Welfare 30, 419–425 (2008)

37. Konieczny, S., Lang, J., Marquis, P.: Distance-based merging: a general framework and some complexity results. In: Proceedings of the Eighth International Conference on Principles of Knowledge Representation and Reasoning (KR 2002), pp. 97–108 (2002)
38. Konieczny, S., Pino-Perez, R.: Merging Information under Constraints: A Logical Framework. Journal of Logic and Computation 12, 773–808 (2002)
39. Kornhauser, L.A., Sager, L.G.: Unpacking the Court. Yale Law Journal 96, 82–117 (2006)
40. Lamboray, C.: A comparison between the prudent order and the ranking obtained with Borda's, Copeland's, Slater's and Kemeny's rules. Mathematical Social Sciences 54, 1–16 (2007)
41. List, C., Puppe, C.: Judgment aggregation. In: Anan, P., Pattanaik, P.K., Puppe, C. (eds.) The Handbook of Rational and Social Choice: An Overview of New Foundations and Application. Oxford University Press, Oxford (2009)
42. Martinez-Panero, M.: Approaches to consensus: the state of the art. In: ESTYLF 2008-Proceedings, pp. 1–8 (2008)
43. McCabe-Dansted, J.C., Pritchard, G., Slinko, A.: Approximability of Dodgson's rule. Social Choice and Welfare 31, 311–330 (2008)
44. Meskanen, T., Nurmi, H.: Distance from consensus: a theme and variations. In: Simeone, B., Pukelsheim, F. (eds.) Mathematics and Democracy: recent advances in voting systems and collective choice, pp. 117–132. Springer, Berlin (2006)
45. Miller, M.K., Osherson, D.: Methods for distance-based judgment aggregation. Social Choice and Welfare 32, 575–601 (2009)
46. Nurmi, H.: Comparing Voting Systems. D.Reidel, Dordrecht (1987)
47. Nurmi, H.: Discrepancies in the outcomes resulting from different voting schemes. Theory and Decision 25, 193–208 (1988)
48. Nurmi, H.: Voting Paradoxes and How to Deal with Them. Springer, Berlin (1999)
49. Nurmi, H.: Voting Procedures und Uncertainty. Springer, Berlin (2002)
50. Nurmi, H.: A comparison of some distance-based choice rules in ranking environments. Theory and Decision 57, 5–24 (2004)
51. Pigozzi, P.: Belief Merging and the Discursive Dilemma: An Argument-Based Account to Paradoxes of Judgment Aggregation. Synthese 152, 285–298 (2006)
52. Ratliff, T.C.: A comparison of Dodgson's method and Kemeny's rule. Social Choice and Welfare 18, 79–89 (2001)
53. Ratliff, T.C.: A comparison of Dodgson's method and the Borda count. Economic Theory 20, 357–372 (2002)
54. Saari, D.G.: Geometry of Voting. Springer, Berlin (1995)
55. Saari, D.G.: Disposing Dictators, Demystifying Voting Paradoxes. Cambridge University Press, New York (2008)
56. Saari, D.G., Merlin, V.R.: A geometric examination of Kemeny's rule. Social Choice and Welfare 17, 403–438 (2000)
57. Sen, A.: Social choice theory. In: Arrow, K.J., Intrilligator, M.D. (eds.) Handbook of Mathematical Economics, vol. III,22 pp. 1073–1181. North-Holland, Amsterdam (1986)
58. Slater, P.: Inconsistencies in a schedule of paired comparisons. Biometrika 48, 303–312 (1961)

59. Xu, Y.: Non Binary Social Choice: A Brief Introduction. In: Schofield, N. (ed.) Collective Decision-Making: Social Choice and Political Economy. Kluwer, Boston (1996)
60. Young, H.P.: Condorcet's Theory of Voting. American Political Science Review 82(4), 1231–1244 (1988)
61. Young, H.P., Levenglick, A.: A consistent extension of Condorcet's election principle. SIAM Journal of Applied Mathematics 35(2), 285–300 (1978)
62. Zwicker, W.S.: Consistency without neutrality in voting rules: when is a vote an average? Mathematical and Computer Modelling 48, 1357–1373 (2008)

On Penalty-Based Aggregation Functions and Consensus

Gleb Beliakov, Tomasa Calvo, and Simon James

1 Introduction

The problem of aggregating individual preferences in order to arrive at a group consensus arises naturally in elections where a candidate must be chosen that best represents the individuals' differing opinions. Other contexts include the judging of sporting competitions and the fusion of sensor readings. In these applications it makes sense that the aggregated result should be as close as possible to the individual inputs, giving rise to the need for methods that minimize this difference. Penalty-based aggregation functions are precisely those functions that aim to accomplish this, drawing upon various notions of "difference" in varying situations.

The study of how best to aggregate ordinal preferences traces back as far as Ramon Llull (thirteenth century) and Nicholas of Cusa (fifteenth century), whose ideas correspond respectively with those proposed by Condercet [15] and Borda [8] in the eighteenth century [38]. The *Borda count* (also equivalent to the method proposed by Kendall later in 1962 [32]) consists of allocating *Marks*, aggregating the results using a weighted arithmetic mean. Drawing upon the result from statistics that the arithmetic mean is a function which minimizes the Euclidean distance between the inputs and output, Cook and Seiford [17, 18] recognized the potential to frame the problem in terms of this objective. Their minimum variance method extended the work of Kemeny and Snell [31] who proposed a distance measure for analyzing rankings along with a number of desirable axioms in the spirit of Arrow

Gleb Beliakov · Simon James
School of Information Technology, Deakin University
221 Burwood Hwy, Burwood 3125, Australia
e-mail: {gleb, sjames}@deakin.edu.au

Tomasa Calvo
Departamento de Ciencias de la Computación, Universidad de Alcalá
28871-Alcalá de Henares (Madrid), Spain
e-mail: tomasa.calvo@uah.es

E. Herrera-Viedma et al. (Eds.): Consensual Processes, STUDFUZZ 267, pp. 23–40.
springerlink.com   © Springer-Verlag Berlin Heidelberg 2011

(relaxing that of irrelevant alternatives). More recently, Cook *et al* [16] have shown how generalized distance functions can be used.

Similar approaches have been taken by Freimer and Yu [21] who used optimization methods to minimize the "regret" of decision makers in a group and bring about a compromise solution, where regret is defined as the difference between an individual's score and the obtained result. These solutions were shown in [41] to satisfy useful properties such as pareto optimality, uniqueness, symmetry and independence of irrelevant alternatives. Such properties are particularly desirable in election processes. Balinski and Laraki have made great efforts in experimentation for determining voting methods that best serve the voters [2, 3].

One of the more difficult issues when aggregating preferences is how best to assign numerical scores if only ordinal or pairwise preferences are given. Indeed, in terms of penalty-based aggregation, the problem is further complicated by determining how the penalties should be calculated fairly in this context. Aggregation of pair-wise comparison matrices was presented in [25], utilizing a goal programming [24] approach that again minimizes distance. Different norms can be used to measure the disagreement between individuals and the group, with the L_1 norm essentially minimizing the aggregate disagreement, and L_∞ minimizing the maximum disagreement. Here also, the authors account for preferences that are conflicting and sometimes inconsistent.

A different type of penalization was presented in [22] where disagreement between an individual and the overall ranking decreases the weighting given to that individual in order to dissuade decision-makers from making extreme judgements in order to bias the results. Penalty-based aggregation methods are also able to accommodate different weightings, tailored to the specific situation.

Much has been said of voting and aggregating preferences, however the use of penalty-based aggregation functions is not restricted to this domain. Hornik and Meyer [29] used aggregation based on distance measures in order to reach consensus in bench-marking experiments, and a number of current applications are presented in [33].

The rationale behind using certain functions and certain measures of distance or penalty is context dependent. To use arithmetic means in group decision making, for example, implies a faith in the individuals to make accurate and sincere judgements that will not skew the results. Medians are not as susceptible in this way. When using aggregation methods, how consensus is interpreted will affect how it is achieved. For instance, is the disagreement expressed by the inputs equally important, or is it more important to minimize the dissatisfaction of a particular party?

Aggregation functions are often interpreted as providing an output which is representative of the inputs. A measure of distance or deviation from this value, or the imposition of a penalty for not having consensus has been studied in various forms, from aggregating relations in [20], to interpreting the use of the median operator in [30] and [39] and the study of *faithful* penalty functions in [13]. Mesiar *et al* also found some general results for what they called *minimization based aggregation operators*, which were interpreted in terms of a dissimilarity function [36]. An overview of penalty-based aggregation functions was given in [5], highlighting also their natural relation to the maximum likelihood principle.

Linguistic quantifiers which model such concepts as "most" and "majority" allow aggregation to be performed in a framework which is more in line with our intuitive notions of consensus. Examples include consensus models in group decision making [28] and more recently the use of quantifiers with order-induced functions, where inputs are ordered by their implied support before being aggregated [37].

This chapter draws on the results synthesized in [5], providing an introduction to penalty-based aggregation functions and penalty functions in general. In Section 2, the preliminaries of aggregation functions will be provided. Section 3 presents some well used definitions of penalty and shows how some aggregation functions correspond to minimizing the overall penalty associated with a given input vector. In Section 4, we consider some alternative frameworks of penalty and also introduce the idea of aggregating penalties using the OWA function. We give some examples of the resulting aggregation functions which can be obtained, noting where monotonicity of the penalty-based function will be lost, or where it will not be well defined. It will be shown throughout that penalty-based aggregation functions provide a natural framework for mathematical interpretations of consensus.

2 Preliminaries: Aggregation Functions

We restrict ourselves to aggregation functions defined on $[0,1]^n$, although of course values in any closed interval can be used as inputs and outputs by means of a simple linear transformation. Inputs and outputs on $\bar{\mathbb{R}}_+ = [0,\infty]$ or $\bar{\mathbb{R}} = [-\infty,\infty]$ can also be dealt with using many averaging functions. Up to date accounts of aggregation functions are given in [1, 6, 11, 26, 38].

Definition 1. A function $f : [0,1]^n \to [0,1]$ is called an aggregation function if it is monotone non-decreasing in each variable and satisfies $f(0,0,...,0) = 0$, $f(1,1,...,1) = 1$. [1]

Definition 2. An aggregation function f is called averaging if it is bounded by the minimum and maximum of its arguments

$$\min(\mathbf{x}) := \min(x_1,\ldots,x_n) \leq f(x_1,\ldots,x_n) \leq \max(x_1,\ldots,x_n) =: \max(\mathbf{x}).$$

It is immediate (because of monotonicity) that averaging aggregation functions are idempotent (i.e., $\forall t \in [0,1] : f(t,t,\ldots,t) = t$) and vice versa. Then clearly the boundary conditions are satisfied. Often averaging aggregation functions are collectively referred to as means.

---

[1] When considering other intervals, the boundary conditions may need special care, e.g. the product with arguments in $\mathbb{R}_+$. In some studies continuity of f is also added to its definition, we do not impose it generally (we remind that some aggregation function are discontinuous, e.g. the drastic sum, drastic product, discontinuous means and others).

Various properties may be satisfied by averaging aggregation functions depending on their definition. The following will be useful when considering penalty-based functions and consensus.

Definition 3 (Properties). An aggregation function $f : [0,1]^n \to [0,1]$ is:

Homogeneous   if $f(\lambda x_1, \ldots, \lambda x_n) = \lambda f(\mathbf{x})$; whenever $(\lambda x_1, \ldots, \lambda x_n) \in [0,1]^n$ and $\lambda f(\mathbf{x}) \in [0,1]$.

Shift-invariant   if $f(x_1 + \lambda, \ldots, x_n + \lambda) = f(\mathbf{x}) + \lambda$ whenever $(x_1 + \lambda, \ldots, x_n + \lambda) \in [0,1]^n$ and $f(\mathbf{x}) + \lambda \in [0,1]$;

Symmetric   if its value does not depend on the permutation of the arguments, i.e. $f(x_1, x_2, \ldots, x_n) = f(x_{\pi(1)}, x_{\pi(2)}, \ldots, x_{\pi(n)})$ for every $\mathbf{x}$ and every permutation $\pi = (\pi(1), \pi(2), \ldots, \pi(n))$ of $(1,2 \ldots, n)$;

Strictly monotone   if $\mathbf{x} \leq \mathbf{y}$ but $\mathbf{x} \neq \mathbf{y}$ implies $f(\mathbf{x}) < f(\mathbf{y})$.

Homogeneity and shift-invariance are useful for defining functions over different intervals, so the degree of consensus is not biased by the selection of evaluation scale. Where symmetric aggregation functions are used, it is implied that the source of each input is considered equally important. In some situations however, consensus may be better achieved by focusing on some penalties more than others, e.g. when combining sensor readings it may be known that some are more reliable than others. The property of strict monotonicity means that an increase to any of the inputs results in an increase to the consensus value. Where this property is not satisfied, it will be the case that some output minimizes the penalty for multiple input vectors within a region of the domain, e.g. the median value will not change unless the relative order of inputs changes (even though the penalty values may change).

One of the most popularly used averaging aggregation functions is the arithmetic mean (or *average*). It is homogeneous, shift-invariant, symmetric and strictly monotone, minimizing the sum of squared distances between the output and inputs. Weighting vectors $\mathbf{w} \in [0,1]^n$, such that $\sum w_i = 1$, can be used to allocate more importance to certain inputs depending on their source (or as with ordered functions, their relative size). The weighted arithmetic mean is generalized by a very important family of aggregation functions known as weighted quasi-arithmetic means.

Definition 4 (Weighted quasi-arithmetic means). For a given strictly monotone and continuous function $g : [0,1] \to [-\infty, +\infty]$, called a generating function or generator, and a weighting vector $\mathbf{w} = (w_1, \ldots, w_n)$, the weighted quasi-arithmetic mean is the function

$$M_{\mathbf{w},g}(\mathbf{x}) = g^{-1}\left(\sum_{i=1}^{n} w_i g(x_i)\right), \tag{1}$$

where $\sum w_i = 1$ and $w_i \geq 0 \; \forall \, i$.

Special cases include:

Arithmetic means $WAM_{\mathbf{w}} = \sum_{i=1}^{n} w_i x_i,$ $g(t) = t$;

Geometric means $G_{\mathbf{w}} = \prod_{i=1}^{n} x_i^{w_i},$ $g(t) = log(t)$;

Harmonic means $H_{\mathbf{w}} = \left(\sum_{i=1}^{n} \frac{w_i}{x_i} \right)^{-1},$ $g(t) = \frac{1}{t}$;

Power means $M_{\mathbf{w},[r]} = \left(\sum_{i=1}^{n} w_i x_i^r \right)^{\frac{1}{r}},$ $g(t) = t^r$.

The scaling function g is not defined uniquely, but up to an arbitrary linear transformation. Prototypical examples of this class are the arithmetic mean A, the geometric mean G and the harmonic mean H and their weighted analogues. In turn, these mentioned functions are special cases of power means $M_{[r]}$ with $h(x) = x^r, r \in \mathbb{R}$, with $M_{[0]} = G$ by definition. The limiting cases $r \rightarrow \pm\infty$ correspond to the maximum and minimum respectively.

It is known that the only homogeneous quasi-arithmetic means are the power means while only the weighted arithmetic mean is shift-invariant. However, there are other (non quasi-arithmetic) homogeneous and shift-invariant means. The symmetric versions are obtained where $w_i = \frac{1}{n}, \forall i$.

OWA functions and their generalizations are also well known examples of averaging functions.

Definition 5 (OWA). Given a weighting vector $\mathbf{w}$, the OWA function is

$$OWA_{\mathbf{w}}(\mathbf{x}) = \sum_{i=1}^{n} w_i x_{(i)},$$

where the (.) notation denotes the components of $\mathbf{x}$ being arranged in non-increasing order $x_{(1)} \geq x_{(2)} \geq \ldots \geq x_{(n)}$.

Special cases of the OWA operator, depending on the weighting vector $\mathbf{w}$ include:

Arithmetic mean   where all the weights are equal, i.e. all $w_i = \frac{1}{n}$
Maximum   function for $\mathbf{w} = (1,0,...,0)$;
Minimum   function for $\mathbf{w} = (0,...,0,1)$;
Median   function for $w_i = 0$ for all $i \neq k$, $w_k = 1$ if $n = 2k+1$ is odd, and $w_i = 0$
    for all $i \neq k, k+1$, $w_k = w_{k+1} = 0.5$ if $n = 2k$ is even;
k-th order statistic   for $w_k = 1, 0$ otherwise, for all $i \neq k$;

The OWA function is a piecewise linear idempotent aggregation function. It is symmetric, homogeneous, shift-invariant and strictly monotone if $w_i > 0, \forall i$.

As stated, the OWA functions have been useful for modeling the concept of consensus as they can be naturally associated with linguistic quantifiers. For example, suppose $n = 5$, "most" can be modeled with the vector $\mathbf{w} = (1/3, 1/3, 1/3, 0, 0)$ and "80% of" with $\mathbf{w} = (0,0,0,1,0)$. An important generalization of the OWA is the induced OWA [40], where the ordering of the input vector is determined by an auxiliary variable $\mathbf{z}$.

Definition 6. Given a weighting vector $\mathbf{w}$ and an inducing variable $\mathbf{z}$, the Induced Ordered Weighted Averaging (IOWA) function is

$$IOWA_{\mathbf{w}}(\langle x_1, z_1 \rangle, \ldots, \langle x_n, z_n \rangle) = \sum_{i=1}^{n} w_i x_{(i)}, \tag{2}$$

where the $(.)$ notation denotes the inputs $\langle x_i, z_i \rangle$ reordered such that $z_{(1)} \geq z_{(2)} \geq \ldots \geq z_{(n)}$ and the convention that if q of the $z_{(i)}$ are tied, i.e. $z_{(i)} = z_{(i+1)} = \ldots = z_{(i+q-1)}$,

$$x_{(i)} = \ldots = x_{(i+q-1)} = \frac{1}{q} \sum_{j=(i)}^{(i+q-1)} x_j. \tag{3}$$

An important class of functions related to ordered aggregation functions are the weighted medians. Weighted medians associate an auxiliary weight with each input x_i, and then return the output of the k-th highest input $x_{(k)}$ where k represents the mid-way point of the progressive sum of *weights*.

Definition 7. Let $\mathbf{w}$ be a weighting vector, and let $\mathbf{u}$ denote the vector obtained from $\mathbf{w}$ by arranging its components in the order induced by the components of the input vector $\mathbf{x}$, such that $u_k = w_i$ if $x_i = x_{(k)}$ is the k-th largest component of $\mathbf{x}$. The lower weighted median is the function

$$Med_{\mathbf{w}}(x_1, \ldots, x_n) = x_{(k)}, \tag{4}$$

where k is the index obtained from the condition

$$\sum_{j=1}^{k-1} u_j < \frac{1}{2} \text{ and } \sum_{j=1}^{k} u_j \geq \frac{1}{2}.$$

The upper weighted median is the function (4) where k is the index obtained from the condition

$$\sum_{j=1}^{k-1} u_j \leq \frac{1}{2} \text{ and } \sum_{j=1}^{k} u_j > \frac{1}{2}.$$

Many notions of penalty lead naturally to weighted medians. These functions are expressible in terms of the IOWA if $\mathbf{z}$ is chosen to correspond with the cumulative weights when the x_i are ordered by size.

We will lastly consider the family of Choquet integrals. It is well known that weighted arithmetic means and the OWA are generalized by the Choquet integral, which is defined by a fuzzy measure. In the context of consensus, the weights in arithmetic means and OWA functions can be interpreted as allocating importance to an input by its source or relative size respectively. The values of the fuzzy measure used in the calculation of the Choquet integral, however represent the importance of each group (or coalition) of inputs [27].

Discrete fuzzy measures (or capacities [14]) are set functions which are interpreted as assigning a weight to all possible groups of criteria, and thus offer a much greater flexibility for modeling aggregation.

Definition 8. Let $\mathbb{N} = \{1, 2, \ldots, n\}$. A discrete fuzzy measure is a set function $v : 2^{\mathbb{N}} \to [0, 1]$ which is monotonic (i.e. $v(S) \le v(T)$ whenever $S \subset T$) and satisfies $v(\emptyset) = 0, v(\mathbb{N}) = 1$.

Consider now the components of $\mathbf{x}$ in non-decreasing order: $x_{(1)} \le x_{(2)} \le \ldots \le x_{(n)}$ and assume, by convention, that $x_{(0)} = 0$ and $x_{(n+1)} = \infty$. Denote by $H_i = \{(i), \ldots, (n)\}$ the subset of elements corresponding with the $n - i + 1$ largest components of $\mathbf{x}$. By convention, $H_{n+1} = \emptyset$.

Definition 9. Let v be a discrete fuzzy measure. The discrete Choquet integral with respect to v, $C_v : X^n \to X$ is defined by

$$C_v(x_1, \ldots, x_n) = \sum_{i=1}^{n} x_{(i)}[v(H_i) - v(H_{i+1})]. \qquad (5)$$

The OWA, weighted medians and Choquet integral can also be generalized by means of generating functions g, as is done with the arithmetic mean to form the family of quasi-arithmetic means (see [12, 6]). In these cases, they are respectively referred to as the generalized OWA (GenOWA), the generalized Choquet integral and weighted quasi-medians.

3   Penalty Functions

Penalty-based aggregation functions have been studied by several authors. The results that arithmetic means minimize the squared differences and medians the absolute differences were already known to Laplace (from [38], see also [23]). The main motivation is the following. Let $\mathbf{x}$ be the inputs and y the output. If all the inputs coincide $x = x_1 = \ldots = x_n$, then the output is $y = x$, since we have a unanimous vote. If some input $x_i \ne y$, then we impose a "penalty" for this disagreement. The larger the disagreement, and the more the inputs disagree with the output, the larger (in general) the penalty. We look for an aggregated value which minimizes the penalty; in some sense we look for a consensus which minimizes the disagreement. This is easily interpreted when aggregating experts' opinions.

Thus we need to define a suitable measure of disagreement, or dissimilarity. We start with a very broad definition of penalties, and then particularize it and obtain many known aggregation functions as special cases. Let us consider a vector of inputs $\mathbf{x}$ and the vector $\mathbf{y} = (y, y, \ldots, y)$.

Definition 10. The function $P : [0, 1]^{n+1} \to [0, \infty)$ is a penalty function if and only if it satisfies:

i) $P(\mathbf{x}, y) \geq 0$ for all $\mathbf{x}, y$;

ii) $P(\mathbf{x}, y) = 0$ if $\mathbf{x} = \mathbf{y}$;

iii) For every fixed $\mathbf{x}$, the set of minimizers of $P(\mathbf{x}, y)$ is either a singleton or an interval.

The penalty-based function is

$$f(\mathbf{x}) = \arg\min_y P(\mathbf{x}, y),$$

if y is the unique minimizer, and $y = \frac{a+b}{2}$ if the set of minimizers is the interval (a, b) (open or closed).

The first two conditions have useful interpretations: no penalty is imposed if there is full agreement, and no negative penalties are allowed. However, since adding a constant to P does not change its minimizers, technically they can be relaxed: P just needs to reach its absolute minimum when $\mathbf{x} = \mathbf{y}$. Condition iii) ensures that the function f is well defined. If P is quasiconvex[2] in y, then iii) is automatically satisfied. We should also note that a penalty-based function is necessarily idempotent, but it is not always monotone.

Definition 11. A penalty-based function f, which is monotone increasing in all components of $\mathbf{x}$ is called a penalty-based aggregation function.

It was shown in [5] that any idempotent function (and hence any averaging aggregation function) can be expressed as a penalty-based function, for instance, using the penalty $P(\mathbf{x}, y) = (f(\mathbf{x}) - y)^2$. This result leads to the potential for existing families of aggregation functions to be interpreted in terms of the penalties they minimize.

There are many perceptions of penalty and consensus that lead to functions which may not be well defined or monotone. The mode is one such function, which in some cases corresponds with a *majority* vote. We will briefly address such functions in Section 4. For some types of penalty-based functions, whether or not monotonicity holds can be determined from the partial derivatives, i.e.

$$\frac{P_{yx_i}}{P_{yy}} \leq 0,$$

is the necessary and sufficient condition for monotonicity where P is twice differentiable [5].

In the next subsections we will present some special types of penalty-based aggregation families.

3.1 Faithful Penalty Functions

Intuitively, it makes sense to take the sum of penalties, applying different weights where the disagreement from each source is not considered to be equally important.

---

[2] A function f is quasiconvex, if its level sets $S = \{\mathbf{x} : f(\mathbf{x}) \leq C\}$ are convex for any $C \in Ran(f)$.

The special class of *faithful* penalty functions was considered in [13], expressed as

$$P(\mathbf{x},y) = \sum_{i=1}^{n} w_i p(x_i,y),$$ (6)

where $\mathbf{w}$ is a weighting vector and $p : [0,1]^2 \rightarrow [0,\infty)$ is the dissimilarity function with the properties:

1) $p(t,s) = 0$ if and only if $t = s$, and
2) p can be represented as $p(t,s) = K(g(t),g(s))$ where K is convex and g is continuous and monotone.

These conditions ensure the resulting penalty-based function $f(\mathbf{x}) = \arg\min_{y} P(\mathbf{x},y)$, will be well defined (with the minimizer y^* either unique or an interval) and monotone. In this case, f is an aggregation function [34] and falls into the class of faithful penalty-based aggregation functions.

There are several well known aggregation functions that are faithful penalty-based aggregation functions:

1. Let $p(t,s) = (t-s)^2$. The corresponding faithful penalty-based aggregation function is a weighted arithmetic mean;
2. Let $p(t,s) = |t-s|$. The corresponding faithful penalty-based aggregation function is a weighted median;
3. Let $P(\mathbf{x},y) = \sum_{i=1}^{n} w_i p(x_{(i)},y)$, where $x_{(i)}$ is the i-th largest component of $\mathbf{x}$. We obtain the ordered weighted counterparts of the means in the previous examples, namely the OWA and weighted medians. Where the order is induced by some auxiliary variable $\mathbf{z}$, we obtain the IOWA;
4. Let $P(\mathbf{x},y) = \sum_{i=1}^{n} w_i(\mathbf{x})(x_{(i)}-y)^2$, where $x_{(i)}$ is the i-th smallest component of $\mathbf{x}$, and $w_i(\mathbf{x}) = v(H_i) - v(H_{i+1})$, as in Definition 9. We obtain the Choquet integral with respect to v. Note that the weights depend on the ordering of the components of $\mathbf{x}$;
5. Let $p(t,s) = (g(t)-g(s))^2$. The corresponding faithful penalty-based aggregation function is a weighted quasi-arithmetic mean with the generator g. The generalized OWA, generalized Choquet and quasi-medians can also be obtained by making the analogous substitutions.

The following examples illustrate how these penalty functions can be interpreted in terms of consensus.

Example 1. The geometric mean is a function which cannot have an aggregated value above zero unless all of its arguments are greater than zero[3]. In the context of consensus it can be interpreted as modeling a veto, ensuring that some degree of satisfaction is met by all inputs. The geometric mean falls under the class of quasi-arithmetic means with generator $g = -log(t)$ and hence each argument x_i is associated with the penalty,

---

[3] This property is also referred to as hard partial conjunction [19].

$$p(x_i, y) = (-log(x_i) + log(y))^2 = (log(x_i/y))^2.$$

We see clearly that the penalty where any input $x_i = 0$ approaches ∞, forcing the resulting minimizer to zero.

Example 2. In [37], the induced OWA was used to model the *majority opinion* by means of a quantifier and support function. Suppose we have expert ratings $\mathbf{x} = (0.6, 0.7, 0, 0.9, 0.1)$. We could say that an overall score of 0.7 would be more representative than say 0.5, since it is within 0.2 of most of the ratings. We can use the "support" for each value, i.e. how close other values are and how many, to induce the order, which provides the reordering $\mathbf{x} = (0.7, 0.6, 0.9, 0.1, 0)$. A weighting vector modeling the concept of *most*, e.g. $\mathbf{w} = (1/3, 1/3, 1/3, 0, 0)$, can then be used to aggregate the inputs. The penalty applied to each value is hence weighted according to its *support*, rather than its source or input size. In the context of group decision making, such aggregation favors those actively trying to achieve consensus rather than those who would skew the results by making extreme or uncompromising decisions[4].

Other interesting faithful penalty-based aggregation functions are:

- Let $p(t, s) = |t - s|^r, r \geq 1$. Then in general no closed form solution exists, but the minimum in (6) can be found numerically. Interestingly, the limiting case $r \to 1$ does not correspond to the median when $n = 2k$ (see [30]), but to a solution of the following equation,

$$(y - x_{(1)})(y - x_{(2)}) \ldots (y - x_{(k)}) = (y - x_{(k+1)}) \ldots (y - x_{(n)}).$$

  For example, for $n = 4$ we have

$$f(\mathbf{x}) = \frac{x_{(1)}x_{(2)} - x_{(3)}x_{(4)}}{(x_{(1)} + x_{(2)}) - (x_{(3)} + x_{(4)})},$$

  whereas the standard definition of median gives $f(\mathbf{x}) = \frac{x_{(2)} + x_{(3)}}{2}$. Weighted medians were considered in [4].
- Let $c \geq 0$ and

$$p(t, s) = \begin{cases} t - s & \text{if } s \leq t, \\ c(s - t) & \text{if } s > t. \end{cases}$$

  Then f is the α-quantile operator, with $\alpha = c/(1 + c)$ [13]. To obtain the i-th order statistic, we take $c = \frac{i - 1/2}{n - i + 1/2}$.

3.2   Other Means Based on Penalties

Most conceptions of penalty, difference or dissimilarity are expressed similarly to Eq. (6). We will briefly discuss some families of averaging aggregation functions

---

[4] It should be noted that the IOWA may not be monotone, depending on the inducing variable. For instance, the mode could be considered in this framework.

which are based on minimizing some type of penalty. More detailed information can be found in [5], in particular, the special cases corresponding to quasi-arithmetic means etc.

Deviation Means. Given a function $d : [0,1]^2 \to \mathbb{R}$ which is continuous, strictly increasing with respect to the second argument and satisfies $d(t,t) = 0, \forall t \in [0,1]$, the equation

$$\sum_{i=1}^{n} w_i d(x_i, y) = 0$$

has the unique solution $y^* = f(\mathbf{x})$ called the deviation mean (see [10]). Quasi-arithmetic means constitute a special case where $d(t,s) = g(s) - g(t)$ and g is the generating function.

Entropic Means. Let $\phi : \mathbb{R}_+ \to \mathbb{R}$ be a strictly convex differentiable function with $(0,1] \subset dom\, \phi$ such that $\phi(1) = \phi'(1) = 0$, and $\mathbf{w}$ a weighting vector. The penalty d_ϕ is defined as

$$d_\phi(x,y) = x\phi(y/x).$$

The entropic mean ([7]) is the function

$$f(\mathbf{x}) = y^* = \arg\min_{y \in \mathbb{R}_+} P_\phi(\mathbf{x}, y) = \arg\min_{y \in \mathbb{R}_+} \sum_{i=1}^{n} w_i d_\phi(x_i, y).$$

The penalty $d_\phi(\alpha, \cdot)$ is strictly convex for any $\alpha > 0$, and $d_\phi(\alpha, \beta) \geq 0$ with equality if and only if $\alpha = \beta$. The resulting means, hence always satisfy the defining properties of aggregation functions. It should be noted, however that $d_\phi(x,y) = x\phi(y/x)$ can be zero even if $x \neq y$ so it does not satisfy the definition of a faithful penalty-based aggregation function.

Bregman loss functions. Let $\psi : \mathbb{R}^n \to \mathbb{R}$ be a strictly convex differentiable function. Then the Bregman penalty-based aggregation function is

$$f(\mathbf{x}) = y^* = \arg\min_{y \in \mathbb{R}_+} \sum_{i=1}^{n} w_i D_\psi(y, x_i),$$

where $D_\psi(x,y) = \psi(x) - \psi(y) - (x-y)\psi'(y)$. The function $D_\psi(x,y)$ is a univariate version of the Bregman loss function [9].

Weighted quasi-arithmetic means can be recovered in special cases for many of these expressions of penalty. In fact, many of these functions can be related back to faithful penalty-based aggregation functions with an appropriate choice of p, each capturing a slightly different notion of difference. Although the penalty used to define entropic means does not satisfy the requirements of faithful penalty functions, the minimizers may coincide.

Since disagreement can be interpreted in terms of distance, it is reasonable to consider norms and similar concepts for capturing the notion of penalty. One example is the Minkowski gauge, which was used in [5] to construct new penalty-based aggregation functions.

4   Other Penalties and Notions of Consensus

In this section we consider alternative frameworks of penalty which could be useful
for achieving consensus in various contexts.

4.1   Heterogeneous Penalties

In constructing penalty-based functions using Eq. (6), one makes the assumption
that a single notion of penalty is sufficient in describing all variables in the given
context. One extension of the general form of this penalty equation is to allow the
penalty of each of the i-th inputs to be modeled differently.

$$P(\mathbf{x},y) = \sum_{i=1}^{n} p_i(x_i,y). \tag{7}$$

Here, in addition to the weights (absorbed in p_i) with their usual interpretation
of the relative importance of the i-th input (or i-th largest input), we can vary contribution of the i-th input based on the functional form of the corresponding penalty
$p_i(x_i,y)$.

Example 3. [34, 35] Consider the inputs of different sensors, which need to be aver-
aged (e.g., temperature sensors). The inputs from sensors are random variables with
different distributions (e.g., normal, Laplace or another member of the exponential
family). Then taking the weighted arithmetic mean or median is not appropriate, because the sensors are heterogeneous. We can take into account the diversity of input
error distributions by means of different penalty functions. Suppose we have one
sensor whose distribution is Laplace, and a second whose is normal. An appropriate
penalty function is then,

$$P(\mathbf{x},y) = |x_1 - y| + (x_2 - y)^2.$$

Solving the equation of the necessary condition for a minimum, and taking into
account that P is convex, we obtain

$$f(x_1,x_2) = Med(x_1,x_2 - \frac{1}{2},x_2 + \frac{1}{2}).$$

For a weighted penalty function

$$P(\mathbf{x},y) = w_1|x_1 - y| + w_2(x_2 - y)^2,$$

the solution is
$$f(x_1,x_2) = Med(x_1,x_2 - \frac{w_1}{2w_2},x_2 + \frac{w_1}{2w_2}).$$

The minimization of Eq. (7) may not always be possible analytically, however
usually a numerical solution will be easily obtained. In cases where it is known

that the function is convex with respect to y, the resulting function can satisfy the properties of an aggregation function. Consider the following example.

Example 4. Let $P(\mathbf{x},y) = \sum_{i=1}^{n-1} w_i(x_i - y)^2 + w_n \max(0, y - x_n)^2$. The meaning of the last term is the following. Suppose the n-th input (e.g., the n-th expert) usually underestimates the result y. Then we wish to penalize $y > x_n$ but not $y < x_n$. So the n-th input is discarded only if $y < x_n$. The resulting penalty P is a piecewise quadratic function whose minimum is easily found: it is the minimum of the weighted arithmetic means of the first $n-1$ and of all components of $\mathbf{x}$, $f(\mathbf{x}) = \min(A(x_1, \ldots, x_{n-1}), A(x_1, \ldots, x_n))$.

This example can be extended by changing more terms of the sum to some asymmetric functions of $y - x_i$, in which case the solution can be found numerically. The interpretation is similar: positive and negative deviations from x_i are penalized differently.

4.2 OWA Based Penalties

Here we consider penalties weighted by the relative *penalty size* by using the OWA function, i.e.

$$P(\mathbf{x},y) = \sum_{i=1}^{n} w_i p(x_{(i)}, y), \qquad (8)$$

where (.) in this case denotes the arguments of $\mathbf{x}$ arranged such that $p(x_{(1)}, y) \geq p(x_{(2)}, y) \geq \ldots \geq p(x_{(n)}, y)$. As we will see, the minimizer y^* may fail to be unique (or a unique interval) and hence may result in functions which are non-monotone and sometimes not well defined.

4.2.1 k-th Projections

Consider a penalty described by Eq. (8) with $w_k = 1, w_i = 0$, otherwise. Provided the dissimilarity function p satisfies

1. $|a - y| = |b - y| \rightarrow p(a,y) = p(b,y)$;
2. $p(a,y) > p(b,y)$ whenever $a > b > y$ or $a < b < y$,

the minimizer will be given by

$$y^* = \frac{x_{(j)} + x_{(j+n-k)}}{2}$$

where $\{(j), (j+1), \ldots, (j+n-k)\} = \mathscr{S}$ is the set of $n - k + 1$ sequential inputs (ordered by input size, not the associated penalty) such that $x_{(j)} - x_{(j+n-k)}$ is minimized.

Special cases of $P(\mathbf{x},y)$ in terms of k include:

- Where $k = n$, $P(\mathbf{x},y) = \min p(x_i, y)$, i.e. we want to minimize the smallest penalty. Since $|\mathscr{S}| = 1$, so any value $y = x_i$ returns the minimum (not unique).

- Where $k = 1$, $P(\mathbf{x}, y) = \max p(x_i, y)$, i.e. we want to minimize the maximum penalty. In this case, $|\mathscr{S}| = n \to x_{(j)} = x_{(1)}, x_{(j+n-k)} = x_{(n)}$ and hence

$$y^* = \frac{x_{(1)} + x_{(n)}}{2}.$$

  This function is known as the *mid-range*.
- If n is odd and $k = \frac{n+1}{2}$, $P(\mathbf{x}, y) = Med(p(x_i, y))$ – i.e. we define a function such that the median penalty is minimized.

For this type of function, determining $f(\mathbf{x})$ can be achieved easily by first calculating $x_{(i)} - x_{(i+n-k)}$ and taking the minimum to determine j. In general, there will be k local minima (provided no inputs are equal), a global minimum only occurs if there is a set $\mathscr{S}_{min}$ with a range strictly lower than all sequential sets $|\mathscr{S}| = n - k + 1$.

4.2.2   Sum of the Largest k Residuals

Consider now Eq. (8) with the weighting vector $w_i = \frac{1}{k}, \forall i \leq k, 0$ otherwise. In this case, the associated function which minimizes the penalty will be influenced by both the form of p, e.g. whether the penalties we consider are based on squared or absolute difference, and the parity of k.

To illustrate the latter case, consider the input vector $\mathbf{x} = (x_1, x_2, ..., x_5)$ such that $x_1 < x_2 < ... < x_5$ and suppose we have a value y^* which minimizes the sum of absolute differences of the k highest penalties. For $k = 3$, it is clear that the sum of the largest two penalties will be $|x_5 - x_1|$ with y^* somewhere between these two values. Minimizing the third highest penalty will be achieved by placing y^* midway between x_2 and x_4.

If k is **odd**, the unique minimizer is given by

$$y = \frac{x_{(\frac{k+1}{2})} + x_{(n - \frac{k-1}{2})}}{2}.$$

Special cases

- The max function has $k = 1$ (as seen before, it results in the mid-range),
- $k = n$ results in the median function, i.e. $x_{(\frac{k+1}{2})} = x_{(n - \frac{k-1}{2})} = Med(\mathbf{x})$.

It is clear that the value of y^* has no effect on the two highest penalties, as long as it is somewhere between. If we then take $k = 4$, we will have a total penalty $|x_5 - x_1| + |x_4 - x_2|$ achieved for a number of values. We can not take *any* value over the interval $[x_2, x_4]$, however, we need to ensure that $|x_3 - y^*|$ is not greater than $|x_2 - y^*|$ or $|x_4 - y^*|$. If this were to happen, the resulting sum of the third and fourth highest penalties would be greater than $|x_4 - x_2|$.

So, if k is **even**, the minimizer can be any point

$$y_* = \frac{x_{(\frac{k}{2}+1)} + x_{(n - \frac{k}{2} + 1)}}{2} \text{ to } y^* = \frac{x_{(\frac{k}{2})} + x_{(n - \frac{k}{2})}}{2}$$

and we can take the mid-point so that the function is well defined.

- $k = n$ results in the Median function, i.e. $\frac{k}{2} = n - \frac{k}{2}$, $y_* = x_{(\frac{n}{2}+1)}$ and $y^* = x_{(\frac{n}{2})}$.

It is worth noting that the resulting value for y^* here is dependent on the relative order of the input values, not on the spacing between them as with k-th projections. The penalty-based aggregation function can then be considered as a standard OWA function (induced by input size only).

The use of penalties based on squared difference or some other form results in more complicated penalty-based functions, which can be piece-wise and perhaps easier to calculate numerically than analytically. For example, minimizing the largest three penalties of the form $(x_i - y)^2$ results in the following function:

$$ y = \begin{cases} Avg(x_{(1)}, x_{(2)}, x_{(n)}), & \text{if } Avg(x_{(1)}, x_{(2)}, x_{(n)}) < Avg(x_{(2)}, x_{(n-1)}) \\ Avg(x_{(1)}, x_{(n-1)}, x_{(n)}), & \text{if } Avg(x_{(1)}, x_{(n-1)}, x_{(n)}) > Avg(x_{(2)}, x_{(n-1)}) \\ Avg(x_{(2)}, x_{(n-1)}), & \text{otherwise.} \end{cases} $$

In other words, if the average of 3 outer values exists such that these are the furthest values, we take this average, otherwise we need the mid-point. Analogously for $k = 4$, if an average of 4 outside values exists such that these are the four furthest values, this will be the minimizer, however otherwise we need to consider mid-points of the potential fourth and fifth largest penalties (corresponding to $x_{(2)}, x_{(3)}, x_{(n-2)}, x_{(n-1)}$ depending on the data).

One could also consider sums of the smallest k penalties, however these result in multiple minimizers and will often not be well defined (for instance, taking any argument of $\mathbf{x}$ gives a 0-valued penalty if we only consider minimizing the smallest penalty. For the smallest k penalties, the minimizing values will be found by considering sequential sets $\mathscr{S}$ of size k and the corresponding $P(\mathbf{x}, y)$.

5 Conclusion

This chapter has shown some of the frameworks that are employed in the study of aggregation functions for modeling consensus. The definitions of many aggregation functions lead naturally from conceptions of consensus which may arise in practical situations. The study of such functions and their relationships to penalties is important, since their properties, their reliability and wide use, have been studied extensively. The desire for well-defined and monotone functions is not artificial, but rather is likely to arise naturally in many contexts. If a function is not well defined, there is no systematic way of dealing with ties and hence the behavior of such rules becomes unreliable. If monotonicity is lacking, anomalies in the decision process can occur, for instance, if an expert's evaluation is increased, it may not make sense that the evaluation representing the group of experts' consensus decreases.

An understanding of the relationship between expressions of penalty and penalty-based functions also allows the use of certain functions to be better interpreted. Functions such as the arithmetic mean and median are often used based on superficial considerations of extreme values or symmetric distributions, however the penalties each are associated with might be a better guiding principle in some contexts.

If a value which should represent consensus is sought, one can specify the penalties to be associated with each input, and then allow the resulting function to be determined either analytically or by means of an optimization process. Some examples where such practices are likely to be useful are in the fusion of sensor readings and the aggregation of experts' opinions. As well as weighting the individual arguments in line with some measure of importance, one can also model the penalty with a function which captures the distribution or deviation from consensus, e.g. when experts consistently over-estimate or over-evaluate.

The flexibility of aggregation functions to model such notions as interaction, importance, support and majority provides many opportunities and potential directions for formal conceptions and studies of consensus.

References

1. Alsina, C., Frank, M.J., Schweizer, B.: Associative Functions: Triangular Norms And Copulas. World Scientific, Singapore (2006)
2. Balinski, M., Laraki, R.: A theory of measuring, electing and ranking. In: Ecole Polytechnique, Centre National de la Rechereche Scientifique, Cahier (November 2006)
3. Balinski, M., Laraki, R.: Election by majority judgement: Experimental evidence. Ecole Polytechnique, pp. 2007–2028. Centre National de la Rechereche Scientifique, Cahier (2007)
4. Barral Souto, J.: El modo y otras medias, casos particulares de una misma expresion matematica. Boletin Matematico 11, 29–41 (1938)
5. Beliakov, G., Calvo, T.: Aggregation functions based on penalties. Fuzzy Sets and Systems 161(10), 1420–1436 (2010)
6. Beliakov, G., Pradera, A., Calvo, T.: Aggregation Functions: A Guide for Practitioners. Springer, New York (2007)
7. Ben-Tal, A., Charnes, A., Teboulle, M.: Entropic means. J. Math. Anal. Appl. 139, 537–551 (1989)
8. de Borda, A.C.: Mémoire sus les élections au scrutin. Historie de l'Academie Royale des Sciences(1784)
9. Bregman, L.M.: The relaxation method of finding the common point of convex sets and its application to the solution of problems in convex programming. USSR Comput. Math. Phys. 7, 200–217 (1967)
10. Bullen, P.S.: Handbook of Means and Their Inequalities. Kluwer, Dordrecht (2003)
11. Bustince, H., Herrera, F., Montero, J.: Fuzzy Sets and Their Extensions: Representation, Aggregation and Models. Springer, Heidelberg (2008)
12. Calvo, T., Kolesárová, A., Komorníková, M., Mesiar, R.: Aggregation operators: properties, classes and construction methods. In: Calvo, T., Mayor, G., Mesiar, R. (eds.) Aggregation Operators. New Trends and Applications, pp. 3–104. Physica-Verlag, Heidelberg (2002)
13. Calvo, T., Mesiar, R., Yager, R.: Quantitative weights and aggregation. IEEE Trans. on Fuzzy Systems 12, 62–69 (2004)
14. Choquet, G.: Theory of capacities. Ann. Inst. Fourier 5(1953–1954) (1953)
15. de Condercet, M.: Essai sur l'application de l'analyse à la probabilité des décisions rendues à la pluralité des voix. L'Imrimierie Royale (1785)

16. Cook, W.D., Kress, M., Seiford, L.M.: A general framework for distance-based consensus in ordinal ranking models. European Journal of Operational Research 96, 392–397 (1996)
17. Cook, W.D., Seiford, L.M.: Priority ranking and consensus formation. Management Science 24, 1721–1732 (1978)
18. Cook, W.D., Seiford, L.M.: The Borda-Kendall consensus method for priority ranking problems. Management Science 28, 621–637 (1982)
19. Dujmovic, J.J.: Characteristic forms of generalized conjunction/disjunction. In: Proc. of IEEE Intl. Conf. on Fuzzy Systems, pp. 1075–1080. Hong Kong (June 2008)
20. Fodor, J., Roubens, M.: Fuzzy Preference Modelling and Multicriteria Decision Support. Kluwer, Dordrecht (1994)
21. Freimer, M., Yu, P.L.: Some new results on compromise solutions for group decision problems. Management Science 22(6), 688–693 (1976)
22. García-Lapresta, J.L.: Favoring consensus and penalizing disagreement in group decision making. Journal of Advanced Computational Intelligence and Intelligent Informatics 12(5), 416–421 (2008)
23. Gini, C.: Le Medie. Unione Tipografico-Editorial Torinese, Milan (Russian translation, Srednie Velichiny, Statistica, Moscow, 1970) (1958)
24. González-Pachón, J., Romero, C.: Distance-based consensus methods: A goal programming approach. Omega, The International Journal of Management Science 27, 341–347 (1999)
25. González-Pachón, J., Romero, C.: Inferring consensus weights from pairwise comparison matrices without suitable properties. Annals of Operations Research 154, 123–132 (2007)
26. Grabisch, M., Marichal, J.-L., Mesiar, R., Pap, E.: Aggregation Functions (Encyclopedia of Mathematics and its Applications). Cambridge Univeristy Press, Cambridge (2009)
27. Grabisch, M., Murofushi, T., Sugeno, M. (eds.): Fuzzy Measures and Integrals. Theory and Applications. Physica-Verlag, Heidelberg (2000)
28. Herrera, F., Herrera-Viedma, E., Verdegay, J.L.: Direct approach processes in group decision making using linguistic OWA operators. Fuzzy Sets and Systems 79, 175–190 (1996)
29. Hornik, K., Meyer, D.: Deriving consensus rankings from benchmarking experiments. In: Department of Statistics and Mathematics. Research Report Series, vol. 33, Wirtschaftsuniversität Wien (2006)
30. Jackson, D.: Note on the median of a set of numbers. Bulletin of the Americam Math. Soc. 27, 160–164 (1921)
31. Kemeny, J.G., Snell, L.J.: Preference Ranking: An Axiomatic Approach. In: Mathematical Models in the Social Sciences, Ginn, New York (1962)
32. Kendall, M.: Rank Correlation Methods, 3rd edn. Hafner, New York (1962)
33. Martínez-Panero, M.: Consensus perspectives: recent theoretical advances and applications. Dep. de Economía Aplicada, PRESAD Research Group, Universidad de Valladolid, Valladolid, 47011, Spain (2006)
34. Mesiar, R.: Fuzzy set approach to the utility, preference relations, and aggregation operators. Europ. J. Oper. Res. 176, 414–422 (2007)
35. Mesiar, R., Komornikova, M., Kolesarova, A., Calvo, T.: Aggregation functions: A revision. In: Bustince, H., Herrera, F., Montero, J. (eds.) Fuzzy Sets and Their Extensions: Representation, Aggregation and Models. Springer, Heidelberg (2008)
36. Mesiar, R., Špirková, J., Vavríková, L.: Weighted aggregation operators based on minimization. Inform. Sci. 178, 1133–1140 (2008)

37. Pasi, G., Yager, R.R.: Modeling the concept of majority opinion in group decision making. Inform. Sci. 176, 390–414 (2006)
38. Torra, Y., Narukawa, V.: Modeling Decisions. Information Fusion and Aggregation Operators. Springer, Heidelberg (2007)
39. Yager, R., Rybalov, A.: Understanding the median as a fusion operator. Int. J. General Syst. 26, 239–263 (1997)
40. Yager, R.R., Filev, D.P.: Induced ordered weighted averaging operators. IEEE Transactions on Systems, Man, and Cybernetics – Part B: Cybernetics 20(2), 141–150 (1999)
41. Yu, P.L.: A class of solutions for group decision making. Management Science 19(8), 936–946 (1973)

Ranking Alternatives in Group Decision-Making with Partial Information: A Stable Approach

I. Contreras, M.A. Hinojosa, and A.M. Mármol

Abstract. The objective of the paper is to propose procedures to construct global rankings of alternatives in situations in which each member of a group is able to provide imprecise or partial information on his/her preferences about the relative importance of the criteria that have to be taken into account.

We first propose an approach based on the assumption that the final evaluation depends on the complete group since no possibility exists that the group might split into coalitions that search for more favorable solutions for the the coalitions members. To this end, the partial information on criteria weights provided by each individual is transformed into ordinal information on alternatives, and then the aggregation of individual preferences is addressed within a distance-based framework.

In a second approach, the possibility of coalition formation is considered, and the goal is to obtain rankings in which the disagreements of all the coalitions are taken into account. These rankings will exhibit an additional property of collective stability in the sense that no coalition will have the incentive to abandon the group and begin a separate evaluation process. This last approach may be of interest in political decisions where different sectors have to be incorporated into a joint evaluation process with the desire to obtain a consensus across all possible subgroups.

Keywords: Group decision-making, multiple criteria, imprecise information.

I. Contreras
Pablo de Olavide university
e-mail: iconrub@upo.es

M.A. Hinojosa
Pablo de Olavide university
e-mail: mahinram@upo.es

A.M. Mármol
Sevilla University
e-mail: amarmol@us.es

E. Herrera-Viedma et al. (Eds.): Consensual Processes, STUDFUZZ 267, pp. 41–52.
springerlink.com   © Springer-Verlag Berlin Heidelberg 2011

1 Introduction

In many group decision problems a set of alternatives must be evaluated on the basis of different and conflicting criteria which have to be taken into account in the final decision. In general, each of the group members has a particular view about the relative importance of the different criteria. The aim of the collective decision-making process is either to identify the best or most preferred alternative(s) from a set or to generate a ranking of alternatives in accordance with these individual preferences about criteria.

Recent research in the field of group decision making incorporates the possibility of dealing with imprecise preference information and permits the procedures to be applied in contexts where the group members are unable or unwilling to provide a precise representation of their preferences over alternatives. See for instance [3], where a class of flexible weight indices for ranking alternatives is proposed, and [16], where a preference aggregation method based on the estimation of utility intervals is presented.

The case of imprecise information has also been addressed for collective decisions in the multicriteria framework. A detailed revision of group decision models with imprecise information can be found, for instance, in [6]. Recent contributions are also [12], [8], [9], [10], [13], [15], [2] and [4].

In this paper we propose multicriteria collective decision procedures which consist of the construction of compromise solutions by using a distance function on the set of rankings when the group wants to rank a set of alternatives. These procedures are especially well suited for group settings where each member of the group individually provides partial information about his/her preferences with respect to the criteria under consideration, whilst not having information about the preferences of the other agents.

The final objective is the construction of a global ranking of the alternatives that combines as accurately as possible the different evaluations of the alternatives with respect to criteria by taking into account the partial information sets provided by the agents.

The first approach proposed is based on the assumption that the final evaluation depends on the complete group since no possibility exists that the group might split into coalitions in order to seek more favorable solutions for the coalition members.

In a second approach, the possibility of coalition forming is considered, and the goal is to obtain rankings in which the disagreements of all the coalitions are taken into account.

This latter approach might be of interest in political decisions where different sectors have to be incorporated into a joint evaluation process with the desire to obtain a consensus across all possible subgroups.

The rest of the paper is organized as follows. In Section 2, we introduce the collective decision-making model to be addressed. In Section 3, the procedure to obtain a final ranking of alternatives is presented, and an illustrative example is provided. In Section 4 the set of group rankings obtained by the procedure described in Section 3 is refined by taking into account the disagreement of all the coalitions. Section 5 is devoted to conclusions.

2   The Model

Let us consider a multicriteria group decision problem in which M alternatives, $X = \{x_1, \ldots, x_M\}$, have been evaluated with respect to N criteria. The evaluations of each alternative with respect to each criterion are represented by a matrix $A \in R^{N \times M}$, whose elements are denoted a_{ij} with $i = 1, \ldots, N$, $j = 1, \ldots, M$. These evaluations are assumed to be objective, in the sense that they do not depend on the assessment of the agents, but are measured independently. Hence, a_{ij} represents the cardinal value or the score given to alternative x_j with respect to the i-th criterion.

There are K Decision Makers (DMs), each of whom offers some information about his/her preferences with respect to the relative importance of the criteria.

We assume that DMs' preferences can be represented by means of an additive function. Thus, the k-th decision makers aggregated value associated to alternative x_h is given by,

$$V^k(x_h) = \sum_{i=1}^{N} w_i^k a_{ih}, \tag{1}$$

where w^k denotes a vector of weights that represents the relative importance of the criteria for agent k.

In contrast to classic approaches which consist of the elicitation of weights, these parameters need not be completely determined beforehand. We allow imprecision by permitting the values of the criteria weights for each agent to vary in partial information sets, $\Phi^k \subseteq R^N$, $k = 1, \ldots, K$.

A partial information set for an agent consists of those vectors of weights that the agent will accept as reasonable for the importance of the criteria. By convention the criteria weights are normalized to add up to one, hence, $\Phi^k \subseteq \{w^k \in R^N, \sum_{i=1}^{N} w_i^k = 1, w_i^k \geq 0, i = 1, \ldots, N\}$, for $k = 1, \ldots, K$. In particular, we will explore the cases where preference information is given by means of linear relations between the weighting coefficients. In this case, Φ^K are polyhedral sets described by linear constraints on the criteria weights.

The process of constructing the information set for each DM can be carried out in a sequential way (see [11]). The DMs can provide information by stating linear relations on the weights. For instance, they can provide partial or complete ordinal information on the importance of the criteria. Another example of representation of information by means of linear relations, which is also easily interpretable by the DM, is when the DM declares a preference of alternative x_h over x_j. This implies that x_h should not be ranked below x_j in any ranking of alternatives induced by the individual preferences of this particular decision-maker. Hence, a relation $\sum_{i=1}^{N} w_i^k a_{ih} - \sum_{i=1}^{N} w_i^k a_{ij} > 0$ must be incorporated into Φ^k.

3 Rankings Minimizing Global Disagreement

The main idea of the procedure proposed here is the achievement of a final consensus or compromise solution between DMs from the individual rankings of alternatives induced by the partial information sets. Implicit in this problem is the existence of a measure of global agreement or disagreement between rankings. Therefore, the approach implies the introduction of a distance function on the set of rankings in order to determine the ranking that minimizes the total distance across DMs. A detailed study of models based on distance functions can be seen in [5] and [7].

A priority vector or a ranking on the set of alternatives $X = \{x_1 \ldots, x_M\}$ is represented by a vector $R = (r_1, \ldots, r_M)$, where r_j denotes the ordinal position assigned to alternative x_j. As standard, the first category is assigned to the most preferred alternative and when ties occur, the average of the values corresponding to the tied alternatives is assigned. Let R_k and R_s be two priority vectors that represent the rankings of the alternatives corresponding to DMs k and s.

The following distance, which is based on the L_1-metric, is considered on the set of priority vectors:

$$d(R_k, R_s) = \sum_{j=1}^{M} |r_{kj} - r_{sj}| \tag{2}$$

In this setting, a consensus vector or a group ranking, R_G, is a priority vector that satisfies

$$\begin{aligned} min \ &\sum_{k=1}^{K} d(R_k, R_G) \\ s.t. \quad &R_G \in \mathscr{R} \end{aligned} \tag{3}$$

where $\mathscr{R}$ is the set of vectors that represent priority vectors or rankings. Notice that the consensus vector, R_G, is not necessarily unique.

It is important to point out that the consideration of partial information sets to represent DMs' preferences implies that, from an individual point of view, agent k would accept an evaluation of alternative x_h consisting of $V^k(x_h) = \sum_{i=1}^{N} w_i^k a_{hi}$ if $w^k \in \Phi^k$. Hence, every ranking of alternatives that can be achieved from the dominance relations induced by any $w^k \in \Phi^k$ is considered acceptable by the k-th DM. As a consequence, different rankings of alternatives (at least one, otherwise $\Phi^k = \emptyset$) can be induced by each DM's preferences. Therefore, the procedure has to include not only an objective for the group in order to determine the ranking that best agrees with the individual preferences, but also a selection criterion to choose a ranking for each DM which represents the individual preferences (those rankings that best agree with the group order).

A dominance relationship between alternatives can be derived from (1). For a fixed vector of weights, $w^k \in \Phi^k$, we can say that alternative x_j strictly dominates alternative x_h under the preference structure of the k-th DM, if $V^k(x_j) - V^k(x_h) > 0$. The consensus ranking is induced here by the ordinal positions of the alternatives, determined by comparing the aggregated values $V^k(x_i)$ for the different values of $w^k \in \Phi^k$.

Hence, an intermediate step consisting of the elicitation of an individual ranking for each DM is required. To this end, we consider the following set of constraints, where δ_{ij} and γ_{ij} are binary variables:

$$
\begin{aligned}
V^k(x_h) - V^k(x_j) + \delta_{hj}^k B \geq 0, \forall h \neq j, \\
V^k(x_h) - V^k(x_j) + \gamma_{hj}^k B \geq \varepsilon, \forall h \neq j.
\end{aligned}
\tag{4}
$$

Here B is a large number and ε is a discriminating factor between alternatives, such that we say alternative x_h strictly dominates x_j if $V^k(x_h) - V^k(x_j) \geq \varepsilon$.

The values of variables δ_{hj}^k will be equal to one each time that x_h is strictly dominated by x_j, i.e., whenever $V^k(x_j) > V^k(x_h)$. In contrast, γ_{hj}^k will be one whenever $V^k(x_j) \geq V^k(x_h)$, that is, each time x_h is not preferred to x_j in the k-th DM preferences. These variables are included in the inequalities in order to reflect the ties between alternatives.

The ranking position induced by the aggregate values $V^k(x_h)$ can be obtained as the sum of the binary variables divided by 2 plus one. That is,

$$
r_{kh} = \sum_{h \neq j} \frac{\delta_{hj}^k + \gamma_{hj}^k}{2} + 1, \forall h = 1, \dots, M.
\tag{5}
$$

To guarantee that the vectors $R_k = (r_{k1}, \dots, r_{kM})$ represent priority vectors, we incorporate the following set of constraints

$$
\delta_{hj}^k + \delta_{jh}^k \leq 1, \forall h \neq j,
\tag{6}
$$

$$
\delta_{hj}^k + \gamma_{jh}^k = 1, \forall h \neq j.
\tag{7}
$$

Constraints (6) and (7) are necessary in order to induce the ranking of alternatives from the values $V^k(x_i)$. These constraints guarantee that the values of the binary variables δ_{hj}^k and γ_{kj}^k are correct, in the sense that only when x_j strictly dominates x_h, then $\delta_{hj}^k = 1$ holds, and consequently $\delta_{jh}^k = 0$. The values of γ_{hj}^k also depend on the value assigned to δ_{hj}^k and represent ties between alternatives. It is worth noting that in each constraint we have to consider, not only the dominance relation of x_h over x_j, but also the relation of x_j over x_h.

Finally, the following requirement is included

$$
\sum_{j=1}^{M} r_{kj} = \frac{M(M+1)}{2}, \forall k = 1, \dots, K.
\tag{8}
$$

In addition to the above constraints, the condition in (8) assures that vector R_k is contained in set $\mathscr{R}$, that is to say, represents a priority vector.

In order to obtain the compromise solution, a set of variables corresponding to the group has to be considered: a weighting vector, aggregate values and a ranking of alternatives for the group. Hence, a new agent labelled as the subindex G in the set of DMs[1] is included in the model.

To determine the group solution we have to deal with the distance defined in (2). The minimization of the sum of these nonlinear functions, as stated in (3) can be reduced to a linear programming model by considering a Goal Programming formulation. By taking into account the following change of variables proposed in [1] and [7],

$$
\begin{aligned}
\alpha_{kj} &= \tfrac{1}{2}\left[|r_{kj} - r_{Gj}| - (r_{kj} - r_{Gj})\right] \\
\beta_{kj} &= \tfrac{1}{2}\left[|r_{kj} - r_{Gj}| + (r_{kj} - r_{Gj})\right],
\end{aligned}
\tag{9}
$$

we will include the following set of constraints in the model to measure the distance between individual priority vectors and the compromise ranking.

$$
r_{kj} - r_{Gj} + \alpha_{kj} - \beta_{kj} = 0, \forall k = 1,\dots,K; j = 1,\dots,N.
\tag{10}
$$

Therefore, the distance between R_k and R_G can be represented equivalently by the following expression

$$
d(R_k, R_G) = \sum_{j=1}^{M} (\alpha_{kj} + \beta_{kj}).
\tag{11}
$$

All the specifications defined above yield the following model

$$
\begin{aligned}
min \ & \sum_{k=1}^{K} \sum_{j=1}^{M} (\alpha_{kj} + \beta_{kj}) \\
s.t. \ & V^k(x_h) = \sum_{i=1}^{N} w_i^k a_{ih}, & \forall h,k \\
& V^k(x_h) - V^k(x_j) + \delta_{hj}^k B \geq 0, & \forall h \neq j, \forall k \\
& V^k(x_h) - V^k(x_j) + \gamma_{hj}^k B \geq \varepsilon, & \forall h \neq j, \forall k \\
& \delta_{hj}^k + \delta_{jh}^k \leq 1, & \forall h \neq j, \forall k \\
& \delta_{hj}^k + \gamma_{jh}^k = 1, & \forall h \neq j, \forall k \\
& r_{kh} = \sum_{h \neq j} \frac{\delta_{hj}^k + \gamma_{hj}^k}{2} + 1, & \forall h,k \\
& \sum_{h=1}^{M} r_{kh} = \frac{M(M+1)}{2} & \forall k \\
& r_{kh} - r_{Gh} + \alpha_{kh} - \beta_{kh} = 0, & \forall h, k \neq G \\
& w^k \in \Phi^k, & \forall k \\
& \delta_{hj}^k, \gamma_{hj}^k \in \{0,1\}, & \forall h,j,k,
\end{aligned}
\tag{12}
$$

where $B, \varepsilon \geq 0$.

By solving (12), the compromise ranking R_G is obtained, together with a ranking of alternatives for each DM. The group ranking is such that the total disagreement is minimized from among those disagreements induced by the individual preferences. This ranking is determined by means of the aggregate values $V^k(x_i)$, hence the vector of weights which best agrees with the group solution is selected for each DM.

---

[1] Subindexes k, that represent DMs, will now vary in the set $\{1,\dots,K,G\}$.

The aggregated values for the group, $V^G(x_h)$, are only considered for computational purposes in order to induce a ranking of alternatives for the group in the same format in which individual rankings have been constructed and have no interpretation from a cardinal point of view.

Some desirable properties of our procedure in the context of social choice processes which are a direct consequence of its construction are: feasibility, anonymity, neutrality and no dictatorship. That is to say, a compromise solution is always obtainable, all the group members are treated equally, all the alternatives are treated equally, and no DM exists whose individual preferences determine the consensus ranking.

3.1 *Illustrative Example*

This section illustrates the proposed procedure for group decision-making problems. We have considered an example in which four decision-makers want to decide between five alternatives denoted by $\{x_1, \ldots, x_5\}$. The alternatives have been evaluated with respect to four different criteria. Table 1 shows the scores of the alternatives with respect to each criterion. These evaluations have been normalized so that the sum with respect to each criterion add up to one.

Table 1 Matrix of scores

Alternative	Crit. 1	Crit. 2	Crit. 3	Crit. 4
x_1	0.5776	0.1692	0.0481	0.1612
x_2	0.2054	0.2900	0.0340	0.2070
x_3	0.1143	0.1692	0.3107	0.2200
x_4	0.0717	0.3052	0.4489	0.1765
x_5	0.0310	0.0664	0.1583	0.2353

Four agents are to evaluate this information. They have provided partial information about the relative importance they assign to each criteria.

Denote the set of criteria weights as

$$\Lambda^+ = \{w \in R^4, \ \sum_{k=1}^{4} w_k = 1, w_k \geq 0, k = 1, \ldots, 4\}.$$

Agent 1 considers that the importance of the criteria is ranked in order of decreasing magnitude. In addition, this agent states that the weight of criterion 4 is not less than 1% of the total. Therefore, the partial information set for agent 1, can be formalized as

$$\Phi^1 = \{w^1 \in \Lambda^+, w_1^1 \geq w_2^1 \geq w_3^1 \geq w_4^1 \geq 0.01\}$$

The preference information about the criteria corresponding to the remaining agents are represented in the following information sets and can be interpreted in a similar way to that above.

$$\Phi^2 = \{w^2 \in \Lambda^+, w_2^2 \geq w_1^2 + w_3^2, w_4^2 \geq w_1^2 + w_3^2, w_1^2 \geq 0.01, w_3^2 \geq 0.01\},$$
$$\Phi^3 = \{w^3 \in \Lambda^+, w_4^3 \geq 2w_3^3 \geq 4w_2^3 \geq 8w_1^3 \geq 0.08\},$$
$$\Phi^4 = \{w^4 \in \Lambda^+, w_3^4 \geq w_1^4 + w_2^4 + w_4^4, w_1^4 \geq 0.01, w_2^4 \geq 0.01, w_3^4 \geq 0.01\}.$$

The solution to problem (12) yields a level of disagreement equal to 8 units. However, the collective ranking which generates this level of disagreement is not unique. Table 2 shows the group solutions that minimize the total disagreement, represented by their respective priority vectors, and the individual disagreements each solution generates.

Table 2 Group rankings

Priority vector (R_G)	$d(R_1,R_G)$	$d(R_2,R_G)$	$d(R_3,R_G)$	$d(R_4,R_G)$
(5, 4, 2, 1, 3)	6	2	0	0
(2, 4, 3, 1, 5)	0	2	6	0
(5, 3, 2, 1, 4)	6	0	2	0
(4.5, 3, 2, 1, 4.5)	5	0	3	0
(5, 3.5, 2, 1, 3.5)	6	1	1	0
(3.5, 3.5, 2, 1, 5)	3	1	4	0

It is interesting to note that the individual disagreement of each solution can vary from one collective ranking to another. Note that we only fix the total disagreement at its minimum level which, in this case, is 8.

4 Stable Rankings across Coalitions

Unfortunately, as shown in the previous example, the procedure described in Section 3 does not always provide a unique group ranking which minimizes the level of disagreement of the DMs. In this section we present a procedure to refine the set of group rankings which is inspired by cooperative game theory.

The individual disagreement associated to the agents for each collective ranking can be seen as an allocation of the total disagreement. The procedure described in Section 3 may provide several collective rankings which minimize the total disagreement and, in addition, there may be different sets of individual rankings associated to the same collective ranking, therefore several allocations among the DMs of the minimum level of disagreement may exist.

Let $y = (y_1, y_2, \ldots, y_K)$ denote any of such allocations, that is to say, for each $k = 1, 2, \ldots, K$, $y_k = d(R_k, R_G)$, where R_G is one of the group rankings that minimizes the total disagreement and R_k is one of the associated individual rankings. Thus, when we choose a collective ranking, R_G, and a set of associated individual rankings, $R_k, k = 1, 2, \ldots, K$, we are assigning a disagreement $y_S = \sum_{k \in S} y_k$ to each subgroup (or coalition) $S \subseteq \mathcal{K} = \{1,\ldots,K\}$. On the other hand, the procedure described in Section 3 can be applied, not only to the whole set of DMs, but also to each coalition $S \subset \mathcal{K}$. In this way a minimum level of disagreement, D_S^*, is obtained for each coalition $S \subseteq \mathcal{K}$, by taking into account the preferences of the DMs in S. The level D_S^* represents the minimum disagreement with a collective ranking that the members of S can achieve if they are in their own.

For each coalition $S \subseteq \mathcal{K}$, we consider the difference $y_S - D_S^*$, which is a measure of the dissatisfaction of coalition S with the collective ranking R_G which provides allocation y. The idea is to choose from among all the collective rankings for which the the minimum level of disagreement for the whole set of DMs is achieved, those that lexicographically minimize the maximum dissatisfaction of the coalitions. If any vector $y = (y_1, y_2, \ldots, y_K)$, $y_k \geq 0$, $y = 1, 2, \ldots, K$, such that $y_{\mathcal{K}} = D_{\mathcal{K}}^*$ could be chosen as an allocation of the collective disagreement, then the problem to solve would be (13).

$$lex - min \ max_{S \subset \mathcal{K}} \ \{y_S - D_S^*\} \qquad (13)$$
$$s.t. \qquad y_{\mathcal{K}} = D_{\mathcal{K}}^*$$

This problem has a unique solution, the pre-nucleolus of the cooperative coalitional game defined by assigning the minimum disagreement, D_S^*, to each coalition $S \subseteq \mathcal{K}$ (see [14]).

Unfortunately, in general, a collective ranking, R_G, and a set of associated individual rankings, R_k, $k = 1, 2, \ldots, K$, for which the corresponding allocation $y = (d(R_k, R_G))_{k \in \mathcal{K}}$ coincides with the pre-nucleolus of the game, do not always exist. Nevertheless, it is always possible to find rankings for which allocation $y = (d(R_k, R_G))_{k \in \mathcal{K}}$ approximates y^* by solving Problem (14).

$$lex - min \ max_{S \subset \mathcal{K}} \ \{y_S - D_S^*\}$$
$$s.t. \qquad d(R_k, R_G) = y_k, \quad \forall k \in \mathcal{K}$$
$$y_{\mathcal{K}} = D_{\mathcal{K}}^* \qquad (14)$$
$$R_k \in \mathcal{R}(\phi^k)$$
$$R_G \in \mathcal{R}(\phi^G)$$

Moreover, in many cases the collective ranking obtained from this problem is unique.

4.1 Example (continued from Section 3.1)

The values of the minimum disagreement of the coalitions are shown in Table 3.

In order to obtain the solution of Problem (13) for this example, in a first step, the following linear problem is solved:

$$min \ \mu$$
$$s.t. \ y_S - D_S^* \leq \mu \quad \forall S \subset \mathcal{K} \qquad (15)$$
$$y_{\mathcal{K}} = D_{\mathcal{K}}^*.$$

Table 3 Minimum disagreements of the coalitions

Coalition (S)	Disagreement (D_S^*)	Coalition (S)	Disagreement (D_S^*)
$\mathcal{K} = \{1,2,3,4\}$	8	$\{2,3\}$	2
$\{1,2,3\}$	8	$\{2,4\}$	0
$\{1,2,4\}$	2	$\{3,4\}$	0
$\{1,3,4\}$	6	$\{1\}$	0
$\{2,3,4\}$	2	$\{2\}$	0
$\{1,2\}$	0	$\{3\}$	0
$\{1,3\}$	6	$\{4\}$	0
$\{1,4\}$	0		

The solution of this problem is $\mu = 4$ and the constraints corresponding to coalitions $\{1,2\}$ and $\{3,4\}$ are active for each allocation y associated to the solution (it is worth noting that for at least one of the coalitions this is always true). The next step consists of solving Problem (16).

$$
\begin{aligned}
&min \ \mu \\
&s.t. \ y_1 + y_2 = 4 \\
&\quad\ \ y_3 + y_4 = 4 \\
&\quad\ \ y_S - D_S^* \leq \mu \quad \forall S \subset \mathcal{K}, S \neq \{1,2\}, \{3,4\} \\
&\quad\ \ y_{\mathcal{K}} = D_{\mathcal{K}}^*.
\end{aligned}
\tag{16}
$$

The optimal value for this problem is $\mu = \frac{10}{3}$, and there exists a unique allocation associated to the optimal solution, $y^* = (\frac{8}{3}, \frac{4}{3}, \frac{10}{3}, \frac{2}{3})$. The allocation y^* is called the pre-nucleolus of the cooperative coalitional game induced by the minimum disagreements of the coalitions.

Since this allocation may not correspond to the disagreement between a collective ranking and a set of individual rankings of the agents, Problem (14) has to be solved.

In this case, the solution to this last problem provides a vector of individual disagreements $y = (3,1,4,0)$, which is as close as possible to the pre-nucleolus. Note that in Problem (14) only distances between rankings are considered.

Table 4 and Table 5 summarize the results obtained. In Table 4 the weighting vector selected for each individual partial information set is shown. Table 5 includes the individual rankings induced from these weighting vectors (note that this ranking is constructed throughout the aggregated values $V^k(x_i)$), the group ranking obtained, and the individual distances from the group ranking R_G.

The final consensus ranking is unique, $R_G^* = (3.5, 3.5, 2, 1, 5)$. Therefore, in this case, the group solution establishes that alternative x_4 is ranked at the first position, followed by x_3 in the second position. Alternatives x_1 and x_2 are tied in the third position and the least-valued alternative is x_5.

Table 4 Individual weighting vectors

	Crit 1	Crit 2	Crit 3	Crit 4
w^1	0,330	0,330	0,329	0,010
w^2	0,050	0,438	0,231	0,281
w^3	0,010	0,020	0,323	0,647
w^4	0,108	0,392	0,500	0,000

Table 5 Individual rankings

	Rank positions				
	DM 1	DM 2	DM 3	DM 4	Group
x_1	2	4	5	3.5	3.5
x_2	4	3	4	3.5	3.5
x_3	3	2	2	2	2
x_4	1	1	1	1	1
x_5	5	5	3	5	5
Disagreement	3	1	4	0	

5 Conclusions

We have proposed a compromise method for collective decision problems which is especially suited for situations where the members of the group provide partial or imprecise information about their preferences with respect to the criteria, and there is no flow of information between these members. In this context the group members are not encouraged to misrepresent their true preferences in order to manipulate the group decision. Hence, the result derived could be considered a fair representation of the group evaluation of all the alternatives.

Imprecise information is formalized by means of linear relations between criteria weights. This way of providing preferential information is one of the most easily interpreted by the DMs, and includes interesting particular cases such as those in which they only provide ordinal information on the weights.

The basic procedure relies on the transformation of the partial information about criteria into ordinal information about alternatives, expressed through rankings. A model is constructed that provides a set of compromise rankings for the group and an additional step enables the achievement of a unique compromise ranking which is stable across coalitions.

The approach presented here uses a measure of agreement based upon the L_1-metric and, therefore, emphasizes the sum of individual disagreements with respect to every alternative. However, the use of other metrics could also be considered in order to analyze this class of collective decision making problems.

Acknowledgements. This research has been partially financed by the Spanish Ministry of Education and Science project SEJ2007-62711/ECON and by Consejería de Innovación, Ciencia y Tecnología (Junta de Andalucía) project P06-SEJ-01801.

References

1. Charnes, A., Cooper, W.W.: Goal Programming and multiple objective optimization. Part 1. European Journal of Operational Research 1, 39–54 (1977)
2. Climaco, J.N., Dias, L.C.: Negotiation Processes with Imprecise Information on Multi-criteria Additive Models. Group Decision and Negotiation 15, 171–184 (2006)
3. Contreras, I., Hinojosa, M.A., Mármol, A.M.: A class of flexible weight indices for ranking alternatives. IMA Journal of Management Mathematics 16, 71–85 (2005)
4. Contreras, I., Mármol, A.M.: A lexicographical compromise method for multiple criteria group decision problems with imprecise information. European Journal of Operational Research 181(3), 1530–1539 (2007)
5. Cook, W.D., Kress, M., Seiford, L.M.: A general framework for distance-based consensus in ordinal ranking models. European Journal of operational Research 96, 392–397 (1996)
6. Dias, L.C., Climaco, J.N.: Dealing with imprecise information in group multicriteria decisions: A methodology and a GDSS architecture. European Journal of Operational Research 160(2), 291–307 (2005)
7. González-Pachón, J., Romero, C.: Distanced based consensus methods: a goal programming approach. Omega 27, 341–347 (1999)
8. Kim, S.H., Ahn, B.S.: Group decision making procedure considering preference strength under incomplete information. Computers and Operations Research 12, 1101–1112 (1997)
9. Kim, S.H., Choi, S.H., Kim, J.K.: An interactive procedure for multi-attribute group decision making with incomplete information: Range-based approach. European Journal of Operational Research 118, 139–152 (1999)
10. Malakooti, B.: Ranking and Screening Multiple Criteria Alternatives with Partial Information and Use of Ordinal and Cardinal Strengths of Preferences. IEEE Transactions on Systems, Man, and Cybernetics Part A 30, 355–367 (2000)
11. Mármol, A.M., Puerto, J., Fernández, F.R.: Sequential incorporation of imprecise information in multiple criteria decision processes. European Journal of Operational Research 137, 123–133 (2002)
12. Salo, A.A.: Interactive decision aiding for group decision support. European Journal of Operational Research 84, 134–149 (1995)
13. Salo, A., Hämäläinen, R.P.: Preference ratios in multiattribute evaluation (PRIME)elicitation and decision procedures under incomplete information. IEEE Transactions on Systems, Man and Cybernetics 31, 533–545 (2001)
14. Schmeidler, D.: The nucleolus of a characteristic function game. SIAM Journal of Applied Mathematics 17, 1163–1170 (1969)
15. Valadares Tavares, L.: A model to support the search for consensus with conflicting rankings: Multitrident. International Transactions in Operational Research 11, 107–115 (2004)
16. Wang, Y.M., Yang, J.B., Xu, D.L.: A preference aggregation method through the estimation of utility interval. Computers and Operations Research 32, 2027–2049 (2005)

Opinion Changing Aversion Functions for Group Settlement Modeling

Matteo Brunelli, Robert Fullér, and József Mezei

Abstract. Opinion changing aversion (OCA) functions are used to quantify the decision makers' resistance to opinion changing. By introducing OCA functions of polynomial form we will show that if each expert has a quadratic opinion changing aversion function then the minimum-cost solution is nothing else but the weighted average of the individual optimal solutions where the weights are the relative importances of the decision makers. We will consider minimum-cost solutions for group settlements under crisp and fuzzy budget constraints.

Introduction

The challenge of group decision is deciding what action a group should take. Decision makers are invited to express their opinions/preferences on a set of alternatives. Their preferences can be elicited either by asking them to pairwise compare alternatives using suitable semantic scales [17, 19, 20], or by means of priority vectors, namely vectors whose components are scores of alternatives. Priority vectors are cardinal rankings (ratings) and, to our aim, we do not need to set any constraint about their interpretation and their admissible values. Therefore, given a rating, say

Matteo Brunelli
IAMSR and TUCS, Åbo Akademi University, Joukahainengatan 3-5A,
FIN-20520 Åbo, Finland
e-mail: matteo.brunelli@abo.fi

Robert Fullér
IAMSR, Åbo Akademi University, Joukahainengatan 3-5A,
FIN-20520 Åbo, Finland
e-mail: robert.fuller@abo.fi

József Mezei
IAMSR and TUCS, Åbo Akademi University, Joukahainengatan 3-5A,
FIN-20520 Åbo, Finland
e-mail: jmezei@abo.fi

E. Herrera-Viedma et al. (Eds.): Consensual Processes, STUDFUZZ 267, pp. 53–63.
springerlink.com   © Springer-Verlag Berlin Heidelberg 2011

$w = (w_1, w_2, \ldots, w_k)$, on a set of k alternatives, the scale of admissible values can be either unipolar or bipolar, but also ratio, interval, absolute and ordinal as we only require that a real number representing a score or an utility level is associated to each alternative, i.e. $w \in \mathbb{R}^k$. Note that, in the case of an ordinal scale, scores can be associated to alternatives, for instance, by using the Borda count method [2].

An important step in obtaining a collective representation of preferences of a group of decision makers is the aggregation process. A large number of papers has been devoted to discuss this issue and various proposals have been made. Normally, preference aggregation can be performed in two different ways: (i) by aggregating individual judgments (AIJ) and (ii) by aggregating individual preferences (AIP) [11]. The difference between them, is that the first one includes all those models performing the aggregation of preferences expressed by means of pairwise comparison matrices whereas the second one includes all those methods that apply priority vectors.

In this paper we will solely focus on AIP models. Our choice can be justified by the fact that, if the decision maker is perfectly consistent, then a pairwise comparison matrix and its associated priority vector are just two different ways for expressing the same opinions over a set of alternatives and also when the decision maker is not fully consistent but reasonably close to being such, the priority vector is assumed to be a sufficiently coherent estimation of his/her preferences. Last but not least, the handling of the ratings is easier and computationally less demanding than performing similar operations on pairwise comparison matrices.

It might be useful to divide preference aggregation models into two families, with respect to the nature of the underlying model. The first one is to employ *aggregation functions* like the weighted arithmetic and geometric means as suggested in [18, 11] or the centroid, or center of mass of the preferences as suggested in [13]. To this class belongs also [1] where the authors suggested the middlemost value instead of the average values. The second family consists of all those methods where the collective representation of preferences is obtained by solving a usually non-linear *optimization problem* (see [16] as examples). As a rule, this optimization method is used to minimize the total level of disagreement and an optimal – or *consensual* [14] solution – can be said to be the less disagreed one.

Whenever such a consensual solution has to be found, a single valued, nonnegative cost function is assigned to each decision maker and then the overall cost is minimized. These cost functions are called opinion changing aversion (OCA) functions [15]. Several different cost functions have been proposed in the literature, e.g. Hamming distance, Euclidean distance, sigmoid functions. In this paper we will focus on quadratic OCA functions and show that the group decision (or settlement) will be nothing else but the center of gravity of the opinions of the decision makers.

1 Opinion Changing Aversion Functions

In the early nineties, Mich et al. [15] introduced a new measure of consensus depending on a function estimated for each expert according to her/his aversion to opinion changing. These functions have become known in the literature as opinion changing aversion (OCA) functions, and have been further developed in [3, 7, 8, 12]. More often than not, the group decision is heavily influenced by the degree of importance of participants. For example, the opinions of executives should be more reflected in the final conclusion of group decision. Yet, it could be also reasonable to assume that the importance of decision makers should be somehow related with their degrees of consistency [9, 10]. Therefore, we are now considering the relative importance weight of each expert. Let the degree of importance of i-th expert be β_i, then his/her relative degree of importance is

$$\frac{\beta_i}{\beta_1 + \beta_2 + \cdots + \beta_n}.$$

Let (a_i, b_i, c_i) denote the best possible settlement (or ideal solution) for the j-th expert. For simplicity, we will consider only three alternatives/criteria, but our results can easily be derived for the general case. We will use the following family of polynomial OCA functions,

$$\mathscr{C}_i^k(x,y,z) = \frac{\beta_i}{\beta_1 + \beta_2 + \cdots + \beta_n} \times \frac{(x-a_i)^{2k} + (y-b_i)^{2k} + (z-c_i)^{2k}}{M}, \quad (1)$$

where $k \in \mathbb{N}$ and M is defined by,

$$M = \max_{i,j}\{(a_i - a_j)^{2k} + (b_i - b_j)^{2k} + (c_i - c_j)^{2k}\}.$$

It is clear that $\mathscr{C}_i^k$ satisfies the conditions of an OCA function since $\mathscr{C}_i^k(a_i, b_i, c_i) = 0$ and the directional derivative of $\mathscr{C}_i^k$ is positive in any direction at point (a_i, b_i, c_i). However, for computational purposes we shall consider only the simplest polynomial OCA function, the quadratic one, defined by

$$\mathscr{C}_i(x,y,z) = \frac{\beta_i}{\beta_1 + \beta_2 + \cdots + \beta_n} \times \frac{(x-a_i)^2 + (y-b_i)^2 + (z-c_i)^2}{M}, \quad (2)$$

where M is defined by,

$$M = \max_{i,j}\{(a_i - a_j)^2 + (b_i - b_j)^2 + (c_i - c_j)^2\}.$$

Here $\mathscr{C}_i(x,y,z)$ measures the aversion of the i-th expert to changing her/his opinion from (a_i, b_i, c_i) to (x,y,z), for $i = 1, \ldots, n$.

2   Allocation of the Optimal Unconditional Settlement

The optimal strategy minimizes the overall aversion to opinion changing, and it can be computed by solving the following optimization problem,

$$\mathscr{C}(x,y,z) = \sum_{i=1}^{n} \mathscr{C}_i(x,y,z) \to \min; \text{ subject to } x,y,z \in \mathbb{R} \qquad (3)$$

If we use quadratic OCA functions of the form (2) then the minimization problem can be stated as,

$$\sum_{i=1}^{n} \frac{\beta_i}{\beta_1 + \beta_2 + \cdots + \beta_n} \times \frac{(x-a_i)^2 + (y-b_i)^2 + (z-c_i)^2}{M} \to \min,$$

which can be written in the form,

$$\frac{1}{M(\beta_1 + \beta_2 + \cdots + \beta_n)} \times \sum_{i=1}^{n} \beta_i \left[(x-a_i)^2 + (y-b_i)^2 + (z-c_i)^2 \right] \to \min$$

Using the Lagrange multiplier method we can easily find,

$$\frac{\partial \mathscr{C}(x,y,z)}{\partial x} = 2 \sum_{i=1}^{n} \beta_i (x-a_i) = 0 \Rightarrow x^* = \frac{a_1\beta_1 + a_2\beta_2 + \cdots + a_n\beta_n}{\beta_1 + \beta_2 + \cdots + \beta_n}$$

$$\frac{\partial \mathscr{C}(x,y,z)}{\partial y} = 2 \sum_{i=1}^{n} \beta_i (y-b_i) = 0 \Rightarrow y^* = \frac{b_1\beta_1 + b_2\beta_2 + \cdots + b_n\beta_n}{\beta_1 + \beta_2 + \cdots + \beta_n}$$

$$\frac{\partial \mathscr{C}(x,y,z)}{\partial z} = 2 \sum_{i=1}^{n} \beta_i (z-c_i) = 0 \Rightarrow z^* = \frac{c_1\beta_1 + c_2\beta_2 + \cdots + c_n\beta_n}{\beta_1 + \beta_2 + \cdots + \beta_n}$$

That is, the unique optimal solution to minimization problem (3) becomes

$$\begin{pmatrix} x^* \\ y^* \\ z^* \end{pmatrix} = \begin{pmatrix} \dfrac{a_1\beta_1 + a_2\beta_2 + \cdots + a_n\beta_n}{\beta_1 + \beta_2 + \cdots + \beta_n} \\[2ex] \dfrac{b_1\beta_1 + b_2\beta_2 + \cdots + b_n\beta_n}{\beta_1 + \beta_2 + \cdots + \beta_n} \\[2ex] \dfrac{c_1\beta_1 + c_2\beta_2 + \cdots + c_n\beta_n}{\beta_1 + \beta_2 + \cdots + \beta_n} \end{pmatrix}$$

which is nothing else but the center of gravity of the experts' initial opinions. We show now some simple special cases.

Example 1. Suppose that we have a four-expert two-issue problem in which the ideal solutions and, consequently, the quadratic OCA functions for the experts are defined by

First expert: The ideal solution is $(a_1, b_1) = (1,1)$ and

$$\mathscr{C}_1(x,y) = \frac{\beta_1}{\beta_1 + \beta_2 + \beta_3 + \beta_4} \times \frac{(x-1)^2 + (y-1)2}{2}.$$

Second expert: The ideal solution is $(a_2, b_2) = (1,0)$ and

$$\mathscr{C}_2(x,y) = \frac{\beta_2}{\beta_1 + \beta_2 + \beta_3 + \beta_4} \times \frac{(x-1)^2 + y^2}{2}.$$

Third expert: The ideal solution is $(a_3, b_3) = (0,1)$ and

$$\mathscr{C}_3(x,y) = \frac{\beta_3}{\beta_1 + \beta_2 + \beta_3 + \beta_4} \times \frac{x^2 + (y-1)^2}{2}$$

Fourth expert: The ideal solution is $(a_4, b_4) = (0,0)$ and

$$\mathscr{C}_4(x,y) = \frac{\beta_4}{\beta_1 + \beta_2 + \beta_3 + \beta_4} \times \frac{x^2 + y^2}{2}.$$

Then problem (3) collapses into,

$$\mathscr{C}(x,y) = \sum_{i=1}^{4} \mathscr{C}_i(x,y) \rightarrow \min; \text{ subject to } x, y \in [0,1]$$

and we can easily compute that,

$$\frac{\partial \mathscr{C}(x,y)}{\partial x} = 2(\beta_1 + \beta_2)(x-1) + 2(\beta_3 + \beta_4)x = 0 \Rightarrow x^* = \frac{\beta_1 + \beta_2}{\beta_1 + \beta_2 + \beta_3 + \beta_4}$$

$$\frac{\partial \mathscr{C}(x,y)}{\partial y} = 2(\beta_1 + \beta_3)(y-1) + 2(\beta_2 + \beta_4)y = 0 \Rightarrow y^* = \frac{\beta_1 + \beta_3}{\beta_1 + \beta_2 + \beta_3 + \beta_4}$$

and the optimal cost is,

$$\mathscr{C}^* = \frac{1}{2(\beta_1 + \beta_2 + \beta_3 + \beta_4)^3} \Big((\beta_3 + \beta_4)(\beta_1 + \beta_2)^2 + (\beta_1 + \beta_2)(\beta_3 + \beta_4)^2$$

$$+ (\beta_2 + \beta_4)(\beta_1 + \beta_3)^2 + (\beta_1 + \beta_3)(\beta_2 + \beta_4)^2 \Big)$$

Example 2. Suppose that we have an eight-expert three-issue problem in which the ideal solutions and, consequently, the quadratic OCA functions for the experts are defined by

(1,1,1):

$$\mathscr{C}_1(x,y,z) = \frac{\beta_1}{\sum_i \beta_i} \frac{(x-1)^2 + (y-1)^2 + (z-1)^2}{3}$$

(1,1,0):

$$\mathscr{C}_2(x,y,z) = \frac{\beta_2}{\sum_i \beta_i} \frac{(x-1)^2 + (y-1)^2 + z^2}{3}$$

(1,0,1):

$$\mathscr{C}_3(x,y,z) = \frac{\beta_3}{\sum_i \beta_i} \frac{(x-1)^2 + y^2 + (z-1)^2}{3}$$

(0,1,1):

$$\mathscr{C}_4(x,y,z) = \frac{\beta_4}{\sum_i \beta_i} \frac{x^2 + (y-1)^2 + (z-1)^2}{3}$$

(0,0,1):

$$\mathscr{C}_5(x,y,z) = \frac{\beta_5}{\sum_i \beta_i} \frac{x^2 + y^2 + (z-1)^2}{3}$$

(0,1,0):

$$\mathscr{C}_6(x,y,z) = \frac{\beta_6}{\sum_i \beta_i} \frac{x^2 + (y-1)^2 + z^2}{3}$$

(1,0,0):

$$\mathscr{C}_7(x,y,z) = \frac{\beta_7}{\sum_i \beta_i} \frac{(x-1)^2 + y^2 + z^2}{3}$$

(0,0,0):

$$\mathscr{C}_8(x,y,z) = \frac{\beta_8}{\sum_i \beta_i} \frac{x^2 + y^2 + z^2}{3}$$

Then problem (3) collapses into,

$$\mathscr{C}(x,y,z) = \sum_{i=1}^{8} \mathscr{C}_i(x,y,z) \to \min; \text{ subject to } x,y,z \in [0,1]$$

and we can easily compute that,

$$\frac{\partial \mathscr{C}(x,y,z)}{\partial x} = 2(\beta_1 + \beta_2 + \beta_3 + \beta_7)(x-1) + 2(\beta_4 + \beta_5 + \beta_6 + \beta_8)x = 0,$$

$$x^* = \frac{\beta_1 + \beta_2 + \beta_3 + \beta_7}{\sum_i \beta_i},$$

and,

$$\frac{\partial \mathscr{C}(x,y,z)}{\partial y} = 2(\beta_1 + \beta_2 + \beta_4 + \beta_6)(y-1) + 2(\beta_3 + \beta_5 + \beta_7 + \beta_8)y = 0$$

$$y^* = \frac{\beta_1 + \beta_2 + \beta_4 + \beta_6}{\sum_i \beta_i},$$

and

$$\frac{\partial \mathscr{C}(x,y,z)}{\partial z} = 2(\beta_1 + \beta_3 + \beta_4 + \beta_5)(z-1) + 2(\beta_2 + \beta_6 + \beta_7 + \beta_8)z = 0$$

$$z^* = \frac{\beta_1 + \beta_3 + \beta_4 + \beta_5}{\sum_i \beta_i}.$$

and the optimal cost is,

$$\mathscr{C}^* = \frac{1}{3(\beta_1+\beta_2+\cdots+\beta_8)^3} \times [(\beta_1+\beta_2+\beta_3+\beta_7)(\beta_4+\beta_5+\beta_6+\beta_8)^2+$$

$$(\beta_1+\beta_2+\beta_3+\beta_7)^2(\beta_4+\beta_5+\beta_6+\beta_8)+(\beta_1+\beta_2+\beta_4+\beta_6)(\beta_3+\beta_5+\beta_7+\beta_8)^2+$$

$$(\beta_1+\beta_2+\beta_4+\beta_6)^2(\beta_3+\beta_5+\beta_7+\beta_8)+(\beta_1+\beta_3+\beta_4+\beta_5)(\beta_2+\beta_6+\beta_7+\beta_8)^2+$$

$$(\beta_1+\beta_3+\beta_4+\beta_5)^2(\beta_2+\beta_6+\beta_7+\beta_8)].$$

3 Allocation of the Optimal Settlement under Budget Constraints

If there exist some budget constraints then the optimal strategy minimizes the overall aversion to opinion changing under these constraints, and it can be computed by solving the following optimization problem,

$$\mathscr{C}(x,y,z) = \sum_{i=1}^{n} \mathscr{C}_i(x,y,z) \to \min; \text{ subject to } W(x,y,z)^T \leq q \qquad (4)$$

where $W \in \mathbb{R}^{m \times 3}$ and $q \in \mathbb{R}^m$. Then problem (4) can be solved using the method of Lagrange multipliers. We will show a simple example.

Example 3. Suppose that we have a four-expert two-issue problem in which the ideal solutions and, consequently, the quadratic OCA functions for the experts are defined in the same manner as in Example 1. Suppose further that we have the following budget constraint $1/4 \leq x+y \leq 3/4$. Then problem (4) collapses into,

$$\mathscr{C}(x,y) = \sum_{i=1}^{4} \mathscr{C}_i(x,y) \to \min$$

$$\text{subject to} \quad x+y \leq 3/4$$

$$x+y \geq 1/4$$

and its Lagrangian function is defined by

$$\mathscr{L}(x,y,\lambda_1,\lambda_2) = \mathscr{C}(x,y) + \lambda_1(x+y-3/4) + \lambda_2(1/4-x-y).$$

Then the Karush-Kuhn-Tucker conditions can be written as

$$\frac{\partial \mathscr{L}(x,y,\lambda_1,\lambda_2)}{\partial x} = 2(\beta_1+\beta_2)(x-1) + 2(\beta_3+\beta_4)x + (\beta_1+\beta_2+\beta_3+\beta_4)(\lambda_1-\lambda_2) = 0$$

$$\frac{\partial \mathscr{L}(x,y,\lambda_1,\lambda_2)}{\partial y} = 2(\beta_1+\beta_3)(y-1) + 2(\beta_2+\beta_4)y + (\beta_1+\beta_2+\beta_3+\beta_4)(\lambda_1-\lambda_2) = 0$$

$$\lambda_1(x+y-3/4) = 0$$

$$\lambda_2(1/4-x-y) = 0$$

Then the optimal solution to problem (4) is $(1/8, 1/8)$ if $\beta_1 - \beta_2 - \beta_3 - 3\beta_4 > 0$ and $(3/8, 3/8)$ if $\beta_1 - \beta_2 - \beta_3 - 3\beta_4 < 0$. If $\beta_1 - \beta_2 - \beta_3 - 3\beta_4 = 0$, then $(1/8, 1/8)$ and $(3/8, 3/8)$ are both optimal solutions, and the optimal cost is,

$$\mathscr{C}^* = \frac{1}{32}\beta_1 + \frac{25}{32}\beta_2 + \frac{25}{32}\beta_3 + \frac{49}{32}\beta_4.$$

4 Allocation of the Optimal Settlement under Flexible Budget Constraints

Sometimes, in real-world decision making processes, e.g. project management, constraints cannot be defined precisely or, even if they are precise they are not required to be a totally tight restriction for the region of feasible solutions. Let us consider the optimal settlement problem with quadratic opinion changing aversion functions when budget constraints are flexible (fuzzy) and the group has a target level for the overall cost, denoted by $q_0 > 0$. For other approaches the reader can consult [5, 6, 7, 8].

Definition 1. A fuzzy set of the real line given by the membership function

$$A(t) = \begin{cases} 1 - \dfrac{|a - t|}{d} & \text{if } |a - t| \le d, \\ 0 & \text{otherwise,} \end{cases}$$

where $d > 0$ will be called a symmetrical triangular fuzzy number with center $a \in \mathbb{R}$ and tolerance level d and we shall refer to it by the pair (a, d).

If the budget constraints are fuzzy and the group has a target level for the overall cost then we can state the following fuzzy mathematical programming (FMP) problem [21]: Find $(x^*, y^*, z^*) \in \mathbb{R}^3$ such that it satisfies the following inequalities as much as possible

$$\mathscr{C}(x, y, z) \le (q_0, d_0)$$
$$w_{11}x + w_{12}y + w_{13}z \le (q_1, d_1)$$
$$\vdots \tag{5}$$
$$w_{m1}x + w_{m2}y + w_{m3}z \le (q_m, d_m)$$

where q_0 is the target level for the overall cost, (q_i, d_i) are fuzzy numbers of symmetric triangular form with center q_i and tolerance level $d_i > 0$, for $i = 0, 1, \ldots, m$, and the inequalities are understood in a possibilistic sense. That is, the degree of satisfaction of the i-th constraint by a point $(x, y, z) \in \mathbb{R}^3$, denoted $\mu_i(x, y, z)$, is defined by

$$\mu_i(x,y,z) = \begin{cases} 1 & \text{if } w_{i1}x + w_{i2}y + w_{i3}z \le q_i, \\ 1 - \dfrac{w_{i1}x + w_{i2}y + w_{i3}z - q_i}{d_i} & \text{if } q_i < w_{i1}x + w_{i2}y + w_{i3}z \le q_i + d_i \\ 0 & \text{if } w_{i1}x + w_{i2}y + w_{i3}z > q_i + d_i \end{cases}$$

for $i = 1, \ldots, m$. The degree of satisfaction of the fuzzy goal (q_0, d_0) by a point $(x, y, z) \in \mathbb{R}^3$ is defined by

$$\mu_0(x,y,z) = \begin{cases} 1 & \text{if } \mathscr{C}(x,y,z) \le q_0, \\ 1 - \dfrac{\mathscr{C}(x,y,z) - q_0}{d_0} & \text{if } q_0 < \mathscr{C}(x,y,z) \le q_0 + d_0 \\ 0 & \text{if } \mathscr{C}(x,y,z) > q_0 + d_0 \end{cases}$$

If for a vector $(x, y, z) \in \mathbb{R}^3$ the value of $w_{i1}x + w_{i2}y + w_{i3}z$ is less or equal than q_i then (x, y, z) satisfies the i-th budget constraint with the maximal conceivable degree: one. If $q_i < w_{i1}x + w_{i2}y + w_{i3}z < q_i + d_i$ then (x, y, z) is not feasible in classical sense, but the group can still tolerate the violation of the crisp budget constraint, and accept (x, y, z) as a solution with a positive degree, however, the bigger the violation the less is the degree of acceptance. Finally, if $w_{i1}x + w_{i2}y + w_{i3}z > q_i + d_i$ then the violation of the i-th constraint is intolerable by the group, that is, $\mu_i(x, y, z) = 0$.

Furthermore, if for a vector $(x, y, z) \in \mathbb{R}^3$ the value of overall cost, $\mathscr{C}(x, y, z)$, is less or equal to q_0 then (x, y, z) satisfies the target level of overall cost with the maximal conceivable degree: one. If $q_0 < \mathscr{C}(x, y, z) < q_0 + d_0$ then (x, y, z) is not feasible in classical sense, but the group can still tolerate the exceeded overall cost, and accept (x, y, z) as a solution with a positive degree, however, the bigger the overstepping the less is the degree of acceptance. Finally, if $\mathscr{C}(x, y, z) > q_i + d_i$ then the exceed of the target level is intolerable by the group, that is, $\mu_0(x, y, z) = 0$.

Then the (fuzzy) solution of the FMP problem (5) is defined as a fuzzy set on $\mathbb{R}^3$ whose membership function is given by

$$\mu(x,y,z) = \min\{\mu_0(x,y,z), \mu_1(x,y,z), \ldots, \mu_m(x,y,z)\},$$

In this setup $\mu(x, y, z)$ denotes the degree to which all inequalities are satisfied at point $(x, y, z) \in \mathbb{R}^3$. To maximize μ on $\mathbb{R}^3$ we have to solve the following crisp quadratic programming problem

$$\max \lambda$$
$$\lambda d_0 + \mathscr{C}(x,y,z) \le q_0 + d_0,$$
$$\lambda d_1 + w_{11}x + w_{12}y + w_{13}z \le q_1 + d_1,$$
$$\cdots$$
$$\lambda d_m + w_{m1}x + w_{m2}y + w_{m3}z \le q_m + d_m,$$
$$0 \le \lambda \le 1, \; x, y, z \in \mathbb{R}.$$

that is,

$$\max \lambda$$

$$\lambda d_0 + \sum_{i=1}^{n} \frac{\beta_i}{\beta_1 + \beta_2 + \cdots + \beta_n} \times \frac{(x - a_i)^2 + (y - b_i)^2 + (z - c_i)^2}{M} \leq q_0 + d_0,$$

$$\lambda d_1 + w_{11}x + w_{12}y + w_{13}z \leq q_1 + d_1,$$

$$\cdots$$

$$\lambda d_m + w_{m1}x + w_{m2}y + w_{m3}z \leq q_m + d_m,$$

$$0 \leq \lambda \leq 1, \ x, y, z \in \mathbb{R}.$$

where M is defined by,

$$M = \max_{i,j} \{ (a_i - a_j)^2 + (b_i - b_j)^2 + (c_i - c_j)^2 \}.$$

5 Summary

We showed that if each expert has a quadratic opinion changing aversion function then the minimum-cost solution is nothing else but the weighted average of the individual optimal solutions where the weights are the relative importances of the experts. In order to present the validity and applicability of this property we introduced some minimum-cost solutions for group settlements under crisp and fuzzy budget constraints.

References

1. Balinski, M., Laraki, R.: A Theory of Measuring, electing and ranking. PNAS 104(21), 8720–8725 (2007)
2. de Borda, J.-C.: Histoire de l'Academie Royale des Sciences, pp. 657–665 (1754)
3. Bordogna., G., Fedrizzi, M., Pasi, G.: A linguistic approach to modeling of consensus in GDSS. In: Proceedings EUROXIII/OR, Glasgow, vol. 36 (1994)
4. Fedrizzi, M., Kacprzyk, J., Zadrozny, S.: An interactive multiuser decision support system for consensus reaching processes using fuzzy logic with linguistic quantifiers. Decision Support Systems 4(3), 313–327 (1988)
5. Fedrizzi, M., Mich, L.: Decision using production rules. In: Proc. of Annual Conference of the Operational Research Society of Italy (AIRO 1991), September 18–10, pp. 118–121. Riva del Garda, Italy (1991)
6. Fedrizzi, M., Fullér, R.: On stability in group decision support systems under fuzzy production rules. In: Trappl, R. (ed.) Proceedings of the Eleventh European Meeting on Cybernetics and Systems Research, pp. 471–478. World Scientific Publisher, London (1992)
7. Fedrizzi, M.: Fuzzy approach to modeling consensus in group decisions. In: Proceedings of First Workshop on Fuzzy Set Theory and Real Applications, Milano, Automazione e strumentazione, Supplement to, issue, May 10, pp. 9–13 (1993)

8. Fedrizzi, M.: Fuzzy consensus models in GDSS. In: Proceedings of the 2nd New Zealand Two-Stream International Conference on Artificial Neural Networks and Expert Systems, pp. 284–287 (1995)
9. Fedrizzi, M., Fedrizzi, M., Marques Pereira, R.A.: On the issue of consistency in dynamical consensual aggregation. In: Bouchon Meunier, B., Gutierrez Rios, J., Magdalena, L., Yager, R.R. (eds.) Technologies for constructing intelligent systems, vol. 89, Studies in Soft Computing, Berlin (1990)
10. Fedrizzi, M., Brunelli, M.: Fair consistency evaluation in reciprocal relations and in group decision making. New Mathematics and Natural Computation 5(2), 407–420 (2009)
11. Forman, E., Peniwati, K.: Aggregating individual judgments and priorities with the Analytic Hierarchy Process. European Journal of Operational Research 108(1), 165–169 (1998)
12. Fullér, R., Mich, L.: Fuzzy reasoning techniques for GDSS. In: Proceedings of EUFIT 1993 Conference, Aachen, Germany, September 7-10, pp. 937–940. Verlag der Augustinus Buchhandlung, Aachen (1993)
13. Jabeur, K., Martel, J.-.M.-.: An agreement index with respect to a consensus preorder. Group Decision and Negotiation, doi:10.1007/s10726-009-9160-3
14. Meskanen, T., Nurmi, H.: Distance from Consensus: A Theme and Variations. In: Simeone, B., Pukelsheim, F. (eds.) Mathematics and Democracy: recent Advances in Voting Systems and Collective Choice, Springer, Heidelberg (2006)
15. Mich, L., Fedrizzi, M., Gaio, L.: Approximate Reasoning in the Modeling of Consensus in Group Decisions. In: Klement, E.-P., Slany, W. (eds.) FLAI 1993. LNCS, vol. 695, pp. 91–102. Springer, Heidelberg (1993)
16. Mikhailov, L.: Group prioritization in the AHP by fuzzy preference programming method. Computers & operations research 31(2), 293–301 (2004)
17. Orlovsky, S.A.: Decision-making with a fuzzy preference relation. Fuzzy Sets and Systems 1(3), 155–167 (1978)
18. Ramanathan, R., Ganesh, L.S.: Group preference aggregation methods employed in AHP: An evaluation and an intrinsic process for deriving members' weightages. European Journal of Operational Research 79(2), 249–265 (1994)
19. Saaty, T.: A scaling method for priorities in hierarchical structures. Journal of Mathematical Psychology 15(3), 234–281 (1977)
20. Tanino, T.: Fuzzy preference orderings in group decision making. Fuzzy Sets and Systems 12(2), 117–131 (1984)
21. Zimmermann, H.-J.: Fuzzy programming and linear programming with several objective functions. Fuzzy Sets and Systems 1(1), 45–55 (1978)

Statistical Preference as a Tool in Consensus Processes

I. Montes, D. Martinetti, S. Díaz, and S. Montes

Abstract. In a consensus process, the intensities of preference can be expressed by means of probability distributions instead of single values. In that case, it is necessary to compare, in a simple way, pairs of probability distributions. Since classical methods do not assure the possibility of comparing any pair of distributions, a modern method is considered in this paper. It is called statistical preference. One of its most remarkable advantages is that it allows to compare any pair of probability distributions.

In this contribution we study in depth some properties of this method and the relationship between most usual stochastic dominance and statistical preference. We also consider some of the most important families of distributions and analyze statistical preference among probability distributions in the same family.

1 Introduction

Consensus and selection processes are essential in order to obtain a final solution in social choice or decision making problems. In this context, we have a number of experts who have expressed their opinions (intensities of preferences) on a set of alternatives. Usually, these intensities are summarized in a representative value by means of some aggregation method (see, for instance, [11, 12, 18]). However, some recent papers (see, for instance, [17, 19]) have worked in a more general framework. Thus, they have considered the resulting probability distributions of intensities as preferences as long as possible. If we deal with probability distributions instead of simple values, it is clear that we increase the information we have about the process. But, as it is logical, we have to pay a price for this extra information. This price is paid

I. Montes · D. Martinetti · S. Díaz · S. Montes
Department of Statistics and Operational Research of University of Oviedo
e-mail: imontes@spi.uniovi.es,
    {martinettidavide.uo,diazsusana,montes}@uniovi.es

E. Herrera-Viedma et al. (Eds.): Consensual Processes, STUDFUZZ 267, pp. 65–92.
springerlink.com   © Springer-Verlag Berlin Heidelberg 2011

by means of a more difficult interpretation and final comparison. A simple method to deal with probability distributions is proposed in [17]. In that context, the probability distributions are compared by means of stochastic dominance (for a complete study, see [13, 15]). This method of comparison of probability distributions has been used in many areas with good results. However, it has some drawbacks. The most important in this context is that not every pair of probability distributions can be compared. This means that not every pair of alternatives can be ordered. An example of this problem was already shown in [17]. There, it was solved by adjusting the probability distributions, but it could be not necessary if we consider a method to order the probability distributions different from stochastic dominance.

This solution was proposed in a general framework by De Schuymer et al. ([6, 9]). They defined a method, called statistical preference, that gives a preference between every pair of probability distributions. Moreover, it has some other advantages with respect to stochastic orders: it uses the joint distribution and it is based on a graded relation. Our contribution is devoted to this method. We study some of its properties and the possible relation between statistical preference and stochastic dominance. Note that some of the results collected here can be found, without proof, in [14]. This method can be applied not only for ordinal intensities of preference (as it was considered in [17]), but also for numerical intensities, which includes the case of fuzzy preferences.

The paper is organized as follows: first of all, we present some concepts necessary to follow the work. Particularly, we recall some known results about stochastic dominance and the definition and some properties of statistical preference. Sect. 3 includes a characterization of statistical preference. We also show here a practical interpretation that helps to understand the idea of statistical preference. In Sect. 4 we compare first and second degree stochastic dominance with statistical preference as methods for ordering independent random variables. In Sect. 5 we consider some of the most known families of distributions. We characterize statistical preference between independent random variables of the same family. In Sect. 6 we briefly discuss the results obtained and we set out some open problems.

2   Previous Concepts

In this section we recall some fundamental concepts in the framework of stochastic dominance and statistical preference. We also introduce the notation.

2.1   Stochastic Dominance

We have already explained in the introduction that stochastic dominance is one of the most usual methods to compare probability distributions, and it has been used successfully in some areas. The many works that can be found in the literature devoted to the analysis and application of this concept show its importance. A complete study can be seen in [13, 15].

Now, we will collect the most relevant definitions and some results that will be necessary to the development of subsequent sections. Furthermore, we will lay emphasis on the intuitive interpretations of the concepts introduced.

In order to avoid any ambiguity, we will represent any probability distribution by means of a random variable modelized by means of this distribution.

Definition 1. Let X and Y be two random variables, and let F_1 and F_2 be their respective cumulative distribution functions (CDF). It is said that X stochastically dominates by first degree Y if

$$F_1(t) \leq F_2(t) \quad \forall t \in I\!R.$$

We will denote it by $X \geq_{\text{FSD}} Y$.

By definition, this concept only takes into account the marginal cumulative distribution functions. Its interpretation may be: if $X \geq_{\text{FSD}} Y$, then it holds that $F_1(t) \leq F_2(t)$, or equivalently $\Pr(X > t) \geq \Pr(Y > t)$, i.e., for every real value that we fix, the probability that X takes a value greater than that fixed value is greater than the probability that Y is greater than the same fixed value.

To see its geometric meaning it suffices to look at Fig. 1. Here we can see that the cumulative distribution function of the random variable X is always under the cumulative distribution function of Y.

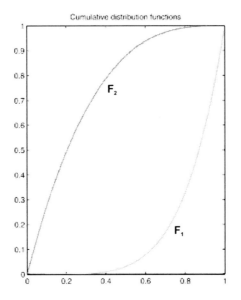

Fig. 1 Example of first degree stochastic dominance.

However, this definition is too restrictive. Only very few pairs of random variables can be ordered by this definition. In order to obtain a less restrictive definition, second degree stochastic dominance appeared.

Definition 2. Let X and Y be two random variables with cumulative distribution functions F_1 and F_2, respectively. It is said that X stochastically dominates by second degree Y if

$$G_1(x) = \int_{-\infty}^{x} F_1(t)dt \leq \int_{-\infty}^{x} F_2(t)dt = G_2(x), \quad \forall x \in I\!R.$$

It will denote it by $X \geq_{SSD} Y$.

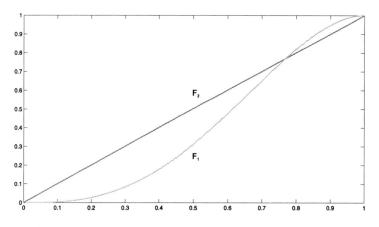

Fig. 2 Example of second degree stochastic dominance.

Let us focus on Fig. 2. The cumulative distribution functions intersect. Therefore, first degree stochastic dominance is not satisfied. However, we can see that the area below the cumulative distribution function of Y is greater than the area below the cumulative distribution function of X and second degree stochastic dominance holds.

Some authors (see for example [13]) consider that a random variable X stochastically dominates by first degree random variable Y if it holds that

$$\forall t \in I\!R, F_1(t) \leq F_2(t) \quad \wedge \quad \exists t_0 \in I\!R \text{ such that } F_1(t_0) < F_2(t_0),$$

and X stochastically dominates by second degree random variable Y if it holds that

$$\forall t \in I\!R, G_1(t) \leq G_2(t) \quad \wedge \quad \exists t_0 \in I\!R \text{ such that } G_1(t_0) < G_2(t_0).$$

However, following [15], in these two last cases we talk about *strong first* and *strong second* degree stochastic dominance. We will use the notation $\geq_{FSD}$ and $\geq_{SSD}$.

However, first and second degree stochastic dominance are also denoted $\succeq_1$ and $\succeq_2$, respectively (see for example [6]).

First and second degree stochastic dominances can be interpreted from the economic standpoint (see [13]):

- The interpretation provided for first degree stochastic dominance is that the person that execute the action is rational, in the sense that he/she always prefers a higher gain.
- Second degree stochastic dominance can be interpreted as follows: the decision maker prefers a higher gain without being prone to gambling.

The concept of stochastic dominance can be extended to any order n, but it has the drawback that its interpretation gets complicated for $n \geq 3$ (see for example [13]).

From here on we focus on first and second degree stochastic dominances, and we present some known results that will be necessary for our study. To begin, we recall a well known result concerning the connection between first and second degree stochastic dominances (see for example [13]).

Proposition 1. *For every pair of random variables X and Y , it holds that*

$$X \geq_{\text{FSD}} Y \Rightarrow X \geq_{\text{SSD}} Y.$$

While the converse is not true in general.

The proof is trivial. If $F_1(t) \leq F_1(t)$ at every point, the area that F_1 encloses at every point is smaller than the area that F_2 encloses at the same point.

Example 1. As an example of random variables for which second degree stochastic dominance holds but first degree stochastic dominance does not, we can consider X a degenerate random variable at the point 2, and Y a random variable with uniform distribution on the interval $(0,3)$. It holds that

$$F_1(x) = \begin{cases} 0, & \text{if } x < 2, \\ 1, & \text{if } x \geq 2, \end{cases}$$

and

$$F_2(y) = \begin{cases} 0, & \text{if } y < 0, \\ \frac{y}{3}, & \text{if } y \in [0,3), \\ 1, & \text{if } y \geq 3. \end{cases}$$

Since $F_1(2) = 1 > \frac{2}{3} = F_2(2)$, $X \not\geq_{\text{FSD}} Y$. However, we can see the graphics of G_1 and G_2 in Fig. 3. There we can see that G_2 is always over G_1, and thus $X \geq_{\text{SSD}} Y$.

Remark 1. Proposition 1 is a particular case of a more general result than can be found for example in [13]. It states that given $k \leq l$, then if X stochastically dominates Y by k^{th} degree, then X also stochastically dominates Y by l^{th} degree.

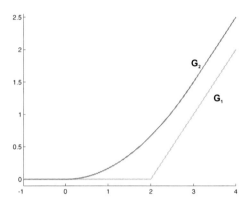

Fig. 3 Graphics of functions G_1 and G_2 corresponding to Example 1.

Next we recall some known characterizations of first and second degree stochastic dominance.

Proposition 2. *Let X and Y be two random variables. It holds that*

- $X \geq_{\text{FSD}} Y \Leftrightarrow E(u(X)) \geq E(u(Y))$, *for every non-decreasing function.*
- $X \geq_{\text{SSD}} Y \Leftrightarrow E(u(X)) \geq E(u(Y))$, *for every non-decreasing and concave function.*

Proposition 1 can be seen as a corollary of this result, because if $E(u(X)) \geq E(u(Y))$ for every non-decreasing function, then in particular it also holds for every non-decreasing and concave function.

Stochastic dominance, in particular first and second degree, have a proven efficiency. However, it is known that they also have some drawbacks. They lead to a crisp order. They do not allow degrees of dominance. Moreover, the order is partial: there are many pairs of random variables that cannot be ordered with this definition (for an example, see [17]). In addition to this, in some cases it becomes difficult to be calculated.

2.2 Statistical Preference

Stochastic dominance has shown to be a very useful method to compare random variables. However we have just seen that it shows some drawbacks. These problems led De Schuymer et al. ([6, 7, 9]) to introduce a natural way to compare random variables. It is based on a probabilistic relation, defined from other relations. We will start introducing all the concepts necessary to establish such probabilistic relation.

Definition 3. Let A be a set of alternatives. A probabilistic relation Q is a mapping

$$Q : A \times A \to [0, 1]$$

such that

$$Q(a,b) + Q(b,a) = 1 \quad \forall a,b, \in A.$$

The value $Q(a,b)$ can be interpreted as the degree of preference between the alternative a and the alternative b. If it is close to 1, a will be clearly preferred to b. If it is close to $\frac{1}{2}$, a and b will be indifferent. And finally, if it is close to 0, b will be preferred to alternative a.

The probabilistic relation used to the pairwise comparison of random variables arises from two probabilities: $\Pr(X > Y)$ and $\Pr(X = Y)$. They can be interpreted as the degree in which X is preferred to Y and the degree in which X and Y are indifferent. Using these probabilities, a probabilistic relation is defined:

Definition 4. Let A be a set of random variables defined on the same probabilistic space. For any two random variables X and Y in A, the degree of statistical preference of X over Y is given by:

$$Q(X,Y) = \Pr(X > Y) + \frac{1}{2}\Pr(X = Y).$$

It is very easy to see that Q is a probabilistic relation:

$$Q(X,Y) + Q(Y,X) = \left(\Pr(X > Y) + \tfrac{1}{2}\Pr(X = Y)\right) + \left(\Pr(Y > X) + \tfrac{1}{2}\Pr(X = Y)\right)$$
$$= \Pr(X > Y) + \Pr(Y > X) + \Pr(X = Y) = 1.$$

This relation Q is consistent with the interpretation given above, in the sense that if the value $Q(X,Y)$ is close to 1, the random variable X will be preferred to Y, if it is close to $\frac{1}{2}$ the random variables will be indifferent and if it is close to 0, the preferred random variable will be Y.

From this valued relation we can introduce a new concept to compare probability distributions: statistical preference.

Definition 5. [7] Let A be a set of random variables defined on the same probabilistic space. Let Q be the probabilistic relation given in Definition 4. For every pair of random variables X and Y on A, it is said that:

- X is statistically preferred to Y if $Q(X,Y) \geq \frac{1}{2}$. It is denoted by $X \geq_{SP} Y$.
- X and Y are indifferent if $Q(X,Y) = \frac{1}{2}$.
- X is strict statistically preferred to Y if $Q(X,Y) > \frac{1}{2}$. It is denoted by $X >_{SP} Y$.

One of the advantages of statistical preference is that it is possible to compare any pair of alternatives with this method.

Example 2. If we consider again the example proposed in [17]. For two alternatives a and b the two cumulative distribution function are:

$$F_{(a,b)}(x) = \begin{cases} 0, & \text{if } x < l_{-3}, \\ \frac{1}{5}, & \text{if } l_{-3} \leq x < l_1, \\ 1, & \text{if } l_1 \leq x, \end{cases} \quad \text{and} \quad F_{(b,a)}(x) = \begin{cases} 0, & \text{if } x < l_{-1}, \\ \frac{4}{5}, & \text{if } l_{-1} \leq x < l_3, \\ 1, & \text{if } l_3 \leq x, \end{cases}$$

where $\{l_{-3}, l_{-2}, l_{-1}, l_0, l_1, l_2, l_3\}$ is the set of ordinal intensities of preference with $l_i < l_{i+1}$.

It is clear that no one of the distributions stochastically dominates the other (first or second degree). However, the value of Q is $16/25$ and therefore, $F_{(a,b)} \geq_{SP} F_{(b,a)}$. This means that a statistical majority for $a \succ b$ is present.

Another of the advantages of statistical preference is that it uses all the available information, because it considers the relation between the variables.

Definition 6. A copula is an application $C : [0,1]^2 \to [0,1]$ such that

- $C(x,0) = C(0,x) = 0 \quad \forall x \in [0,1]$.
- $C(x,1) = C(1,x) = x \quad \forall x \in [0,1]$.
- It holds the moderate-increasing property:

$$C(x_1,y_1) + C(x_2,y_2) \geq C(x_1,y_2) + C(x_2,y_1) \quad \forall (x_1,x_2,y_1,y_2) \in [0,1]^4.$$

For every copula C and every $(x,y) \in [0,1]^2$, it holds that $W(x,y) \leq C(x,y) \leq M(x,y)$. Where $M(x,y) = \min(x,y)$ is the Fréchet-Hoeffding upper bound and $W(x,y) = \max(x+y-1,0)$ is the Fréchet-Hoeffding lower bound. Both of them are copulas. Another important copula is the product, $\Pi(x,y) = x \cdot y$. A complete study on copulas can be seen in [16].

It is well known in Probability that for every pair of random variables X and Y, with respective cumulative distribution functions F_1 and F_2, any joint CDF $F_{1,2}$ obtained from them is built as a copula defined over the marginal cumulative distribution functions:

$$F_{1,2}(x,y) = C(F_1(x),F_2(y)).$$

Given two random variables X and Y, if their joint CDF is obtained as the Fréchet-Hoeffding upper bound (lower bound) over the partial cumulative distribution functions of X and Y, the random variables are said comonotonic (countermonotonic). If the joint CDF is obtained as the product of the marginal cumulative distribution functions, the random variables are said independent:

$$F_{1,2}(x,y) = F_1(x) \cdot F_2(y).$$

This paper is focused on statistical preference between independent random variables. Recall that when both random variables are discrete,

$$Q(X,Y) = \Pr(X > Y) + \frac{1}{2}\Pr(X = Y)$$

$$= \sum_i \Pr(X = i) \sum_{j<i} \Pr(Y = j) + \frac{1}{2}\sum_i \Pr(X = i)\Pr(Y = i).$$

When both random variables are continuous, $\Pr(X = Y) = 0$ and $X \geq_{SP} Y$ if and only if $\Pr(X > Y) \geq \frac{1}{2}$. Equivalently, $X \geq_{SP} Y$ if and only if $\Pr(X > Y) \geq \Pr(Y > X)$. Also,

$$Q(X,Y) = \int_{-\infty}^{\infty} f_1(x)dx \int_{-\infty}^{x} f_2(y)dy.$$

Remark 2. Observe that statistical preference does not depend on the support of the random variables, but on the way those supports are ordered. This can be seen easily for discrete random variables. If we consider the random variables

$$
\begin{array}{c|cc}
X & a & b \\
\hline
\mathrm{Pr}_1 & 0.4 & 0.6
\end{array}
\qquad
\begin{array}{c|c}
Y & c \\
\hline
\mathrm{Pr}_2 & 1
\end{array}
$$

with $a < c < b$, and we assume they are independent, we can compute $Q(X,Y)$ without knowing the values of a, b and c: $Q(X,Y) = 0.6$. Thus, $X \geq_{\mathrm{SP}} Y$ independently of the values of a, b or c.

Remark 3. After Remark 2 one could think that statistical preference is related to the order of the medians of the variables. However, it is not.
Consider on the one hand, the random variables

$$
\begin{array}{c|ccc}
X & -1 & 0 & 1 \\
\hline
\mathrm{Pr}_1 & 0.4 & 0.2 & 0.4
\end{array}
\qquad
\begin{array}{c|ccc}
Y & -10 & 0.1 & 0.5 \\
\hline
\mathrm{Pr}_2 & 0.4 & 0.2 & 0.4
\end{array}
$$

and assume they are independent. It holds that $\mathrm{Me}(X) = 0 < 0.1 = \mathrm{Me}(Y)$. However, $X \geq_{\mathrm{SP}} Y$ since $Q(X,Y) = 0.64$.
Consider on the other hand, two random variables X and Y with uniform distributions, $U(2,3)$ and $U(0,1)$ respectively, and assume they are independent, then clearly $X \geq_{\mathrm{SP}} Y$ and $\mathrm{Me}(X) > \mathrm{Me}(Y)$.

However, our intuition about the relationship between statistical preference and median was right, as it will be proven in Corollary 3. By now, let us notice that the median of $X - Y$ is 0.5.

Statistical preference admits an easy geometric interpretation: the joint density function of the two random variables X and Y must cover a bigger volume in the semiplane where $x \geq y$ than in the other one, $x \leq y$. As an example, consider Fig. 4. There we can see the joint density function of two random variables X and Y with exponential distributions $Exp(1/20)$ and $Exp(1/2)$ respectively. We can see there that $X >_{\mathrm{SP}} Y$ since the function covers a bigger volume in the right-hand side of the graphic.

3 Characterizing Statistical Preference

3.1 Necessary and Sufficient Conditions

In Subsect. 2.1 we recall two characterizations of first and second degree stochastic dominance in terms of the expected values. We can provide a similar result for statistical preference.

Theorem 1. *Let X and Y be two independent random variables with CDF F_1 and F_2 respectively. Then, if X' is a random variable identically distributed as X and independent of X and Y,*

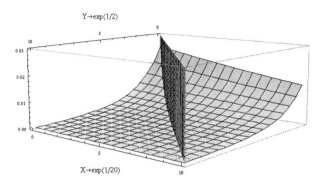

Fig. 4 Graphical representation of the statistical preference.

$$X \geq_{\mathrm{SP}} Y \Leftrightarrow E(F_2(X)) - E(F_1(X)) \geq \frac{1}{2}(\Pr(X = Y) - \Pr(X = X')), \qquad (1)$$

*where given a random variable X with cumulative distribution function F_1, $E(h(X))$
stands for the expected value of the function h with respect to the variable X, this is,
$E(h(X)) = \int_{\mathbb{R}} h(x)dF_1(x)$.*

Proof. It holds that $X \geq_{\mathrm{SP}} Y$ if and only if $\Pr(Y > X) + \frac{1}{2}\Pr(X = Y) \leq \frac{1}{2}$. On the
other hand let us recall (see for example [1]) that $E(F_1(X)) = \frac{1}{2} + \frac{1}{2}\Pr(X = X')$,
then
$$\Pr(Y > X) = 1 - \Pr(Y \leq X) = 1 - E(F_2(X))$$
$$= \frac{1}{2} + E(F_1(X)) - \frac{1}{2}\Pr(X = X') - E(F_2(X)).$$

Whereas, $X \geq_{\mathrm{SP}} Y$ if and only if

$$\frac{1}{2} + E(F_1(X)) - \frac{1}{2}\Pr(X = X') - E(F_2(X)) + \frac{1}{2}\Pr(X = Y) \leq \frac{1}{2}$$

or equivalently,

$$E(F_2(X)) - E(F_1(X)) \geq \frac{1}{2}(\Pr(X = Y) - \Pr(X = X')). \qquad \square$$

The general expression presented in Equation (1) can be simplified for the particular
case of continuous random variables.

Corollary 1. *Let X and Y be two independent continuous random variables with
CDF F_1 and F_2 respectively. Then,*

$$X \geq_{\mathrm{SP}} Y \Leftrightarrow E(F_2(X)) \geq E(F_1(X)). \qquad (2)$$

Proof. If X and Y are continuous random variables, X' is identically distributed as X and all of them are independent, $\Pr(X = Y) = 0 = \Pr(X = X')$. Therefore, Equation (1) simplifies into Equation (2). $\qquad\square$

Remark 4. Note that $X \geq_{\text{FSD}} Y \Leftrightarrow F_1(t) \leq F_2(t)$ for all $t \in I\!R$. Thus, if X and Y are continuous, first stochastic dominance force the inequality $F_1(t) \leq F_2(t)$ for all t, while statistical preference force the inequality in mean (with respect to X), $E(F_1(X)) \leq E(F_2(X))$.

Equation (1) cannot be simplified into (2) in general. In fact, no one of the implications involved in Equation (2) hold in general.

If X and Y are two independent random variables,

- in general
$$X \geq_{\text{SP}} Y \not\Rightarrow E(F_1(X)) \leq E(F_2(X)).$$

It suffices to consider X a Bernoulli distribution with parameter 0.75 and Y a degenerate random variable at $c \in (0,1)$. It holds that $Q(X,Y) = \Pr(X > Y) = 0.75$ and $X \geq_{\text{SP}} Y$. However,

$$E(F_1(X)) = \Pr(X = 0)F_1(0) + \Pr(X = 1)F_1(1) = 0.25^2 + 0.75 = 0.8125.$$
$$E(F_2(X)) = \Pr(X = 0)F_2(0) + \Pr(X = 1)F_2(1) = 0.25 \cdot 0 + 0.75 = 0.75.$$

Then, $E(F_1(X)) > E(F_2(X))$.

- Moreover,
$$E(F_1(X)) \leq E(F_2(X)) \not\Rightarrow X \geq_{\text{SP}} Y.$$

Assume that X and Y are independent random variables distributed as follows

X	0	1	2
$\Pr_X$	0.3	0.4	0.3

Y	0	1	2
$\Pr_Y$	0.2	0.57	0.23

It holds that
$$E(F_2(X)) = 0.06 + 0.4 \cdot 0.77 + 0.3 = 0.668,$$
$$E(F_1(X)) = 0.09 + 0.28 + 0.3 = 0.67,$$

but $X \not\geq_{\text{SP}} Y$ since

$$Q(X,Y) = \Pr(X > Y) + \frac{1}{2}\Pr(X = Y) = 0.311 + \frac{1}{2}0.357 = 0.4895.$$

Remark 5. It seems quite intuitive that if X, Y and X' are three random variables and X and X' are identically distributed, $\Pr(X = Y) \leq \Pr(X = X')$. However it follows from this last counterexample that this is not necessarily true in general.

Let us recall that in the previous counterexample $X \not\geq_{\text{SP}} Y$. Then, it follows from Theorem 1 that $E(F_2(X)) - E(F_1(X)) < \frac{1}{2}(\Pr(X = Y) - \Pr(X = X'))$. Since in that example $E(F_2(X)) - E(F_1(X)) > 0$, it follows that $\Pr(X = Y) > \Pr(X = X')$. In fact, $\Pr(X = Y) = 0.357$ and $\Pr(X = X') = 0.34$.

Lemma 1. *Let X and Y be two random variables. Then it holds that*

$$X \geq_{SP} Y \Rightarrow \Pr(X < Y) \leq \frac{1}{2}.$$

The converse implication only holds if $\Pr(X = Y) \leq 1 - 2\Pr(X < Y)$*. In particular, the equivalence*

$$X \geq_{SP} Y \Leftrightarrow \Pr(X < Y) \leq \frac{1}{2}$$

holds for continuous random variables.

Proof. It holds that $Q(X,Y) = \Pr(X > Y) + \frac{1}{2}\Pr(X = Y) \geq \frac{1}{2}$. Then,

$$\Pr(X < Y) = 1 - \Pr(X > Y) - \Pr(X = Y) \leq \frac{1}{2} - \frac{1}{2}\Pr(X = Y) \leq \frac{1}{2}.$$

The inequality $\Pr(X = Y) \leq 1 - 2\Pr(X < Y)$ is equivalent to $\Pr(X < Y) + \frac{1}{2}\Pr(X = Y) \leq \frac{1}{2}$. Since $\Pr(X < Y) + \Pr(X = Y) + \Pr(X > Y) = 1$, the implication follows.

In particular, for continuous random variables, $\Pr(X = Y) = 0$ and trivially $\Pr(X = Y) \leq 1 - 2\Pr(X < Y)$.   □

Proposition 3. *Let X, Y and Z be three random variables defined on the same probabilistic space and let* λ *and* μ *be two real numbers. It holds that*

1. $X \geq_{SP} Y \Leftrightarrow X + Z \geq_{SP} Y + Z.$

2. $\lambda X \geq_{SP} \mu \Leftrightarrow \begin{cases} X \geq_{SP} \frac{\mu}{\lambda}, & \text{if } \lambda > 0, \\ \frac{\mu}{\lambda} \geq_{SP} X, & \text{if } \lambda < 0, \\ 0 \geq \mu, & \text{if } \lambda = 0. \end{cases}$

Proof. 1. It holds that

$$Q(X,Y) = \Pr(X > Y) + \frac{1}{2}\Pr(X = Y)$$
$$= \Pr(X + Z > Y + Z) + \frac{1}{2}\Pr(X + Z = Y + Z)$$
$$= Q(X + Z, Y + Z).$$

Then, $Q(X,Y) \geq \frac{1}{2}$ if and only if $Q(X + Z, Y + Z) \geq \frac{1}{2}$.

2. First of all, suppose that $\lambda > 0$:

$$\lambda X \geq_{SP} \mu \Leftrightarrow Q(\lambda X, \mu) \geq \frac{1}{2} \Leftrightarrow \Pr(\lambda X > \mu) + \frac{1}{2}\Pr(\lambda X = \mu) \geq \frac{1}{2}$$
$$\Leftrightarrow Pr\left(X > \frac{\mu}{\lambda}\right) + \frac{1}{2}Pr\left(X = \frac{\mu}{\lambda}\right) \geq \frac{1}{2}$$
$$\Leftrightarrow Q\left(X, \frac{\mu}{\lambda}\right) \geq \frac{1}{2} \Leftrightarrow X \geq_{SP} \frac{\mu}{\lambda}.$$

Now, suppose that $\lambda < 0$. Reasoning by analogy:

$$\lambda X \geq_{SP} \mu \Leftrightarrow Q(\lambda X, \mu) \geq \frac{1}{2}$$

$$\Leftrightarrow Q\left(\frac{\mu}{\lambda}, X\right) \geq \frac{1}{2} \Leftrightarrow \frac{\mu}{\lambda} \geq_{SP} X.$$

Finally, if $\lambda = 0$, it is obvious that

$$\lambda X \geq_{SP} \mu \Leftrightarrow 0 \geq \mu. \qquad \square$$

Note that in addition to the equivalence between the statistical preference, we have seen that X is preferred to Y in the same degree as $X + Z$ is preferred to $Y + Z$. Some new equivalences can be deduced from the previous ones.

Corollary 2. *Let X and Y be a pair of random variables and α a constant. Then it holds that*

1. $X \geq_{SP} Y \Leftrightarrow X - Y \geq_{SP} 0$.
2. $X + Y \geq_{SP} Y \Leftrightarrow X \geq_{SP} 0$.
3. $X \geq_{SP} X + \alpha \Leftrightarrow \alpha \leq 0$.
4. $X \geq_{SP} \alpha X \Leftrightarrow \begin{cases} 0 \geq_{SP} X, & \text{if } \alpha > 1, \\ X \geq_{SP} 0, & \text{if } \alpha < 1. \end{cases}$

Proof. Points 1, 2 and 3 are immediate from the first point of Proposition 3. Consider the last one. Applying previous Proposition,

$$X \geq_{SP} \alpha X \Leftrightarrow (1 - \alpha)X \geq_{SP} 0.$$

Applying Point 2 of Proposition 3,

$$(1 - \alpha)X \geq_{SP} 0 \Leftrightarrow \begin{cases} 0 \geq_{SP} X, & \text{if } \alpha > 1, \\ X \geq_{SP} 0, & \text{if } \alpha < 1. \end{cases} \qquad \square$$

Corollary 3. *Let X and Y be a pair of continuous random variables. Then it holds that*

$$X \geq_{SP} Y \; iff \; Me_{X-Y} \geq 0.$$

Proof. Let us denote $Z = X - Y$. By the previous corollary, we will prove the equivalence between $Z \geq_{SP} 0$ and $M_Z \geq 0$.

If $Z \geq_{SP} 0$, then

$$\Pr(Z > 0) \geq \frac{1}{2} \Rightarrow Me_Z \geq 0.$$

Conversely, if $Me_Z \geq 0$, then

$$\frac{1}{2} = \Pr(Z > Me_Z) \leq \Pr(Z > 0). \qquad \square$$

3.2 Intuitive Interpretation of Statistical Preference

In this last part of the section we try to clarify the meaning of statistical preference by means of a gambling example.

Suppose that we have two random variables X and Y defined over the same probabilistic space such that $X >_{SP} Y$, i.e., such that $Q(X,Y) > \frac{1}{2}$. Consider the following game: *We obtain a pair of random values of X and Y simultaneously. For example, if X and Y model the results of the dice, we would roll them simultaneously; otherwise, they can be simulated by a computer. Player 1 bets 1 euro on Y to take a value greater than X. If this holds, Player 1 wins 1 euro, he loses 1 euro if the value of X is greater, and he does not lose anything if the values are equal.*

Denote by Z_i the random variable "reward of Player 1 in the i-th iteration of the game". Then it holds that

$$Z_i = \begin{cases} 1, & \text{if } Y > X \\ 0, & \text{if } Y = X \\ -1, & \text{if } Y < X \end{cases}$$

Then, applying the hypothesis $\Pr(X > Y) + \frac{1}{2}\Pr(X = Y) > \frac{1}{2}$, it holds that

$$\Pr(X > Y) > \frac{1}{2}(1 - \Pr(X = Y)) = \frac{1}{2}(\Pr(X > Y) + \Pr(Y > X))$$
$$\Rightarrow \Pr(X > Y) > \Pr(Y > X),$$

or equivalently, $q > p$, if we consider the notation $p = \Pr(X < Y)$ and $q = \Pr(X > Y)$. Thus

$$E(Z_i) = \Pr(Y > X) - \Pr(Y < X) = p - q < 0.$$

$\{Z_1, Z_2, ...\}$ is an infinite sequence of independent random variables with the same mean and variance. Then, applying the weak law of big numbers, it holds that

$$\overline{Z_n} \xrightarrow{p} p - q,$$

or equivalently,

$$\forall \varepsilon > 0, \lim_{n \to \infty} \Pr\left(|\overline{Z_n} - (p - q)| > \varepsilon\right) = 0.$$

Denote the reward of Player 1 after n iterations of the game by S_n. It holds that $S_n = Z_1 + ... + Z_n$. Then, Player 1 wins the game after n iterations with probability:

$$\Pr(S_n > 0) = \Pr(Z_1 + ... + Z_n > 0) = \Pr(\overline{Z_n} > 0) = \Pr(\overline{Z_n} - (p - q) > q - p)$$
$$\leq \Pr(|\overline{Z_n} - (p - q)| > q - p),$$

and thus, if we consider $\varepsilon = q - p > 0$, it holds that

$$\lim_{n \to \infty} \Pr(S_n > 0) \leq \lim_{n \to \infty} \Pr(|\overline{Z_n} - (p - q)| > q - p) = 0.$$

We have proven that the probability of the event: "Player 1 wins before n iterations of the game" goes to 0 when n goes to ∞.

An immediate consequence is the next proposition:

Proposition 4. *Let X and Y be two random variables such that $X >_{SP} Y$. Consider the experiment that consists of obtaining simultaneously random values of X and Y, and let*

$B_n \equiv$ *"In the first n iterations, at least half of the times the*

value obtained by X is greater than or equal to the value obtained by Y".

Then,

$$\lim_{n \to \infty} \Pr(B_n) = 1.$$

This proposition can be also obtained considering it as a random walk (see for example [10]).

Then we can say that if we consider the game consisting of obtaining a random value of X and a random value of Y and we repeat it a large enough number of times, if $X \geq_{SP} Y$, we will obtain that most of the half of the times the variable X will take a value greater than the value obtained by Y. However, this does not guarantee that the mean value obtained by the variable X is greater than the mean value obtained by the variable Y.

Let us consider a new example:

Example 3. ([9]) Let us consider the game consisting of rolling two special dice, denoted A and B. Their faces do not show the classical values but the following numbers:

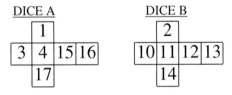

DICE A   DICE B

In each iteration, the dice with the greatest number wins.
In this case the relation Q takes the value:

$$Q(A,B) = \Pr(A > B) + \frac{1}{2}\Pr(A = B) = \Pr(A > B)$$
$$= \Pr(A \in \{3,4\}, B \in \{2\})$$
$$+ \Pr(A \in \{15,16,17\}, B \in \{2,10,11,12,13,14\}) = \frac{21}{36}.$$

Thus, $X \geq_{SP} Y$ and applying the previous result, if we repeat the game indefinitely, it holds that the probability of winning at least half of the times tends to 1.

However, if we calculate the expected value of every dice, we obtain that

$$E(A) = \frac{1}{6}(1+3+4+15+16+17) = \frac{28}{3},$$

$$E(B) = \frac{1}{6}(2+10+11+12+13+14) = \frac{31}{3}.$$

Then, by the criterium of the highest mean, dice B should be preferred. However, if we want to win more times we should choose dice A. Thus, we have an example where the method considered in this paper is better than the classical methods.

4 Statistical Preference versus Stochastic Dominance

In this section we will study the relationships that can be established between first and second degree stochastic dominance and statistical preference, when the random variables are independent.

We recall once more that stochastic dominance only uses the marginal distributions of the variables compared. As we have seen in Subsect. 2.2, every joint cumulative distribution function is the copula of the marginal cumulative distribution functions. For this reason, although we will focus on the product copula, this study can be repeated considering comonotonic or countermonotonic variables, or in general any copula. Some studies have been already published on this topic (see for example [2, 8]).

We first provide a result necessary to prove the connection between first degree stochastic dominance and statistical preference.

Lemma 2. *Let X and Y be two independent random variables with CDF F_1 and F_2 respectively, such that $X \geq_{FSD} Y$, and let h be a positive non-decreasing real upper bounded function. Then,*

$$E(h(X)) \geq E(h(Y)).$$

Proof. Assume first of all that h is a simple function of the form

$$h = x_1 I_{A_1} + x_2 I_{A_2} + \ldots + x_n I_{A_n}$$

where $x_1 \leq x_2 \leq \ldots \leq x_n$ and A_i intervals in $I\!R$ such that $\{A_1, \ldots, A_n\}$ is a partition of $I\!R$ and for all $x \in A_i$ and $y \in A_{i+1}$, $x < y$, this is, the intervals are ordered. Assume $B_0 = \emptyset$ and define $B_i = \cup_{j=1}^{i} A_j$. Then $B_n = I\!R$. Also, $B_i = B_{i-1} \cup A_i$ and $\Pr(A_i) = \Pr(B_i) - \Pr(B_{i-1})$. Then

$$\begin{aligned}
E(h(X)) &= \sum_{i=1}^{n} x_i \Pr(X \in A_i) = \sum_{i=1}^{n} x_i (\Pr(X \in B_i) - \Pr(X \in B_{i-1})) \\
&= \sum_{i=1}^{n} x_i \Pr(X \in B_i) - \sum_{i=1}^{n} x_i \Pr(X \in B_{i-1}) \\
&= \sum_{i=1}^{n-1} (x_i - x_{i+1}) \Pr(X \in B_i) + x_n \Pr(X \in B_n).
\end{aligned}$$

Since $\Pr(X \in B_i) \leq \Pr(Y \in B_i)$ and $x_i - x_{i+1} \leq 0$ for all i,

$$(x_i - x_{i+1}) \Pr(X \in B_i) \geq (x_i - x_{i+1}) \Pr(Y \in B_i) \qquad \forall i \leq n-1.$$

The previous inequality leads to

$$E(h(X)) = \sum_{i=1}^{n-1} (x_i - x_{i+1}) \Pr(X \in B_i) + x_n$$
$$\geq \sum_{i=1}^{n-1} (x_i - x_{i+1}) \Pr(Y \in B_i) + x_n = E(h(Y)).$$

If h is not a simple function, we can build a sequence of simple functions $\{f^m\}_{m \in \mathbb{N}}$ such that $f^m \uparrow h$, then

$$E(h(X)) = \lim_{m \to \infty} E(f^m(X)) \geq \lim_{m \to \infty} E(f^m(Y)) = E(h(Y)). \qquad \Box$$

Theorem 2. *Let X and Y be two independent random variables. It holds that*

$$X \geq_{FSD} Y \quad \Rightarrow \quad X \geq_{SP} Y.$$

Proof. Since

$$\Pr(X > Y) + \frac{1}{2} \Pr(X = Y) + \frac{1}{2} \Pr(X = Y) + \Pr(Y > X) = 1,$$

it suffices to prove that
$$\Pr(X > Y) \geq \Pr(Y > X).$$

Equivalently, $\Pr(X \geq Y) \geq \Pr(Y \geq X)$. It holds that

$$\Pr(X \geq Y) = E(F_2(X)) \qquad \text{and} \qquad \Pr(Y \geq X) = E(F_1(Y)) \leq E(F_2(Y)),$$

where F_1 stands for the CDF of X, F_2 for the CDF of Y and the last inequality follows from the fact that $X \geq_{FSD} Y$. Then, it suffices to prove that

$$E(F_2(X)) \geq E(F_2(Y)).$$

This follows from the previous lemma. $\qquad \Box$

Next we prove that no other implication among first and second degree stochastic dominance and statistical preference holds.

- First of all, we are going to see that the converse implication to the one presented in Theorem 2 does not hold in general.
  For this purpose consider X a continuous random variable with uniform distribution on $(0,1)$, and Y defined as $\frac{3}{4}0 + \frac{1}{4}U(1,2)$, i.e., Y takes the value 0 with probability $\frac{3}{4}$, and the rest of the probability is uniformly distributed in the interval $(1,2)$. It holds that $Q(X,Y) \geq \frac{1}{2}$, because

$$Q(X,Y) = \Pr(X > Y) = \Pr(X \in (0,1)) \Pr(Y = 0) = \frac{3}{4}.$$

Thus $X \geq_{SP} Y$. However, first degree stochastic dominance does not hold because

$$F_1(1) = 1 > F_2(1) = 0.75.$$

Then, we have seen that first degree stochastic dominance is stronger than statistical preference. However, we will see in next section that in some cases they are equivalent properties.

$$\boxed{X \geq_{SP} Y \not\Rightarrow X \geq_{FSD} Y}$$

- Now consider X and Y defined by

X	0	2
Pr_1	$\frac{1}{4}$	$\frac{3}{4}$

Y	1
Pr_2	1

It holds that random variable X is statistically preferred to Y:

$$Q(X,Y) = \Pr(X > Y) + \frac{1}{2}\Pr(X = Y) = \Pr(X > Y)$$

$$= \Pr(X = 2)\Pr(Y = 1) = \frac{3}{4}.$$

Thus $X \geq_{SP} Y$. However, $X \not\geq_{SSD} Y$. The cumulative distribution functions are:

$$F_1(t) = \begin{cases} 0, & \text{if } t < 0, \\ \frac{1}{4}, & \text{if } t \in [0,2), \\ 1, & \text{if } t \geq 2. \end{cases}$$

$$F_2(t) = \begin{cases} 0, & \text{if } t < 1, \\ 1, & \text{if } t \geq 1. \end{cases}$$

and therefore the function G_1 and G_2 are given by:

$$G_1(x) = \int_{-\infty}^{x} F_1(t)dt = \begin{cases} 0, & \text{if } x < 0, \\ \frac{1}{4}x, & \text{if } x \in [0,2), \\ x - \frac{3}{2}, & \text{if } x \geq 2. \end{cases}$$

$$G_2(y) = \int_{-\infty}^{y} F_2(t)dt = \begin{cases} 0, & \text{if } y < 1, \\ y - 1, & \text{if } y \geq 1. \end{cases}$$

If we represent them graphically, we obtain Fig. 5. Second degree stochastic dominance is not satisfied. Then, we have proven that

$$\boxed{X \geq_{SP} \not\Rightarrow X \geq_{SSD} Y}$$

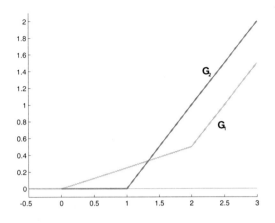

Fig. 5 Graphics of G_1 and G_2.

- Furthermore, the reciprocal implication neither holds. We can see it with the random variables X and Y given by

$$
\begin{array}{c|cc}
X & 1 & 2 \\
\hline
\Pr_1 & \frac{1}{4} & \frac{3}{4}
\end{array}
\qquad
\begin{array}{c|cc}
Y & 0 & 2 \\
\hline
\Pr_2 & \frac{1}{8} & \frac{7}{8}
\end{array}
$$

It holds that X stochastically dominates by second degree random variable Y:

$$
F_1(t) = \begin{cases} 0, & \text{if } t < 1, \\ \frac{1}{4}, & \text{if } t \in [1,2), \\ 1, & \text{if } t \geq 2. \end{cases}
$$

$$
F_2(t) = \begin{cases} 0, & \text{if } t < 0, \\ \frac{1}{8}, & \text{if } t \in [0,2), \\ 1, & \text{if } t \geq 2. \end{cases}
$$

And therefore

$$
G_1(t) = \begin{cases} 0, & \text{if } t < 1, \\ \frac{1}{4}t - \frac{1}{4}, & \text{if } t \in [1,2), \\ t - \frac{7}{4}, & \text{if } t \geq 2. \end{cases}
$$

$$
G_2(t) = \begin{cases} 0, & \text{if } t < 0, \\ \frac{1}{8}t, & \text{if } t \in [0,2), \\ t - \frac{7}{4}, & \text{if } t \geq 2. \end{cases}
$$

In Fig. 6 we can see the graphics of those functions. We can observe that $X \geq_{\text{SSD}} Y$. However, it can be easily proven that $X \ngeq_{\text{SP}} Y$.

$$
\boxed{X \geq_{\text{SSD}} Y \nRightarrow X \geq_{\text{SP}} Y}
$$

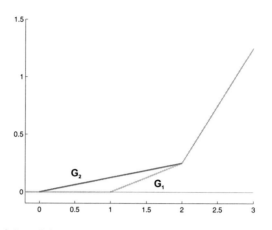

Fig. 6 Graphics of G_1 and G_2.

- We can conclude that there is no relation between second degree stochastic dominance and statistical preference. Furthermore, we have seen that neither second degree stochastic dominance nor statistical preference imply first degree stochastic dominance. It only remains to see if both properties together imply first degree stochastic dominance. The answer is again negative. Consider X and Y two independent random variables given by:

$$\begin{array}{c|cc} X & 1 & 5 \\ \hline \Pr_1 & \frac{1}{2} & \frac{1}{2} \end{array} \qquad \begin{array}{c|cc} Y & 0 & 10 \\ \hline \Pr_2 & \frac{9}{10} & \frac{1}{10} \end{array}$$

For these variables it holds that

$$Q(X,Y) = \Pr(X > Y) + \frac{1}{2}\Pr(X = Y) = \Pr(X > Y)$$
$$= \Pr(Y = 0) = \frac{9}{10} > \frac{1}{2}.$$

Thus $X \geq_{\text{SP}} Y$. Furthermore, since the cumulative distribution functions are

$$F_1(t) = \begin{cases} 0, & \text{if } t < 1, \\ \frac{1}{2}, & \text{if } t \in [1,5), \\ 1, & \text{if } t \geq 5, \end{cases}$$

$$F_2(t) = \begin{cases} 0, & \text{if } t < 0, \\ \frac{9}{10}, & \text{if } t \in [0,10), \\ 1, & \text{if } t \geq 10, \end{cases}$$

the functions G_1 and G_2 are given by

$$G_1(t) = \begin{cases} 0, & \text{if } t < 1, \\ \frac{1}{2}(t-1), & \text{if } t \in [1,5), \\ t-3, & \text{if } t \geq 5, \end{cases}$$

$$G_2(t) = \begin{cases} 0, & \text{if } t < 0, \\ \frac{9}{10}t, & \text{if } t \in [0,10), \\ t-1, & \text{if } t \geq 10. \end{cases}$$

If we look at their graphical representations in Fig. 7, we can see that $X \geq_{\text{SSD}} Y$. However, it holds that

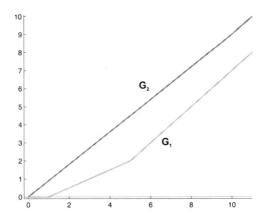

Fig. 7 Graphics of G_1 and G_2.

$$F_1(5) = 1 > \frac{9}{10} = F_2(5),$$

thus X cannot stochastically dominate Y by first degree, i.e., $X \not\geq_{\text{FSD}} Y$. Therefore:

$$\left. \begin{array}{c} X \geq_{\text{SP}} Y \\ X \geq_{\text{SSD}} Y \end{array} \right\} \not\Rightarrow X \geq_{\text{FSD}} Y$$

To summarize the results obtained in this section, we present Fig. 8. Here, we denote by FSD and SSD first and second degree stochastic dominance and by SP, statistical preference.

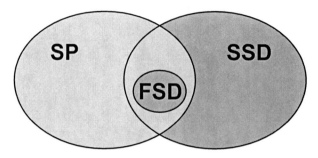

Fig. 8 Connection among statistical preference, first degree and second degree stochastic dominance.

5   Statistical Preference and Some Usual Families of Distributions

In this section we characterize statistical preference for some of the most usual distributions. In some cases, our work will be easy since for some of the distributions we will manage, the expressions that take the relations Q have been calculated in [8].

We begin by the Bernoulli distribution, after that we will focus on some continuous distributions.

5.1   Discrete Distributions

In this first subsection, we study statistical preference between two independent Bernoulli distributions. We recall that the probability distribution of a Bernoulli random variable with parameter $p \in (0, 1)$ is given by

Value	0	1
Probability	$1 - p$	p

Proposition 5. *Let X and Y be two independent random variables Bernoulli distributed with parameters p_1 and p_2 respectively. Then it holds that*

$$X \geq_{SP} Y \Leftrightarrow p_1 \geq p_2.$$

Proof. First of all we are going to calculate the expression of the relation Q:

$$Q(X,Y) = \Pr(X > Y) + \frac{1}{2}\Pr(X = Y)$$

$$= \Pr(X = 1)\Pr(Y = 0) + \frac{1}{2}\left(\Pr(X = 0)\Pr(Y = 0) + \Pr(X = 1)\Pr(Y = 1)\right)$$

$$= p_1(1 - p_2) + \frac{1}{2}\left((1 - p_1)(1 - p_2) + p_1 p_2\right) = \frac{1}{2}(p_1 - p_2 + 1).$$

Then it holds that

$$X \geq_{\text{FSD}} Y \Leftrightarrow Q(X,Y) \geq \frac{1}{2} \Leftrightarrow \frac{1}{2}(p_1 - p_2 + 1) \geq \frac{1}{2} \Leftrightarrow p_1 \geq p_2. \qquad \square$$

As a corollary we can state that for Bernoulli random variables first and second degree stochastic dominance and statistical preference are equivalent properties.

Corollary 4. *Let X and Y be two independent random variables with Bernoulli distributions. Then*

$$X \geq_{\text{FSD}} Y \Leftrightarrow X \geq_{\text{SP}} Y \Leftrightarrow X \geq_{\text{SSD}} Y.$$

5.2 Continuous Distributions

Before beginning the study of statistical preference between continuous random variables, we summarize in Table 1 the density function and the parameters of the distributions that we will analyze trough this section.

Table 1 Density functions and parameters of the continuous distributions studied.

Distribution	Density function	Parameters
$Exp(\lambda)$ Exponential	$\lambda e^{-\lambda x}, \quad x \in [0, \infty)$	$\lambda > 0$
$\beta(p,1)$ Beta	$px^{p-1}, \quad x \in [0, 1]$	$p > 0$
$\mathscr{P}_a(\lambda)$ Pareto	$\lambda x^{-(\lambda+1)}, \quad x \in [1, \infty)$	$\lambda > 0$
$U(a,b)$ Uniform	$\frac{1}{b-a}, \quad x \in [a, b]$	$a, b \in \mathbb{R}, \ a < b$
$N(\mu, \sigma)$ Normal	$\frac{1}{\sqrt{2\pi\sigma^2}} e^{-\frac{(x-\mu)^2}{2\sigma^2}}, \quad x \in \mathbb{R}$	$\mu \in \mathbb{R}, \ \sigma > 0$

Remark 6. In general, the beta distribution depends on two parameters, $p, q > 0$, and its density function is

$$f_{\beta(p,q)}(x) = \frac{1}{B(p,q)} x^{p-1}(1-x)^{q-1}, \quad x \in [0,1].$$

In particular, we consider the case $q = 1$.

Analogously, the Pareto distribution depends on two parameters, a, b, and the density function is given by

$$f(x) = \frac{ab^a}{x^{a+1}}, \qquad x > b.$$

As in [8] we will focus on the case $b = 1$.

Lemma 3. *Let X and Y be two independent random variables with exponential distributions, $X \equiv Exp(\lambda_1)$ and $Y \equiv Exp(\lambda_2)$, respectively. Then, X is statistically preferred to Y if and only if*

$$\lambda_1 \leq \lambda_2.$$

Proof. We first prove that $Q(X,Y) = \dfrac{\lambda_2}{\lambda_1 + \lambda_2}$.

$$Q(X,Y) = \Pr(X > Y) = \int_0^\infty \lambda_1 e^{-\lambda_1 x} dx \int_0^x \lambda_2 e^{-\lambda_2 y} dy = \int_0^\infty \lambda_1 e^{-\lambda_1 x}(1 - e^{-\lambda_2 x}) dx$$

$$= \int_0^\infty \lambda_1 e^{-\lambda_1 x} dx - \int_0^\infty \lambda_1 e^{-(\lambda_1 + \lambda_2)x} dx = 1 - \frac{\lambda_1}{\lambda_1 + \lambda_2} = \frac{\lambda_2}{\lambda_1 + \lambda_2}.$$

Thus, it holds that

$$X \geq_{SP} Y \Leftrightarrow Q(X,Y) \geq \frac{1}{2} \Leftrightarrow \frac{\lambda_2}{\lambda_1 + \lambda_2} \geq \frac{1}{2} \Leftrightarrow \lambda_2 \geq \lambda_1. \qquad \square$$

As for Bernoulli random variables, statistical preference and first and second degree stochastic dominance are equivalent properties for exponential distributions.

Corollary 5. *Let X and Y be two independent random variables with exponential distribution. Then, it holds that*

$$X \geq_{FSD} Y \Leftrightarrow X \geq_{SP} Y \Leftrightarrow X \geq_{SSD} Y.$$

Now we are going to analyze the behavior of uniform distributions.

Proposition 6. *Let X and Y be two independent random variables with uniform distributions, $U(a,b)$ and $U(c,d)$ respectively. Then, $X \geq_{SP} Y$ if and only if:*

1. $a + b \geq c + d$, if $a \leq c \leq d < b$ or $c < a < b \leq d$.
2. always, if $c < a < d \leq b$.

Proof. 1. Suppose that $a \leq c \leq d < b$. Then,

$$Q(X,Y) = \Pr(X > Y) = \int_d^b \frac{1}{b-a} dx + \int_c^d \int_c^x \frac{1}{b-a} \frac{1}{d-c} dy dx$$

$$= \frac{b-d}{b-a} + \int_c^d \frac{1}{b-a} \frac{x-c}{d-c} dx = \frac{b-d}{b-a} + \frac{(d-c)^2}{2(d-c)(b-a)}$$

$$= \frac{2b-c-d}{2(b-a)}.$$

Then, it holds that $X \geq_{SP} Y$ if and only if:

$$\frac{2b-c-d}{2(b-a)} \geq \frac{1}{2} \Leftrightarrow b + a \geq c + d.$$

In the case in which $c < a < b \leq d$, it is similar to see that

$$Q(X,Y) \geq \frac{b+a-2c}{2(d-c)}.$$

Thus, $Q(X,Y) \geq \frac{1}{2}$ if and only if $a + b \geq c + d$.
2. If $c < a < d \leq b$, $X \geq_{FSD} Y$, and then $X \geq_{SP} Y$. $\qquad \square$

For uniform distributions, first degree stochastic dominance and statistical preference are not equivalent in general. In fact, first degree stochastic dominance does not hold when the first case of the proof of the previous proposition holds.

We now study normal distributions.

Proposition 7. *Let X and Y be two independent random variables with normal distributions, $N(\mu_1, \sigma_1)$ and $N(\mu_2, \sigma_2)$, respectively. Then, X will be statistically preferred to Y if and only if*

$$\mu_1 \geq \mu_2.$$

Proof. The relation Q takes the value (see [7]):

$$Q(X,Y) = F_{N(0,1)}\left(\frac{\mu_1 - \mu_2}{\sqrt{\sigma_1^2 + \sigma_2^2}}\right).$$

Then:

$$X \geq_{SP} Y \Leftrightarrow Q(X,Y) \geq \frac{1}{2} \Leftrightarrow F_{N(0,1)}\left(\frac{\mu_1 - \mu_2}{\sqrt{\sigma_1^2 + \sigma_2^2}}\right)$$

$$\Leftrightarrow \frac{\mu_1 - \mu_2}{\sqrt{\sigma_1^2 + \sigma_2^2}} \geq 0 \Leftrightarrow \mu_1 = \mu_2. \qquad \square$$

Given two normal random variables, $X \sim N(\mu_1, \sigma_1)$ and $Y \sim N(\mu_2, \sigma_2)$, it holds that $X \geq_{FSD} Y$ if and only if they are identically distributed, $\mu_1 = \mu_2$ and $\sigma_1 = \sigma_2$, (see [15]). Then, statistical preference is not equivalent to first degree stochastic dominance for normal random variables.

Remark 7. For independent normal distributions, the variance of the variables are not important when studying statistical preference. For this reason, statistical preference is equivalent to the criterium of maximum mean in the comparison of normal random variables.

We next focus on the family of Pareto distribution.

Proposition 8. *Let X and Y be two independent random variables with Pareto distributions, $X \equiv \mathscr{P}_a(\lambda_1)$ and $Y \equiv \mathscr{P}_a(\lambda_2)$, respectively. Then, X is statistically preferred to Y if and only if:*

$$\lambda_2 \geq \lambda_1.$$

Proof. First of all, calculate the expression of Q:

$$Q(X,Y) = \Pr(X > Y) = \int_1^\infty \int_1^x \lambda_1 x^{-\lambda_1 - 1} \lambda_2 y^{-\lambda_2 - 1} \, dy \, dx$$

$$= \int_1^\infty \lambda_1 x^{-\lambda_1 - 1} \left(1 - x^{-\lambda_2}\right) dx = 1 - \frac{\lambda_1}{\lambda_1 + \lambda_2}.$$

Then,

$$X \geq_{SP} Y \Leftrightarrow 1 - \frac{\lambda_1}{\lambda_1 + \lambda_2} \geq \frac{1}{2} \Leftrightarrow \lambda_2 \geq \lambda_1. \qquad \Box$$

As for exponential and Bernoulli distributions, the equivalence between first and second degree stochastic dominance and statistical preference holds for Pareto distributions.

Corollary 6. *Let X and Y be two independent random variables with Pareto distributions. Then*

$$X \geq_{FSD} Y \Leftrightarrow X \geq_{SP} Y \Leftrightarrow X \geq_{SSD} Y.$$

Concerning the beta distribution, we have proven the following result.

Proposition 9. *Let X and Y be two independent random variables with beta distributions, $X \equiv \beta(p_1, 1)$ and $Y \equiv \beta(p_2, 1)$, respectively. Then $X \geq_{SP} Y$ if and only if*

$$p_1 \geq p_2.$$

Proof. We first compute the expression of the relation Q.

$$Q(X,Y) = \Pr(X > Y) = \int_0^1 \int_0^x p_1 x^{p_1-1} p_2 y^{p_2-1} dy\, dx$$

$$= \int_0^1 p_1 x^{p_1-1} p x^{p_2} dx = \frac{p_1}{p_1 + p_2}.$$

Table 2 Characterizations of statistical preference between independent random variables included in the same family of distributions.

Distribution X	Distribution Y		Q(X,Y)	Condition
$B(p_1)$	$B(p_2)$		$\frac{1}{2}(p_1 - p_2 + 1)$	$p_1 \geq p_2$
$Exp(\lambda_1)$	$Exp(\lambda_2)$		$\frac{\lambda_2}{\lambda_1 + \lambda_2}$	$\lambda_2 \geq \lambda_1$
$U(a,b)$	$U(c,d)$			
		$a \leq c \leq d < b$	$\frac{2b-c-d}{2(b-a)}$	$a+b \geq c+d$
		$c < a < b \leq d$	$\frac{a+b-2c}{2(d-c)}$	$a+b \geq c+d$
		$c < a < d \leq b$	$\frac{-d^2-a^2+2bd-2bc+2ac}{2(d-c)(b-a)}$	Always
$\mathscr{P}_a(\lambda_1)$	$\mathscr{P}_a(\lambda_2)$		$\frac{\lambda_2}{\lambda_1 + \lambda_2}$	$\lambda_2 \geq \lambda_1$
$\beta(p_1,1)$	$\beta(p_2,1)$		$\frac{p_1}{p_1 + p_2}$	$p_1 \geq p_2$
$N(\mu_1,\sigma_1)$	$N(\mu_2,\sigma_2)$		$F_{N(0,1)}\left(\frac{\mu_1-\mu_2}{\sqrt{\sigma_1^2+\sigma_2^2}}\right)$	$\mu_1 \geq \mu_2$

The it holds that

$$X \geq_{\mathrm{SP}} Y \Leftrightarrow \frac{p_1}{p_1 + p_2} \geq \frac{1}{2} \Leftrightarrow p_1 \geq p_2. \qquad \square$$

The equivalence between first and second degree stochastic dominances and statistical preference also holds for beta distributions.

Corollary 7. *Let X and Y be two independent random variables with beta distributions. Then,*

$$X \geq_{\mathrm{FSD}} Y \Leftrightarrow X \geq_{\mathrm{SP}} Y \Leftrightarrow X \geq_{\mathrm{SSD}} Y.$$

In Table 2 we have summarized the results that we have obtained in this section.

6 Conclusions

This contribution is an approach to a new way of pairwise comparison of probability distributions: statistical preference. This concept has some advantages over other methods like stochastic dominance or maximum mean: it uses all the available information, every pair of random variables can be compared,.... Therefore it could be more appropriate in consensus problems where distributions of intensities as preferences are employed as long as possible and it is necessary to compare them.

This concept was introduced by De Schuymer et al. ([4]). We have made an analysis of the definition, interpretation and properties of statistical preference. We have also compared it with stochastic dominance, trying to establish general relations between them. We have obtained that in general statistical preference is a weaker condition than first degree stochastic dominance, and it is not related to second degree stochastic dominance. For some of the most usual distributions, we have characterized statistical preference, and we have seen that in some cases first and second degree stochastic dominance and statistical preference are equivalent.

As a future work we consider the following open points:

- Q is a graded relation, and we use its $\frac{1}{2}$-cut to get a crisp order. We would like to study what happens if we choose an α-cut with $\alpha \neq \frac{1}{2}$.
- We will try to generalize the results obtained for the product operator (independent random variables) to any copula. We are particularly interested in comonotonic and countermonotonic variables.
- We also want to continue the analysis of statistical preference in the most usual distributions, trying to characterize the cases in which stochastic dominance and statistical preference are equivalent.
- We would like to be able to apply these studies in real consensus processes.

Acknowledgements

The research reported on in this contribution has been partially supported by Project FEDER-MEC-MTM2007-61193.

The authors would like to thank Enrique Miranda and an anonymous referee for their valuable suggestions.

References

1. Billingsley, P.: Probability and Measure. John Wiley & Sons, Chichester (1986)
2. De Baets, B., De Meyer, H.: On the cycle-transitive comparison of artificially coupled random variables. International Journal of Approximate Reasoning 47, 306–322 (2008)
3. De Baets, B., De Meyer, H., De Schuymer, B.: Transitive comparison of random variables. In: Klement, E.P., Mesiar, R. (eds.) Logical, Algebraic, and Probabilistic Aspects of Triangular Norms. Elsevier, Amsterdam (2005)
4. De Baets, B., De Meyer, H., De Schuymer, B., Jenei, S.: Cyclic evaluation of transitivity of reciprocal relations. Social Choice and Welfare 26, 217–238 (2006)
5. De Meyer, H., De Baets, B., De Schuymer, B.: On the transitivity of the comonotonic and countermonotonic comparison of random variables. Journal of Multivariate Analysis 98, 177–193 (2007)
6. De Schuymer, B., De Meyer, H., De Baets, B.: A fuzzy approach to stochastic dominance of random variables. In: De Baets, B., Kaynak, O., Bilgiç, T. (eds.) IFSA 2003. LNCS (LNAI), vol. 2715, pp. 253–260. Springer, Heidelberg (2003)
7. De Schuymer, B., De Meyer, H., De Baets, B.: Cycle-transitive comparison of independent random variables. Journal of Multivariate Analysis 96, 352–373 (2005)
8. De Schuymer, B., De Meyer, H., De Baets, B.: Extreme copulas and the comparison of ordered lists. Theory and Decision 62, 195–217 (2007)
9. De Schuymer, B., De Meyer, H., De Baets, B., Jenei, S.: On the cycle-transitivity of the dice model. Theory and Decision 54, 261–285 (2003)
10. Feller, H.: An introduction to probability theory and its applications, vol. II. John Wiley & Sons,Inc. Chichester (1973)
11. Herrera, F., Herrera-Viedma, E., Verdegay, J.L.: A model of consensus in group decision making under linguistic assessments. Fuzzy Sets Syst 78, 73–87 (1996)
12. Herrera-Viedma, E., Herrera, F., Chiclana, F.: A Consensus Model for Multiperson Decision Making With Different Preference Structures. IEEE Transactions on systems, man and cybernetics–Part A: Systems and Humans 32, 394–402 (2002)
13. Levy, H.: Stochastic dominance. Kluwer Academic Publishers, Dordrecht (1998)
14. Montes, I., Martinetti, D., Díaz, S.: On the Statistical Preference as a Pairwise Comparison for Random Variables. In: Proceedings of EUROFUSE 2009 Workshop: Preference modelling and decision analysis, Pamplona, Spain (2009)
15. Müller, A., Stoyan, D.: Comparison Methods for Stochastic Models and Risk. John Wiley & Sons, Chichester (2002)
16. Nelsen, R.: An Introduction to Copulas. Lecture Notes in Statistics, vol. 139. Springer, New York (2006)
17. Rademaker, M., De Baets, B.: Voting with intensities, group consensus and stochastic dominance. In: Proceedings of EUROFUSE 2009 Workshop: Preference modelling and decision analysis, Pamplona, Spain (2009)
18. Triantaphyllou, E.: Multi-Criteria Decision Making Methods: A Comparative Study. Kluwer, Norwell (2000)
19. Zendehdel, K., Rademaker, M., De Baets, B., Van Huylenbroeck, G.: Qualitative valuation of environmental criteria though a group consensus based on stochastic dominance. Ecological Economics 67, 253–264 (2008)

Part II: Aspects of Group Decision Making, Social Choice and Voting

Consensus with Oneself: Within-Person Choice Aggregation in the Laboratory

Michel Regenwetter and Anna Popova

Abstract. Unfortunately, the decision sciences are segregated into nearly distinct academic societies and distinct research paradigms. This intellectual isolationism has allowed different approaches to the decision sciences to suffer from different, but important, conceptual gaps. Following earlier efforts to cross-fertilize individual and social choice research, this paper applies behavioral social choice concepts to individual decision making.

Repeated individual choice among identical pairs of choice alternatives often fluctuates dramatically over even very short time periods. Social choice theory usually ignores this because it identifies each individual with a single fixed weak order. Behavioral individual decision research may expose itself to Condorcet paradoxes because it often interprets a decision maker's modal choice (i.e., majority choice) over repeated trials as revealing their "true" preference. We investigate variability in choice behavior within each individual in the research lab. Within that paradigm, we look for evidence of Condorcet cycles, as well as for the famed disagreement between the Condorcet and Borda aggregation methods. We also illustrate some methodological complexities involved with likelihood ratio tests for Condorcet cycles in paired comparison data.

Keywords: Behavioral social choice, Borda score, Condorcet paradox, consensus among consensus methods, voting paradoxes, weak stochastic transitivity.

1 Introduction

The decision sciences are segregated and include two nearly distinct academic constituencies: social choice theorists and individual decision researchers.

Michel Regenwetter · Anna Popova
University of Illinois at Urbana-Champaign, USA

E. Herrera-Viedma et al. (Eds.): Consensual Processes, STUDFUZZ 267, pp. 95–121.
springerlink.com   © Springer-Verlag Berlin Heidelberg 2011

Unfortunately, these two groups engage in very little interaction and cross-fertilization. They meet at separate meetings and publish in separate journals. Yet, each of these research groups has much to offer to the other (see, e.g., Regenwetter et al., 2007a, for a related discussion).

For example, social choice theorists take it for granted that preferences vary across individuals and they agonize over the possibility or impossibility of aggregating such preferences. Yet, behavioral decision research aggregates individual choices as a matter of routine. The most frequent choice made among a pair of objects is called the *modal choice*. Much influential work in behavioral decision research tests theories of individual decision making by focusing on the modal choices among multiple decision makers (e.g., Brandstätter, Gigerenzer, and Hertwig, 2006; and to some degree even the seminal work of Kahneman and Tversky, 1979, and Tversky and Kahneman, 1981). By routinely testing individual decision theory against interindividual modal choice behavior, behavioral decision research may expose itself to aggregation artifacts such as the famous Condorcet paradox (Condorcet, 1785). Many researchers in individual decision making have highlighted that even a single individual can fluctuate substantially in her choice among the exact same choice alternatives when asked to make the same choice repeatedly, even over a short period of time. Some of these researchers then proceed to aggregate an individual's choices by majority rule, i.e., they focus on the decision maker's modal pairwise choices, apparently unconcerned about within-respondent Condorcet paradoxes (e.g., the influential work of Tversky, 1969).

At the same time, by assuming that each individual has a fixed weak order preference, social choice theory may unnecessarily create its own problems. When individual preferences are probabilistic, ballots become random variables. This raises important issues of statistical confidence in election outcomes. The problem is exacerbated when ballot casting is error prone (as exemplified prominently in the 2000 Florida recounts). With the exception of spatial model fitting and some econometric analyses, statistical issues in the analysis and interpretation of empirical data, such as goodness-of-fit, hypothesis testing, inference, confidence, and statistical replicability of social choice outcomes have essentially been a nontopic in social choice theory.

A closely related major distinction between social choice theory and individual decision research is that the latter has developed a full-fledged behavioral program that compares and contrasts rational choice theory with actual choice behavior (e.g., Kahneman and Tversky, 2000). Such a program in social choice is still in its infancy, with very few scholars systematically studying social choice procedures in the laboratory, on survey data or on real ballot data. In addition, much early behavioral social choice research circumnavigated the nontrivial methodological problems that are associated with behavioral research, such as the use of statistical concepts and tools that permit scientifically sound inferences from empirical data (see, e.g., Regenwetter et al., 2006, 2009b; Regenwetter, 2009, for a discussion).

In the empirical part of this paper, we show two applications of social choice concepts to individual decision research. First, we look for Condorcet cycles in data aggregated within a given person who made repeated choices among gambles in the laboratory. This also illustrates a major methodological hurdle for maximum likelihood based testing of Condorcet cycles. Second, we consider the famed disagreement between Condorcet and Borda (going back to Borda, 1770; Condorcet, 1785), again on choice data aggregated within each person.

2 Basic Concepts

DEFINITION. Let C be a finite collection of choice alternatives. A *binary relation* R on C is a collection of ordered pairs of objects in C, that is, $R \subseteq C \times C$. Let $I_C = \{(x, x) |\ x \in C\}$, let $R^{-1} = \{(y, x) | (x, y) \in R\}$, and let $\bar{R} = (C \times C) \setminus R$. A binary relation R on C is

$$\text{asymmetric if } R \subseteq \bar{R}^{-1},$$
$$\text{complete if } R \cup R^{-1} \cup I_C = C \times C,$$
$$\text{strongly complete if } R \cup R^{-1} = C \times C,$$
$$\text{transitive if } RR \subseteq R,$$
$$\text{negatively transitive if } \bar{R}\bar{R} \subseteq \bar{R}.$$

A *weak order* is a transitive and strongly complete binary relation, and a *strict weak order* is an asymmetric, and negatively transitive binary relation. A *strict linear order* is a transitive, asymmetric, and complete binary relation.

In models of preference, it is natural to write $(x, y) \in R$ as xRy and to read the relationship as "x is preferred to (better than) y." For related definitions and classical theoretical work on binary preference representations, see, e.g., Fishburn (1979); Krantz et al. (1971); Roberts (1979).

Throughout this paper, we consider binary paired comparison data from psychological decision making experiments. The main features of the experiments are spelled out in the Appendix (see Regenwetter et al., 2010, 2009a; Regenwetter and Davis-Stober, 2009, for a full description of the experimental paradigms). Participants in these laboratory experiments make hundreds of decisions, many of which are repetitions of a small set of pairwise choices, spread out over time throughout the study. Each individual pairwise decision observation is called a *trial* of the experiment. In some cases, the experiment uses a *two-alternative forced choice paradigm*, where the decision maker is offered two choice alternatives on any given trial and asked (i.e., "forced") to choose one of the two offered alternatives. Figures 1 and 2 provide screen shots of two such trials.

In some cases, the experiment uses a *ternary paired comparison paradigm*, where the decision maker is offered two choice alternatives on any given trial and can either choose one of the two, or indicate indifference. For simplicity

Fig. 1 Example display of a Cash I two-alternative forced choice stimulus.

Fig. 2 Example display of a Noncash two-alternative forced choice stimulus.

and tractability, in this paper, we only analyze data from participants who never used the indifference option in the ternary paired comparison paradigm. In all experiments, each pair of choice alternatives was offered equally many times to the decision maker over the course of the experiment. Hence, in all cases, we observe frequencies N_{xy} with which a person chose x over y, where $N_{xy} + N_{yx} = N$, the total number of times that each pair was presented to the participant.

2.1 The Empirical Sample Space

We now review the statistical assumptions we make about the empirical sample space. We assume that, for each individual, there exists an unknown probability $\bar{P}_{xy}$ that the individual chooses x over y at any given moment. All the experiments we analyze have used decoys between related pairs of choice alternatives so as to minimize the participants' ability to recognize or remember earlier decisions. The decoys allow us to assume that the observed binary choices among nondecoys are statistically independent. Because the experiment typically takes only an hour or two, we further assume that the binary choice probabilities do not change over time. As a consequence of these two assumptions about the data generating process, i.e., as a consequence of *independent and identically distributed (iid) sampling* assumptions, the quantities N_{xy} form a system of independent binomial random variables, each with a known number N of repetitions and with an unknown probability $\bar{P}_{xy}$ of choosing x over y.

In some parts of the paper, we will refer to a situation more closely related to classical social choice theory, in which we sample individuals from a population, with each individual having a deterministic preference among any given pair of choice alternatives. In that case, we denote the probability that a randomly selected decision maker prefers x to y by $\widetilde{P}_{xy}$. Our definitions can be applied to either scenario. Throughout the paper, we use P_{xy} whenever either $\bar{P}_{xy}$ or $\widetilde{P}_{xy}$ may be substituted, depending on the context. In the empirical part, we only study within-person choice probabilities $\bar{P}_{xy}$ in this paper. Population choice probabilities have been the focus in most of our cited prior work.

2.2 The Condorcet Criterion

DEFINITION. Consider a finite set $\mathcal{C}$ of choice alternatives and a system of probabilities P_{xy} for distinct $x, y \in \mathcal{C}$. A choice alternative $x \in \mathcal{C}$ is *strictly majority preferred* (i.e., *strictly Condorcet preferred*) to a choice alternative $y \in \mathcal{C}, y \neq x$, if and only if $P_{xy} > \frac{1}{2}$. A choice alternative x is a *strict Condorcet winner* if and only if

$$P_{xy} > \frac{1}{2}, \qquad \forall y \in \mathcal{C}, y \neq x. \tag{1}$$

A *strict Condorcet cycle* occurs when $P_{xy} > 1/2$, $P_{yz} > 1/2$, $P_{zx} > 1/2$, for some selection of distinct $x, y, z \in \mathcal{C}$.

This definition of majority rule and of a Condorcet cycle is consistent with the more general framework developed in Regenwetter et al. (2002). In social choice theory, where individual preferences are routinely assumed to be deterministic (strict) weak orders, treating P_{xy} as the probability $\widetilde{P}_{xy}$ that a randomly selected voter prefers x to y, the existence of a Condorcet cycle is commonly referred to as a *Condorcet paradox* because individual decision makers have transitive (weak order) preferences, whereas the aggregate preference relation is intransitive. This is often interpreted to mean that rational individuals can make collectively irrational decisions.

2.3 What Can Social Choice Theory and Individual Decision Research Teach Each Other about the Condorcet Paradox?

Suppose that $\mathcal{C} = \{A, B, C\}$. Denote the strict linear order $\{(A, B), (B, C), (A, C)\}$ by ABC and do likewise for all other strict linear orders. Write

$ABCA$ for the cyclical binary relation $\{(A, B), (B, C), (C, A)\}$. Figure 3 gives a geometric illustration of the Condorcet paradox using the unit cube of joint Binomial probabilities $(P_{AB}, P_{AC}, P_{BC}) \in [0, 1]^3$. In the upper left and bottom displays, the vertex labeled ABC with coordinates $(P_{AB}, P_{AC}, P_{BC}) = (1, 1, 1)$ denotes the degenerate distribution where all probability mass is concentrated on the linear order ABC. We consider first a situation in which we are sampling individuals from a population, and where P_{xy} denotes the probability $\widetilde{P}_{xy}$ that such an individual prefers x over y. From that point of view, each of the vertices ABC, BCA and CAB corresponds to a degenerate distribution where the entire electorate is unanimous. The shaded triangle is the convex hull of these three vertices, and this is the collection of all possible joint Binomial probabilities that can occur when the only possible individual preferences are the linear orders ABC, BCA, and CAB.

The upper right display of Figure 3 shows the collection of joint Binomials that lead to the Condorcet cycle $ABCA$. Geometrically, they form a half-unit cube attached to the vertex marked $ABCA$. Behavioral decision researchers have discussed the possibility that individual decision makers may have cyclical preferences, themselves (e.g., Tversky, 1969). If, contrary to standard social choice theoretic assumptions, the entire electorate had the unanimous cyclical preference $ABCA$, then the joint Binomials would be located at that vertex. In that case, the Condorcet cycle would not be a voting paradox, since it would then be representative of the population's unanimously cyclical preferences.

The bottom display of Figure 3 shows the standard example of a Condorcet paradox, where (again substituting $\widetilde{P}_{xy}$ for P_{xy}) one third of the population has strict linear order ABC, one third of the population has preference BCA, and another third has preference order CAB. This yields a Condorcet cycle, because $\widetilde{P}_{AB} = \widetilde{P}_{BC} = \widetilde{P}_{CA} = \frac{2}{3}$: Indeed, the center of gravity of the vertices ABC, BCA, CAB lies in the interior of the half-unit cube associated with the Condorcet cycle $ABCA$.

Figure 4 turns the Condorcet paradox on its head. Here (upper left), if we substitute $\widetilde{P}_{xy}$ for P_{xy}, the population is made up entirely of voters who either have preference orders BAC, ACB, or the cyclical preference $ABCA$. The Binomial probabilities where Condorcet yields the linear order ABC are indicated on the upper right. The star in the bottom display shows a special case of the upper left, where each of $BAC, ACB, ABCA$ is held by one third of the population. Here, the aggregate Condorcet outcome is the linear order ABC, a preference held not even by a single individual. Arguably, one could label this situation a voting paradox.

While these observations may have some implications for social choice theory, we concentrate on the important implications for behavioral individual decision research. Consider the fact that the choices of a single individual fluctuate over repeated decisions. Tversky (1969) set out to show that some individual decision makers sometimes have intransitive individual preferences. Tversky tackled variable choice data from individual decision

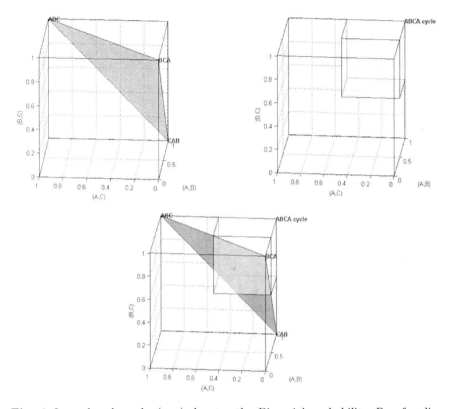

Fig. 3 In each subgraph, (x, y) denotes the Binomial probability $P_{x,y}$ for distinct $x, y \in \mathcal{C} = \{A, B, C\}$. Upper left: Substituting $\widetilde{P}_{xy}$ for P_{xy}, these are the binary choice probabilities if all voters have preferences ABC, BCA, or CAB. The triangle shows all possible binomial probabilities consistent with probability distributions over $\{ABC, BCA, CAB\}$. Again substituting $\widetilde{P}_{xy}$ for P_{xy}, the vertices ABC, BCA, CAB denote the three cases where voters are unanimous. Upper right: Binary choice probabilities that yield the majority cycle $ABCA$. Bottom: Classical Condorcet paradox. The star denotes the binary choice probabilities induced by a uniform distribution on $\{ABC, BCA, CAB\}$. This point, which is the center of gravity of the vertices marked ABC, BCA, CAB, lies inside the half-unit cube associated with the majority cycle $ABCA$.

makers by identifying transitivity of individual preferences with "weak stochastic transitivity," which we define next at the level of the general binary choice probabilities P_{xy}.

DEFINITION. *Weak stochastic transitivity* (see Block and Marschak, 1960; Luce and Suppes, 1965) is the Null Hypothesis in the following test:

$$\begin{cases} H_0 : \forall \text{ (distinct) } x, y, z \in \mathcal{C} : [(P_{xy} \geq 1/2) \wedge (P_{yz} \geq 1/2)] \Rightarrow (P_{xz} \geq 1/2) \\ H_A : \exists \text{ (distinct) } x, y, z \in \mathcal{C} : (P_{xy} \geq 1/2) \wedge (P_{yz} \geq 1/2) \wedge (P_{xz} < 1/2). \end{cases} \quad (2)$$

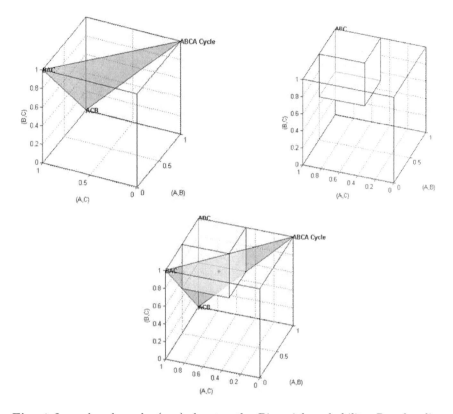

Fig. 4 In each subgraph, (x,y) denotes the Binomial probability $P_{x,y}$ for distinct $x, y \in \mathcal{C} = \{A, B, C\}$. Upper left: Substituting $\tilde{P}_{xy}$ for P_{xy}, these are the binary choice probabilities if all voters have preferences BAC, ACB, or $ABCA$. The triangle shows all possible binomial probabilities consistent with probability distributions over $\{BCA, ACB, ABCA\}$. Again substituting $\tilde{P}_{xy}$ for P_{xy}, the vertices $BCA, ACB, ABCA$ denote the three cases where voters are unanimous. Upper right: Binary choice probabilities that yield the majority order ABC. Bottom: "Reverse" Condorcet paradox. The star denotes the binary choice probabilities induced by a uniform distribution on $\{BAC, ACB, ABCA\}$. This point, which is the center of gravity of the vertices marked $BAC, ACB, ABCA$, lies inside the half-unit cube associated with the transitive majority order ABC.

In other words, by Tversky's criterion, individual preferences are transitive if modal choices are transitive. Up to the difference between strict and weak inequality signs, when substituting $\bar{P}_{xy}$ for P_{xy}, the Alternative Hypothesis in this test states the existence of a *Condorcet cycle* in the individual choice probabilities. Figure 3 shows how a single individual, who fluctuates in his preferences, could generate a Condorcet cycle, and, in fact, a Condorcet paradox. Regenwetter et al. (2009b) gave an example of a decision maker

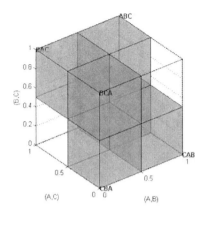

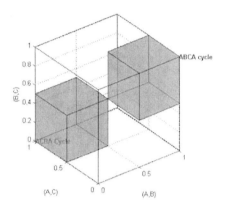

Fig. 5 In each subgraph, (x, y) denotes the Binomial probability $P_{x,y}$ for distinct $x, y \in C = \{A, B, C\}$. Upper left: Joint Binomial choice probabilities consistent with weak stochastic transitivity. Bottom right: Joint Binomial choice probabilities violating weak stochastic transitivity.

who satisfies Cumulative Prospect Theory (Tversky and Kahneman, 1992), but whose probability weighting function and utility function fluctuate. This decision maker has a uniform distribution over instantaneous preference relations ABC, BCA, and CAB. The decision maker's preferences are strict linear orders, but her modal choices form a Condorcet cycle. Figure 5 displays weak stochastic transitivity for the case where $C = \{A, B, C\}$. The Null Hypothesis is displayed on the upper left, the Alternative Hypothesis is shown on the lower right.

Tversky (1969) and many scholars after him have operationalized individual intransitivity of preferences as a violation of weak stochastic transitivity by the probabilities $\bar{P}_{xy}$. Loomes and Sugden (1995) and more recently, Regenwetter et al. (2010) and Regenwetter et al. (2009a), have pointed out that violations of weak stochastic transitivity by an individual could be due to a Condorcet paradox within that respondent, and not necessarily indicate intransitive preferences in that respondent. For more than a hundred data

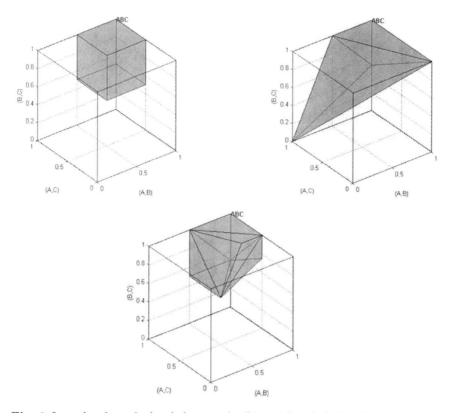

Fig. 6 In each subgraph, (x, y) denotes the Binomial probability $P_{x,y}$ for distinct $x, y \in \mathcal{C} = \{A, B, C\}$. Upper left: The half-unit cube of joint Binomial probabilities that yield Condorcet social order ABC. Upper right: The convex polytope of joint Binomial probabilities that yield Borda social order ABC. Bottom: The convex polytope of joint Binomial probabilities that yield both Condorcet and Borda social order ABC. This polytope is the intersection of the two polytopes above.

sets that use a two-alternative forced choice paradigm, Regenwetter et al. (2010) and Regenwetter et al. (2009a) provided quantitative evidence for a model according to which each individual's preferences follow a (unknown) probability distribution over strict linear orders. Most of these data sets were from experiments that were designed to demonstrate intransitive preferences in individuals. This analysis boils down to testing whether Binomial probabilities are consistent with the 10-dimensional convex polytope formed by the convex hull of the 120 vertices that correspond to linear orders of the five choice alternatives. Likewise, Regenwetter and Davis-Stober (2009) showed that the individual choice behavior of 30 respondents in a ternary paired comparison task is consistent with a model according to which each individual's preferences follow an (unknown) probability distribution over strict

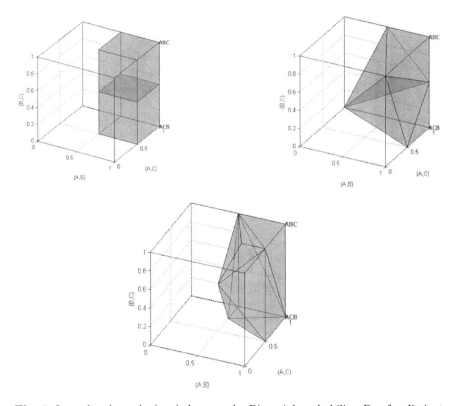

Fig. 7 In each subgraph, (x, y) denotes the Binomial probability $P_{x,y}$ for distinct $x, y \in \mathcal{C} = \{A, B, C\}$. Upper left: The convex polytope of joint Binomial probabilities that yield Condorcet winner A. Upper right: The convex polytope of joint Binomial probabilities that yield Borda winner A. Bottom: The convex polytope of joint Binomial probabilities that yield both Condorcet and Borda winner A. This polytope is the intersection of the two polytopes above.

weak orders. We will check some of these data for Condorcet cycles here. Should we find such evidence, this could be evidence for within participant Condorcet paradoxes, not necessarily for individual intransitive preferences.

2.4 The Borda Score

DEFINITION. Consider a finite set $\mathcal{C}$ of choice alternatives and a system of probabilities P_{xy} for distinct $x, y \in \mathcal{C}$. The *Borda score* of $x \in \mathcal{C}$ is

$$Borda(x) = \sum_{\substack{y \in \mathcal{C} \\ y \neq x}} \left(P_{xy}(\mathcal{T}) - P_{yx}(\mathcal{T}) \right) \tag{3}$$

The *Borda winner* is the choice alternative with the highest Borda score. The *Borda order* is the overall ordering of the choice alternatives by decreasing Borda score. This definition is in line with the general definition in Regenwetter and Rykhlevskaia (2007) that built on an axiomatization by Young (1974).

Figure 6 shows the relationship between the Condorcet and Borda social orders in three dimensions. The upper left display shows the joint Binomials that yield the Condorcet order ABC, whereas the upper right shows the joint Binomials yielding that Borda order. The lower polytope is the intersection of the two polytopes in the top. This is the collection of joint Binomials that yield ABC both by Condorcet and by Borda aggregation. Figure 7 shows the same information, but focussing only on the winner (option A), rather than the entire social order.

3   Behavioral Social Choice

Individual and social choice research areas generally engage in limited cross-fertilization. While social choice theory has relied systematically on individual rational choice theory, e.g., through its wide use of weak orders or of strict weak orders as descriptions of individual preferences, little work has incorporated behavioral approaches. One of the rare fertile areas with active interaction between normative and descriptive approaches is fair division and justice (Balinski and Young, 1982; Schokkaert and Lagrou, 1983; Kahneman et al., 1986; Brams and Taylor, 1996; Schokkaert and Devooght, 2003).

3.1   *The Condorcet Paradox*

Much of social choice theory has focussed on the abstract axiomatic structure of aggregation methods (e.g., Arrow, 1951; Black, 1958; Sen, 1970; Gehrlein and Fishburn, 1976; Riker, 1982; Tangiane, 1991; Saari, 1995; Mueller, 2003). In particular, much of that literature has suggested that Condorcet cycles should be ubiquitous (e.g., DeMeyer and Plott, 1970; Gehrlein and Fishburn, 1976; Gehrlein, 1983; Lepelley, 1993; Jones et al., 1995; McKelvey, 1979; Riker, 1982; Van Deemen, 1999).

Various scholars, including Feld and Grofman (1992) and Mackie (2003) questioned whether these predictions had empirical support. Regenwetter et al. (2006) and its component predecessor papers developed tools to evaluate the mathematical properties of social choice procedures on empirical behavioral data, with a special emphasis on the Condorcet paradox. That project, as well as Regenwetter et al. (2007a) and Regenwetter et al. (2007b) searched a broad range of empirical data sources for evidence of Condorcet paradoxes. The only cases where they could not rule out the paradox were situations with statistical identifiability problems or where statistical replicability was questionable. List and Goodin (2001), Regenwetter et al.

(2006), Tangian (2000) and others also considered theoretical conditions that would eliminate the paradox. Dryzek and List (2003) and List et al. (2007) suggested deliberation among decision makers as a tool to avoid the paradox.

3.2   The Incompatibility of Consensus Methods

The theoretical social choice literature has highlighted impossibility theorems and the mutual incompatibility of social choice procedures that are based on different principles of consensus formation (e.g., Arrow, 1951; Riker, 1982; Mueller, 2003). Saari (1999, 2000a,b, 2001) designed mathematical tools for constructing profiles with nearly any prespecified pattern of disagreements among consensus methods, when mathematically possible. Tangian (2000) discussed theoretical conditions that allow Condorcet and Borda to agree.

   Empirically, Felsenthal et al. (1993), using 37 election data sets, provided evidence that a range of competing social choice methods yielded very similar outcomes. Hastie and Kameda (2005) found dramatic agreement among multiple consensus methods in computer simulations of a hunter-gatherer society. Regenwetter et al. (2006) and its component papers, Regenwetter et al. (2007a), Regenwetter et al. (2007b), and Regenwetter et al. (2009b) compared the outcomes of competing social choice procedures against each other using a range of quantitative methods. In all cases, they found striking agreements between rival social choice methods, especially near perfect consensus among Condorcet and Borda winners, as well as between Condorcet and Borda losers (most of these were elections with five candidates). In some cases, these authors used bootstrap methods to evaluate statistical confidence and usually found the statistical replicability of the agreement to be very high.

4   Consensus with Oneself

For the rest of this paper, we will concentrate on aggregation within persons. Tversky (1969) studied eight individuals who made pairwise choices among five lotteries. Each individual was offered each of the 10 distinct nonordered pairs 20 times over the course of the experiment, with repeated choice being separated by decoys to avoid memory effects. Tversky interpreted a decision maker's modal choice on a given pair of lotteries as indicating that person's "true" binary preference for that pair. He reported that the pattern of modal choices was intransitive for six of the eight participants. We reanalyze Tversky's data using a quantitative maximum likelihood test of weak stochastic transitivity that redresses some methodological problems faced by Tversky (1969) in his original study.

   We repeat the same type of analysis for 54 data sets from 18 participants in three experimental conditions of Regenwetter et al. (2010) and Regenwetter et al. (2009a), where, like in Tversky's study, the respondent had to make each decision 20 times in a two-alternative forced choice paradigm. Finally, we include an analysis for 28 data sets from 13 different participants in Regenwetter and Davis-Stober (2009), where each respondent had to make each decision 45 times over the course of the experiment, using a ternary paired comparison paradigm. Here, we analyze only participants who never used the "indifference" option in the ternary paired comparison task. For the data from Regenwetter et al. (2010) and Regenwetter and Davis-Stober (2009), we also compare Condorcet and Borda outcomes for each individual. Here, we use a bootstrap method similar to that of Regenwetter et al. (2009b) to quantify our confidence in the agreement or disagreement among Condorcet and Borda outcomes.

4.1 Likelihood Ratio Test of Weak Stochastic Transitivity

We now discuss the evaluation of Condorcet cycles in a full-fledged maximum likelihood framework. Figure 5 shows that neither the Null nor the Alternative Hypothesis is a convex set. Furthermore, both are full-dimensional in the empirical outcome space. This means that a maximum likelihood test of weak stochastic transitivity, and hence a maximum likelihood test of Condorcet cycles, is anything but a routine endeavor.

Writing $\mathbf{N} = (N_{xy})_{x,y \in \mathcal{C}, x \neq y}$ for the frequency vector of the number of times each x is chosen over each y in N trials, and $\mathbf{P} = (P_{xy})_{x,y \in \mathcal{C}, x \neq y}$ for the vector of binary choice probabilities, the likelihood function $Lik_{\mathbf{N},\mathbf{P}}$ is

$$Lik_{\mathbf{N},\mathbf{P}} = \kappa \times \prod_{\substack{(x,y) \in \mathcal{C} \times \mathcal{C} \\ x \neq y}} P_{xy}^{N_{xy}}, \tag{4}$$

with κ a constant. (Note that we always have $N_{xy} + N_{yx} = N$ and $P_{xy} + P_{yx} = 1$.)

Figure 5 displays the joint Binomials for three choice alternatives A, B, C. When there are five choice alternatives, we are considering 10 Binomial parameters, i.e., the empirical sample space is a 10-dimensional unit hypercube. Weak stochastic transitivity is a full-dimensional nonconvex union of 10-dimensional hypercubes of length $\frac{1}{2}$ located at those vertices whose coordinates directly translate into linear orders (see previous figures for 3D examples).

This insight goes back to Iverson and Falmagne (1985), who showed that the log-likelihood ratio test statistic, G^2, in a test of weak stochastic transitivity fails to follow an asymptotic χ^2 distribution, because point estimates typically lie at the boundary of the parameter space (namely on a face of a half-unit hypercube inside the unit hypercube.) Tversky (1969) was aware of this problem, but lacked the technical tools to fix it. Using a custom designed conservative test, Iverson and Falmagne (1985) concluded that all but one of Tversky's violations of weak stochastic transitivity were statistically nonsignificant.

Recently, general Bayesian and frequentist methods have become available to deal with such so-called "order constrained inference" problems (Myung et al., 2005; Davis-Stober, 2009). Table 1 shows our analysis summary of Tversky's data for 8 respondents. Tversky (1969, Table 3, p. 36) reported that five participants had p-values below .05. Using the algorithm of Davis-Stober (2009) we find three individuals who violate weak stochastic transitivity significantly. This reflects that the algorithm of Davis-Stober (2009) is not as conservative as that of Iverson and Falmagne (1985) who only found one significant violation. The tests of Iverson and Falmagne (1985) and Davis-Stober (2009) accommodate the nonconvexity of H_0 and H_A, as well as the boundary problem, by leveraging the geometric shape of the parameter space around the maximum likelihood point estimates. This implies that the goodness-of-fit statistic, G^2, has an asymptotic $\bar{\chi}^2$ distribution that is a mixture of χ^2 distributions. We include that distribution in Table 1 for our analysis for each participant.

Regenwetter et al. (2010) found that all but Respondent 3 were consistent with a probability distribution over linear order preferences. This means that Respondents 1 and 6 provide statistically significant evidence for within-participant Condorcet paradoxes. These individuals' choice proportions are consistent with linear order preferences and with a within-respondent Condorcet cycle. Respondents 2, 4, 5, 7, and 8 are consistent with linear order preferences and with linearly ordered modal (Condorcet) outcomes. Respondent 3 violates the linear order model, but the technique of Regenwetter et al. (2010) and Regenwetter et al. (2009a) does not allow us to infer that this person's preferences were intransitive. In other words, we cannot tell, at this point, whether we are dealing with a "reverse" Condorcet paradox like the one illustrated in Figure 4. As Regenwetter et al. (2009a, 2010) discuss in detail, there are some complications in interpreting Tversky's data from a perspective of statistical significance, because Tversky (1969) collected data only on 8 out of 18 participants. The remaining 10 respondents were excluded from the experiment because they did not appear to act sufficiently "intransitively" in a pretest.

Table 2 summarizes our reanalysis of the data collected by Regenwetter et al. (2010). We find a perfect fit of weak stochastic transitivity in 44 out of 54 cases. The column marked "Condorcet Paradox?" indicates whether we have evidence for a Condorcet paradox in the sense that weak stochastic transitivity was violated while choices were nonetheless consistent with linear order preferences: "No" means no evidence at all (because both weak stochastic transitivity and the linear ordering model fit perfectly), whereas "n.s." indicates statistically nonsignificant evidence for a violation of weak stochastic transitivity. "No*" means that there was no evidence for a Condorcet paradox, but that it is possible a (nonsignificant) "reverse Condorcet paradox" occurred, because the linear order model was

Table 1 Reanalysis of all eight respondents in the first experiment of Tversky (1969). For each respondent, we give the log-likelihood ratio (G^2), the χ^2 distribution and the p-value that Tversky (1969) originally reported in his Table 3. We also provide the asymptotic $\bar{\chi}^2$-distribution according to Davis-Stober (2009), the log-likelihood ratio (G^2) at the maximum likelihood estimate, and the p-value resulting from Davis-Stober's state-of-the-art order constrained test. Significant violations at a 5% significance level are marked in bold.

Resp.		Tversky (1969)			Using Davis-Stober's (2009) method	
	G^2	Asym. χ^2 Distr. of G^2	p-value	G^2	Asymptotic $\bar{\chi}^2$ Distribution of G^2	p-value
1	11.82	χ_3	< .01	9.33	$0.13 + .39\chi_1^2 + .37\chi_2^2 + .11\chi_3^2$	< .01
2	7.84	χ_3	< .05	4.46	$0.07 + .24\chi_1^2 + .37\chi_2^2 + .26\chi_3^2 + .06\chi_4^2$	.125
3	6.02	χ_2	< .05	6.03	$.25 + .49\chi_2^2 + .26\chi_3^2$	< .02
4	15.94	χ_3	< .01	2.63	$.13 + .37\chi_1^2 + .35\chi_2^2 + .15\chi_3^2$	.20
5	5.18	χ_2	< .10	4.66	$.06 + .27\chi_1^2 + .37\chi_2^2 + .25\chi_3^2 + .06\chi_4^2$	.11
6	7.36	χ_1	< .01	7.26	$.12 + .38\chi_1^2 + .37\chi_2^2 + .13\chi_3^2$	.02
7	0.4	χ_1	< .75	0.20	$.22 + .51\chi_1^2 + .27\chi_2^2$	.58
8	0	-	perfect fit	0	-	perfect fit

Table 2 Likelihood ratio test of weak stochastic transitivity for the two-alternative forced choice data of Regenwetter et al. (2010) using the algorithm of Davis-Stober (2009) to determine the asymptotic $\bar{\chi}^2$ distribution of G^2. Significant violations at a 5% significance level are marked in bold. See the text for explanations.

Resp.	Cash I (Tversky Replication)			Cash II			Noncash		
	G^2	p-value	Condorcet Paradox?	G^2	p-value	Condorcet Paradox?	G^2	p-value	Condorcet Paradox?
1	0	perfect fit	No	6.03	.04	Maybe	0	perfect fit	No
2	0	perfect fit	No	0	perfect fit	No	0	perfect fit	No*
3	0	perfect fit	No	0	perfect fit	No	0	perfect fit	No*
4	7.91	< .01	Maybe	0	perfect fit	No*	0	perfect fit	No
5	0	perfect fit	No	0	perfect fit	No	0	perfect fit	No
6	0.2	.53	n.s.	1	.45	n.s.	0	perfect fit	No
7	0	perfect fit	No	0	perfect fit	No	0	perfect fit	No*
8	0	perfect fit	No	0	perfect fit	No	0	perfect fit	No
9	0	perfect fit	No	0	perfect fit	No	0	perfect fit	No
10	0	perfect fit	No	0	perfect fit	No*	0	perfect fit	No
11	0	perfect fit	No	0	perfect fit	No*	0	perfect fit	No
12	0.4	.61	n.s.	.2	.32	n.s.	0	perfect fit	No
13	0.6	.68	n.s.	0	perfect fit	No	0	perfect fit	No
14	0	perfect fit	No	0	perfect fit	No	0	perfect fit	No*
15	0	perfect fit	No	0	perfect fit	No	0	perfect fit	No
16	0	perfect fit	Reverse?	2.63	.13	Maybe	0	perfect fit	No*
17	2.42	.31	n.s.	0	perfect fit	No	0	perfect fit	No
18	0	perfect fit	No	3.03	.23	n.s.	0	perfect fit	No*

nonsignificantly violated. "Reverse?" indicates a case where weak stochastic transitivity holds but the linear order model is significantly violated, hence allowing for a potential "reverse Condorcet paradox." "Maybe" denotes inconclusive cases where weak stochastic transitivity is significantly violated, but the linear order model is also (significantly or nonsignificantly) violated.

Note that this study did not prescreen participants as Tversky (1969) did. Hence, two significant violations of weak stochastic transitivity out of 54 data sets, with a significance level of 5%, is just about the number of violations we expect by Type I error. In other words, we have no reason to believe that weak stochastic transitivity was violated in this study. Regenwetter et al. (2009a) analyzed the same data with a slightly different algorithm for determining the appropriate $\bar{\chi}^2$ distributions and found one more significant violation of weak stochastic transitivity.

Table 3 shows a similar analysis of the ternary paired comparison data collected by Regenwetter and Davis-Stober (2009), concentrating only on data where respondents did not use the "indifference" option. That paper used similar choice options as Regenwetter et al. (2010). Here, we find no significant violations of weak stochastic transitivity at all. Overall, for all studies combined, because of the extremely infrequent significant violations of weak stochastic transitivity, the evidence of any within-person Condorcet paradoxes is, consequently, very weak.

Table 3 Likelihood ratio test of weak stochastic transitivity for ternary paired comparison data of Regenwetter and Davis-Stober (2009) using the algorithm of Davis-Stober (2009) to determine the asymptotic $\bar{\chi}^2$ distribution of G^2. Blank cells are omitted cases, where the respondents used the "indifference" response category once or more.

Resp.	Cash I G^2	p-value	Condorcet Paradox?	Cash II G^2	p-value	Condorcet Paradox?	Noncash G^2	p-value	Condorcet Paradox?
1	0	perfect fit	No	0	perfect fit	No			
2	0	perfect fit	No	.78	.50	n.s.	0	perfect fit	No*
3				0	perfect fit	No*			
6	0	perfect fit	No	0	perfect fit	No	0	perfect fit	No
7	0	perfect fit	No	0	perfect fit	No			
8	0	perfect fit	No	0	perfect fit	No	0	perfect fit	No
10				0	perfect fit	No			
11	0	perfect fit	No						
12	0	perfect fit	No						
13	0	perfect fit	No	0	perfect fit	No			
17				0	perfect fit	No			
19				.22	.87	n.s.			
23	0	perfect fit	No*						
24	0	perfect fit	No						
27				0	perfect fit	No			
28				0	perfect fit	No			
29	.02	.45	n.s.	0	perfect fit	Reverse?	0	perfect fit	No

4.2 Boostrap Analysis: Within Person Consensus between Condorcet and Borda Winners/Losers

Bootstrap methods provide a convenient tool for evaluating, through computer simulation, how a quantity computed from empirical data would behave if small perturbations were to occur in the data (see, e.g., Efron and Tibshirani, 1993). This is particularly useful for intractable statistical problems. We use a *nonparametric bootstrap*, in which we sample with replacement from the observed data. For each pair of choice alternatives, we sample the same number of simulated observations as there were observations in the actual experiment. We then recompute the social choice outcomes by Condorcet and by Borda. We use a bootstrap with 1,000 simulated data sets for each participant. For brevity, we concentrate on unique winners and unique losers under Condorcet and Borda.

We do not report on a bootstrap of Tversky's original data. Recall that three respondents led to significant violations of weak stochastic transitivity. In these data sets, there often is no unique Condorcet winner (due to either an intransitivity or a tie) or no unique Borda winner (due to a tie). The same occurs for the losers.

Table 4 Bootstrap analysis of the agreement among Condorcet and Borda for two-alternative forced choice from Cash I of Regenwetter et al. (2010). See the text for explanations.

Resp.	C.W. = B.W.	Confidence of agreement	C.L. = B.L.	Confidence of agreement	C.L. = B.W.	Confidence of disagreement	C.W. = B.L.	Confidence of disagreement
1	Yes	0.91	Yes	0.78	No	1.0	No	1.0
2	Yes	0.96	**No**	0.72	No	1.0	No	1.0
3	Yes	1.00	Yes	0.99	No	1.0	No	1.0
4	Yes	1.00	CC	0.58	No	1.0	No	1.0
5	Yes	1.00	Yes	0.99	No	1.0	No	1.0
6	Yes	0.93	CC	0.70	No	1.0	No	1.0
7	Yes	0.98	Yes	0.97	No	1.0	No	1.0
8	Yes	1.00	Yes	0.95	No	1.0	No	1.0
9	Yes	0.82	Yes	0.81	No	1.0	No	1.0
10	Yes	0.94	Yes	0.99	No	1.0	No	1.0
11	Yes	1.00	Yes	1.00	No	1.0	No	1.0
12	CC	0.82	Yes	0.79	No	1.0	No	1.0
13	CC	0.82	BT	0.73	No	1.0	No	1.0
14	Yes	1.00	Yes	0.99	No	1.0	No	1.0
15	Yes	0.82	Yes	0.91	No	1.0	No	1.0
16	Yes	0.98	Yes	0.91	No	1.0	No	1.0
17	CC	0.64	CC	0.71	No	1.0	No	1.0
18	Yes	0.73	**No**	0.85	No	1.0	No	1.0

Table 5 Bootstrap analysis of the agreement among Condorcet and Borda for two-alternative forced choice from Cash II of Regenwetter et al. (2010). See the text for explanations.

Resp.	C.W. = B.W.	Confidence of agreement	C.L. = B.L.	Confidence of agreement	C.L. = B.W.	Confidence of disagreement	C.W. = B.L.	Confidence of disagreement
1	CC	0.79	**No**	0.60	No	1.0	No	1.0
2	Yes	0.99	Yes	0.83	No	1.0	No	1.0
3	Yes	0.98	Yes	0.90	No	1.0	No	1.0
4	Yes	0.98	Yes	0.82	No	1.0	No	1.0
5	Yes	0.98	Yes	0.95	No	1.0	No	1.0
6	CC	0.79	CC	0.71	No	1.0	No	1.0
7	Yes	1.00	Yes	0.88	No	1.0	No	1.0
8	Yes	0.99	Yes	0.99	No	1.0	No	1.0
9	Yes	0.90	Yes	0.80	No	1.0	No	1.0
10	Yes	0.89	Yes	1.00	No	1.0	No	1.0
11	Yes	1.00	Yes	0.97	No	1.0	No	1.0
12	CC, BT	0.70	Yes	0.80	No	1.0	No	1.0
13	CT	0.71	CT	0.71	No	1.0	No	1.0
14	Yes	1.00	Yes	1.00	No	1.0	No	1.0
15	Yes	0.99	CT	0.72	No	1.0	No	1.0
16	CC, BT	0.73	CC	0.74	No	1.0	No	1.0
17	Yes	0.92	Yes	0.89	No	1.0	No	1.0
18	CC	0.65	CC	0.68	No	1.0	No	1.0

Tables 4-6 summarize our analysis of the agreement among Condorcet and Borda outcomes in the data of Regenwetter et al. (2010). Likewise, Tables 7-9 summarize the corresponding analysis for the data of Regenwetter and Davis-Stober (2009). In all these tables, the first column lists the respondent ID. The next column, marked "C.W. = B.W." reports whether Condorcet and Borda yielded unique and identical (Condorcet and Borda) winners. The column marked "C.L. = B.L." reports whether Condorcet and Borda yielded unique and identical (Condorcet and Borda) losers (by a *loser*, we mean a choice option that loses against all other candidates). Boldfaced entries are cases where we observe a disagreement among Condorcet and Borda. Sometimes the data did not contain a unique winner (or loser) for one or both methods. We indicate with "CC" when a Condorcet cycle prevented the existence of a unique Condorcet winner (or loser) and with "CT" when there was a tie among more than one Condorcet winner (or loser). Likewise, "BT" indicates a tied outcome for Borda for the winner (or loser). The column entitled "Confidence of agreement" gives the bootstrap results for matching unique winners (losers). It reports the proportion of bootstrap samples in which the two outcomes in question matched (even if the outcomes did not match in the data). For example, in the two-alternatives forced choice paradigm of Regenwetter et al. (2010), Respondent 1, Cash I, yielded

Table 6 Bootstrap analysis of the agreement among Condorcet and Borda for two-alternative forced choice from the Noncash condition of Regenwetter et al. (2010). See the text for explanations.

Resp.	C.W. = B.W.	Confidence of agreement	C.L. = B.L.	Confidence of agreement	C.L. = B.W.	Confidence of disagreement	C.W. = B.L.	Confidence of disagreement
1	Yes	1.00	**No**	0.74	No	1.0	No	1.0
2	Yes	0.92	Yes	1.00	No	1.0	No	1.0
3	Yes	1.00	Yes	0.75	No	1.0	No	1.0
4	Yes	1.00	Yes	1.00	No	1.0	No	1.0
5	Yes	1.00	Yes	0.98	No	1.0	No	1.0
6	Yes	1.00	Yes	0.99	No	1.0	No	1.0
7	Yes	1.00	Yes	1.00	No	1.0	No	1.0
8	Yes	0.79	Yes	1.00	No	1.0	No	1.0
9	Yes	0.79	Yes	0.86	No	1.0	No	1.0
10	Yes	1.00	Yes	1.00	No	1.0	No	1.0
11	Yes	0.93	Yes	1.00	No	1.0	No	1.0
12	Yes	1.00	Yes	1.00	No	1.0	No	1.0
13	Yes	0.83	Yes	0.45	No	1.0	No	1.0
14	Yes	1.00	Yes	1.00	No	1.0	No	1.0
15	Yes	1.00	Yes	1.00	No	1.0	No	1.0
16	Yes	1.00	Yes	1.00	No	1.0	No	1.0
17	Yes	0.99	Yes	0.59	No	1.0	No	1.0
18	Yes	1.00	Yes	1.00	No	1.0	No	1.0

Table 7 Bootstrap analysis of the agreement among Condorcet and Borda for ternary paired comparison data from the Cash I condition of Regenwetter and Davis-Stober (2009). See the text for explanations.

Resp.	C.W. = B.W.	Confidence of agreement	C.L. = B.L.	Confidence of agreement	C.L. = B.W.	Confidence of disagreement	C.W. = B.L.	Confidence of disagreement
1	Yes	0.99	Yes	0.73	No	1.0	No	1.0
2	Yes	1.00	Yes	1.00	No	1.0	No	1.0
6	Yes	1.00	Yes	0.86	No	1.0	No	1.0
7	Yes	0.87	Yes	0.91	No	1.0	No	1.0
8	Yes	0.97	**No**	0.75	No	1.0	No	1.0
11	Yes	1.00	Yes	1.00	No	1.0	No	1.0
12	Yes	0.98	Yes	0.99	No	1.0	No	1.0
13	Yes	1.00	Yes	1.00	No	1.0	No	1.0
23	Yes	0.94	**No**	0.77	No	1.0	No	1.0
24	Yes	0.69	Yes	0.72	No	1.0	No	1.0
29	CC	0.70	CC	0.75	No	1.0	No	1.0

Table 8 Bootstrap analysis of the agreement among Condorcet and Borda for ternary paired comparison data from the Cash II condition of Regenwetter and Davis-Stober (2009). See the text for explanations.

Resp.	C.W. = B.W.	Confidence of agreement	C.L. = B.L.	Confidence of agreement	C.L. = B.W.	Confidence of disagreement	C.W. = B.L.	Confidence of disagreement
1	Yes	0.83	Yes	0.95	No	1.0	No	1.0
2	Yes	1.00	Yes	0.71	No	1.0	No	1.0
3	Yes	1.00	Yes	0.79	No	1.0	No	1.0
6	Yes	1.00	Yes	0.79	No	1.0	No	1.0
7	Yes	0.91	No	0.76	No	1.0	No	1.0
8	Yes	0.56	No	0.78	No	1.0	No	1.0
10	Yes	0.99	Yes	0.99	No	1.0	No	1.0
13	Yes	1.00	Yes	1.00	No	1.0	No	1.0
17	Yes	0.99	No	0.72	No	1.0	No	1.0
19	Yes	0.81	CC	0.67	No	1.0	No	1.0
27	Yes	1.00	Yes	0.93	No	1.0	No	1.0
28	Yes	0.84	Yes	0.75	No	1.0	No	1.0
29	Yes	0.92	Yes	0.81	No	1.0	No	1.0

Table 9 Bootstrap analysis of the agreement among Condorcet and Borda for ternary paired comparison data from the Noncash condition of Regenwetter and Davis-Stober (2009). See the text for explanations.

Resp.	C.W. = B.W.	Confidence of agreement	C.L. = B.L.	Confidence of agreement	C.L. = B.W.	Confidence of disagreement	C.W. = B.L.	Confidence of disagreement
2	Yes	0.86	Yes	1.00	No	1.0	No	1.0
6	Yes	0.69	Yes	1.00	No	1.0	No	1.0
8	Yes	0.94	Yes	0.70	No	1.0	No	1.0
29	Yes	0.95	Yes	0.99	No	1.0	No	1.0

unique and identical winners in 91% of bootstrapped samples. Respondent 2, Cash II, yielded unique and identical losers in 72% of bootstrapped samples (even though the observed Condorcet loser and Borda loser did not match in the experiment). The last four columns show whether the winner by either consensus method ever coincided with the loser by the other method. Throughout all our analyses, we never observed the winner by one rule to match the loser by the other rule, either in the original data or in the tens of thousands of bootstrapped samples, hence our confidence of disagreement is 1.0 throughout.

Notice that some violations of weak stochastic transitivity in Table 2 do not involve the winner or loser, i.e., go hand in hand with agreement between Condorcet and Borda for the winner (loser) here. For example, Respondent

4 shows a violation of weak stochastic transitivity in Cash I in Table 2. But there is a unique Condorcet winner, the cycle involves only the other four choice alternatives. Hence, Table 4 shows a unique winner that matches the Borda winner, and a Condorcet cycle that prevents a unique loser from existing in the data.

5   Conclusions

Tversky (1969) reported, what he believed to be statistically significant violations of weak stochastic transitivity within individual decision makers, and he concluded from those results that his Respondents 1 - 6 had intransitive individual preferences. There are two important caveats. 1) Decision makers who violate weak stochastic transitivity could, nonetheless, have transitive preferences. This would mean that these decision makers generate a Condorcet paradox within themselves. 2) Weak stochastic transitivity leads to order constrained inference, where the log-likelihood ratio test statistic does not obey a χ^2 distribution. Regenwetter et al. (2009a, 2010) discuss this problem in detail. After revisiting the literature on intransitive preferences and analyzing large amounts of individual decision making data, they conclude that individual preferences do not appear to be intransitive.

We have considered weak stochastic transitivity from a social choice perspective, but within each person. Our results and those of Iverson and Falmagne (1985), as well as Regenwetter et al. (2009a), suggest that violations of weak stochastic transitivity occur at a rate smaller than permitted by Type I error. In other words, outside Tversky's (1969) study with pre-selected participants, statistically compelling evidence for violations is lacking. As an immediate consequence, the evidence for Condorcet paradoxes (where the decision maker acts in accordance with linear order preferences, but also generates a cycle by modal choice) is statistically weak. However, we would warn the reader not to misinterpret this to mean that modal choice reveals the true (deterministic) preference of individual decision makers. Many of our respondents vary substantially in their choices, often choosing one choice alternative over another only on, say, two thirds of occasions.

We have further concluded that Condorcet and Borda yield the same unique winner and the same unique loser with high statistical confidence, as established through a nonparametric bootstrap procedure. The winner of Condorcet and the loser by Borda coincided not once in our 82,000 bootstrapped samples. The same holds for Condorcet losers and Borda winners. As far as we can tell from these 82 data sets from our laboratory, the famed disagreement among Condorcet and Borda does not appear to occur for choice data that are aggregated within a person.

Social choice theory and behavioral decision research can continue to synergize in the future. Here, we have shown what social choice theory can do for behavioral decision research on individuals. Likewise, laboratory research on individual voting behavior can benefit social choice research. For example, it may provide important insights for electoral reform in an effort to find voting methods that are both ergonomic and strategy-resistant, yet appear to yield outcomes that are typically consistent with normative criteria.

Acknowledgments. Special thanks to Clintin P. Davis-Stober for his advice on countless occasions, to Sergey V. Popov for helping with a computer implementation of the bootstrap algorithm, and to Shiau Hong Lim for implementing the algorithm of Davis-Stober (2009) as a computer program. This work is supported by the *Decision, Risk and Management Science* Program of the National Science Foundation under Award No. SES #08-20009 (to M. Regenwetter, PI) entitled "A Quantitative Behavioral Framework for Individual and Social Choice." Any opinions, findings, and conclusions or recommendations expressed in this publication are those of the authors and do not necessarily reflect the views of the University of Illinois or the National Science Foundation.

References

Arrow, K.J.: Social Choice and Individual Values. Wiley, New York (1951)

Balinski, M.L., Young, H.P.: Fair Representation: Meeting the Idea of One Man, One Vote. Yale University Press, New Haven (1982)

Black, D.: The Theory of Committees and Elections. Cambridge University Press, Cambridge (1958)

Block, H.D., Marschak, J.: Random orderings and stochastic theories of responses. In: Olkin, I., Ghurye, S., Hoeffding, H., Madow, W., Mann, H. (eds.) Contributions to Probability and Statistics, pp. 97–132. Stanford University Press, Stanford (1960)

Borda, J.-C.: On elections by ballot. Paper orally presented to the French Academy (1770)

Brams, S.J., Taylor, A.D.: Fair Division: From Cake Cutting to Dispute Resolution. Cambridge University Press, New York (1996)

Brandstätter, E., Gigerenzer, G., Hertwig, R.: The priority heuristic: Making choices without trade-offs. Psychological Review 113, 409–432 (2006)

Condorcet, M.: Essai sur l'application de l'analyse à la probabilité des décisions rendues à la pluralité des voix. Essai on the application of the probabilistic analysis of majority vote decisions, Imprimerie Royale, Paris (1785)

Davis-Stober, C.P.: Analysis of multinomial models under inequality constraints: Applications to measurement theory. Journal of Mathematical Psychology 53, 1–13 (2009)

DeMeyer, F., Plott, C.R.: The probability of a cyclical majority. Econometrica 38, 345–354 (1970)

Dryzek, J., List, C.: Social choice theory and deliberative democracy: A reconciliation. British Journal of Political Science 33, 1–28 (2003)

Efron, B., Tibshirani, R.J.: An Introduction to the Bootstrap. Chapman and Hall, New York (1993)

Feld, S.L., Grofman, B.: Who is afraid of the big bad cycle? Evidence from 36 elections. Journal of Theoretical Politics 4, 231–237 (1992)

Felsenthal, D.S., Maoz, Z., Rapoport, A.: An empirical evaluation of 6 voting procedures - do they really make any difference? British Journal of Political Science 23, 1–27 (1993)

Fishburn, P.C.: Utility Theory for Decision Making. In: Krieger, R.E. (ed.) Huntington (1979)

Gehrlein, W.V.: Concorcet's paradox. Theory and Decision 15, 161–197 (1983)

Gehrlein, W.V., Fishburn, P.C.: The probability of the paradox of voting: A computable solution. Journal of Economic Theory 13, 14–25 (1976)

Hastie, R., Kameda, T.: The robust beauty of majority rules in group decisions. Psychological Review 112, 494–508 (2005)

Iverson, G.J., Falmagne, J.-C.: Statistical issues in measurement. Mathematical Social Sciences 10, 131–153 (1985)

Jones, B., Radcliff, B., Taber, C., Timpone, R.: Condorcet winners and the paradox of voting: Probability calculations for weak preference orders. The American Political Science Review 89(1), 137–144 (1995)

Kahneman, D., Knetsch, J.L., Thaler, R.: Fairness as a constraint on profit seeking entitlements in the market. American Economic Review 76, 728–741 (1986)

Kahneman, D., Tversky, A.: Prospect theory: An analysis of decision under risk. Econometrica 47, 263–291 (1979)

Kahneman, D., Tversky, A. (eds.): Choices, Values, and Frames. Cambridge University Press, New York (2000)

Konow, J.: Is fairness in the eye of the beholder? an impartial spectator analysis of justice (2008)

Krantz, D.H., Luce, R.D., Suppes, P., Tversky, A.: Foundations of Measurement, vol. 1. Academic Press, San Diego (1971)

Lepelley, D.: Concorcet's paradox. Theory and Decision 15, 161–197 (1993)

List, C., Goodin, R.E.: Epistemic democracy: Generalizing the condorcet jury theorem. Journal of Political Philosophy 9, 277–306 (2001)

List, C., Luskin, R.C., Fishkin, J.S., McLean, I.: Deliberation, single-peakedness, and the possibility of meaningful democracy: Evidence from deliberative polls. In: Working pape, London School of Economics and Stanford University, London (2007)

Loomes, G., Sugden, R.: Incorporating a stochastic element into decision theories. European Economic Review 39, 641–648 (1995)

Luce, R.D., Suppes, P.: Preference, utility and subjective probability. In: Luce, R.D., Bush, R.R., Galanter, E. (eds.) Handbook of Mathematical Psycholog, vol. III, pp. 249–410. Wiley, New York (1965)

Mackie, G.: Democracy Defined. Cambridge University Press, New York (2003)

McKelvey, R.D.: General conditions for global intransitivities in formal voting models. Econometrica 47, 1085–1112 (1979)

Mueller, D.C.: Public Choice III. Cambridge University Press, Cambridge (2003)

Myung, J., Karabatsos, G., Iverson, G.: A Bayesian approach to testing decision making axioms. Journal of Mathematical Psychology 49, 205–225 (2005)

Regenwetter, M.: Perspectives on preference aggregation. Perspectives on Psychological Science 4, 403–407 (2009)

Regenwetter, M., Dana, J., Davis-Stober, C.P.: Testing transitivity of preferences on two-alternative forced choice data. Frontiers in Quantitative Psychology and Measurement 1(148) (2010),
http://www.frontiersin.org/Journal/Abstract.aspx?
f=69\&name=psychology\&ART_DOI=10.3389/fpsyg.2010.00148,
doi:10.3389/fpsyg.2010.00148

Regenwetter, M., Dana, J., Davis-Stober, C.P.: Transitivity of preferences. Psychological Review (in press)

Regenwetter, M., Davis-Stober, C.P.: Choice variability versus structural inconsistency of preferences. In: Manuscript under review (2009)

Regenwetter, M., Grofman, B., Marley, A., Tsetlin, I.: Behavioral Social Choice. Cambridge University Press, Cambridge (2006)

Regenwetter, M., Grofman, B., Popova, A., Messner, W., Davis-Stober, C.P., Cavagnaro, D.R.: Behavioural social choice: A status report. Philosophical Transactions of the Royal Society of London B: Biological Sciences 364, 833–843 (2009b)

Regenwetter, M., Ho, M.-H., Tsetlin, I.: Sophisticated approval voting, ignorance priors, and plurality heuristics: A behavioral social choice analysis in a Thurstonian framework. Psychological Review 114, 994–1014 (2007a)

Regenwetter, M., Kim, A., Kantor, A., Ho, M.-H.: The unexpected empirical consensus among consensus methods. Psychological Science 18, 559–656 (2007b)

Regenwetter, M., Marley, A.A.J., Grofman, B.: A general concept of majority rule. Mathematical Social Sciences: special issue on random utility theory and probabilistic measurement theory 43, 407–430 (2002)

Regenwetter, M., Rykhlevskaia, E.: A general concept of scoring rules: General definitions, statistical inference, and empirical illustrations. Social Choice and Welfare 29, 211–228 (2007)

Riker, W.H.: Liberalism v. Populism. In: Freeman, W.H. (ed.) Liberalism v. Populism, W. H. Freeman and Co, San Fransisco (1982)

Roberts, F.S.: Measurement Theory. Addison-Wesley, London (1979)

Saari, D.G.: Basic Geometry of Voting. Springer, New York (1995)

Saari, D.G.: Explaining all three-alternative voting outcomes. Journal of Economic Theory 87, 313–355 (1999)

Saari, D.G.: Mathematical structure of voting paradoxes 1: Pairwise vote. Economic Theory 15, 1–53 (2000a)

Saari, D.G.: Mathematical structure of voting paradoxes 2: Positional voting. Economic Journal 15, 55–101 (2000b)

Saari, D.G.: Decisions and Elections: Explaining the Unexpected. Cambridge University Press, Cambridge (2001)

Schokkaert, E., Devooght, K.: Responsibility-sensitive fair compensation in different cultures. Social Choice and Welfare 21, 207–242 (2003)

Schokkaert, E., Lagrou, L.: An empirical approach to distributive justice. Journal of Public Economics 21(1), 33–52 (1983)

Sen, A.K.: Collective Choice and Social Welfare. Holden-Day, San Fransisco (1970)

Tangian, A.: Unlikelihood of Condorcet's paradox in a large society. Social Choice and Welfare 17, 337–365 (2000)

Tangiane, A.S.: Aggregation and representation of preferences: Introduction to mathematical theory of democracy. Springer, Berlin (1991)

Tversky, A.: Intransitivity of preferences. Psychological Review 76, 31–48 (1969)

Tversky, A., Kahneman, D.: The framing of decisions and the psychology of choice. Science 211, 453–458 (1981)

Tversky, A., Kahneman, D.: Advances in prospect theory: Cumulative representation of uncertainty. Journal of Risk and Uncertainty 5, 297–323 (1992)

Van Deemen, A.: The probability of the paradox of voting for weak preference orderings. Social Choice and Welfare 16, 171–182 (1999)

Young, H.P.: An axiomatization of Borda's rule. Journal of Economic Theory 9, 43–52 (1974)

A Appendix

We briefly describe the main features of the experiment used in Regenwetter et al. (2009a, 2010). Details can be found in Regenwetter et al. (2010). The experiment was reviewed and approved by the Institutional Review Board of the University of Illinois at Urbana-Champaign. Participants were 18 young adults recruited primarily through fliers on campus at the University of Illinois at Urbana-Champaign. The experiment was designed to closely mimic Tversky's famous (1969) Experiment 1, except that we used a computer interface rather than paper and pencil, participants were not prescreened, and we used three interwoven experimental conditions: "Cash I" was a replication of Tversky's five cash gamble choice options, but with dollar amounts updated to contemporary equivalents, "Cash II" was a variation in which all five gambles had equal expected value ($8.80), whereas "Noncash" denoted a condition with five nonmonetary prizes in the form of gift certificates (e.g., 40 free movie rentals). All participants received a flat payment of $10, plus played one of their chosen gambles for real at the experiment's conclusion. That gamble was randomly selected from the choices they had made during the experiment.

The experiment consisted in presenting gambles two at a time (see Figures 1 and 2) and requiring the decision maker to make a choice on each trial. The software cycled through Cash I, Cash II, Noncash, and a distractor gamble set, in that fixed sequence, in order to keep gambles from any given gamble set maximally separated from each other. To replicate Tversky (1969), we repeated each pairwise choice $N = 20$ times, but we took precautions to separate these repetitions over time in order to eliminate memory effects. Within any gamble sets, on any given trial, the two gambles were offered randomly with the constraint that this pair had not been used for the previous five trials from that set and the constraint that neither gamble had been used in the past trial from that set. Altogether, each decision maker made 818 choices over the course of the roughly one and a half hour long experiment, including 18 warm-up choices at the beginning that were not used for analysis.

The experiment of Regenwetter and Davis-Stober (2009) was similar, but it permitted participants to express indifference among the offered gambles, the experiment involved 30 participants and used $N = 45$ repetitions per pair per person to guarantee ample statistical power. To accommodate the longer duration needed for this experiment, the data collection was split into two sessions held on two different days in close proximity to each other.

The Social Choice Approach to Group Identification

Dinko Dimitrov

Abstract. The chapter gives an overview of selected topics from the theory of group identification. As a starting point serve different axiomatic characterizations of the "libera" rule for group identification whereby the group consist of those and only those individuals each of which views oneself a member of the group. We then focus on consent rules and recursive procedures for collective determination where the opinions of other individuals in the society also count. Finally, we address recent developments in the literature with respect to gradual opinions and group identity functions.

1 Introduction

One possibility to study the process of group formation is to view it as a group identification problem. The latter can be formulated as follows: Given a group of individuals, how to define the extent of a subgroup of it? This problem serves as a background in many social and economic contexts. For example, when one examines the political principle of self-determination of a newly formed country, one would like to define the extension of a given nationality. Or when a newly arrived person in New York chooses where to live, the person is interested in finding out a residential neighborhood that would suit her. In all those contexts, it is typically assumed that there is a well-defined group of people who share some common values, beliefs, expectations, customs, jargon, or rituals. Consequently, questions like "how to define a social group" or "who belongs to the social group" arise. This chapter is devoted to the social choice approach in answering these questions.

Kasher's paper on collective identity ([15], 1993) can be considered as a first, non-formal attempt to look at the group identification problem as an aggregation

Dinko Dimitrov
Chair of Economic Theory, Saarland University,
Campus C3 1, 66123 Saarbrücken, Germany
e-mail: dinko.dimitrov@mx.uni-saarland.de

E. Herrera-Viedma et al. (Eds.): Consensual Processes, STUDFUZZ 267, pp. 123–134.
springerlink.com   © Springer-Verlag Berlin Heidelberg 2011

task. In that paper the author views that each individual of a society has an opinion about every individual, including oneself, whether the latter is a member of a group to be formed. The collective identity of the group to be formed is then determined by aggregating opinions of all the individuals in the society. The formal link between Kasher's approach and the theory of aggregators mainly developed in economic theory was made by Kasher and Rubinstein ([16], 1997) and the corresponding framework is presented in Sect. 2. These authors introduce and axiomatically characterize a "liberal" rule for group identification whereby the group consist of those and only those individuals each of which views oneself a member of the group. Different axiomatic systems leading to this rule are presented and discussed in Sect. 3. Section 4 is devoted to the different ways in which the opinions of other individuals in the society also count for one's group identification with the main focus being on consent rules and recursive procedures for collective determination. In Sect. 5 we address other themes and developments in the literature.

2 The Aggregation Framework

We denote by $N = \{1, 2, \ldots, n\}$ the set of all individuals in the society. Each individual $i \in N$ forms a set $G_i \subseteq N$ consisting of all society members that in the view of i deserve to be accepted as group members. A *profile of views* is an n-tuple $G = (G_1, \ldots, G_n)$ where $G_i \subseteq N$ for every $i \in N$. Let $\mathscr{G}$ be the set of all profiles of views, i.e., $\mathscr{G} = (P(N))^n$ where $P(N)$ is the power set of N. A *collective identity function* (CIF) $F : \mathscr{G} \to P(N)$ assigns to each profile $G \in \mathscr{G}$ a set $F(G) \subseteq N$ of socially accepted group members. We denote by $\mathscr{F}$ the set of all collective identity functions.

3 Characterizing the Liberal Aggregator

Kasher and Rubinstein ([16], 1997) provide axiomatic characterizations of three aggregators: the "dictatorship" aggregator whereby a pre-designated member of the society determines who deserves to became a group member; the "oligarchical" aggregator whereby the decision is taken by consensus among the members of a pre-designated subgroup of the society; and the "liberal" aggregator whereby a person's own opinion about oneself completely determines his social status. The first two characterizations are based on previous results by Fishburn and Rubinstein ([12], 1986) and Rubinstein and Fishburn ([19], 1986) whereas the characterization of the "liberal" aggregator is new.

3.1 The Liberal Principle

Kasher and Rubinstein ([16], 1997) use five axioms as to characterize their *strong liberal* CIF $F^L \in \mathscr{F}$ defined as follows:

For every $G \in \mathscr{G}$,

$$F^L(G) = \{i \in N \mid i \in G_i\}. \tag{1}$$

As shown by Sung and Dimitrov ([21], 2005), these five axioms are not independent and, moreover, a characterization of the "liberal" aggregator can be reached by means of the three of the remaining axioms. These axioms are symmetry (SYM), independence (I), and the liberal principle (L).

We say that a CIF $F \in \mathscr{F}$ satisfies

- *Symmetry* (SYM) if, for every $G \in \mathscr{G}$ and for every $i, j \in N$,

  - $G_i \setminus \{i, j\} = G_j \setminus \{i, j\}$,
  - $i \in G_k \Leftrightarrow j \in G_k$, for every $k \in N \setminus \{i, j\}$,
  - $i \in G_i \Leftrightarrow j \in G_j$,
  - $i \in G_j \Leftrightarrow j \in G_i$,
    imply
  - $i \in F(G) \Leftrightarrow j \in F(G)$.

- *Independence* (I) if, for every $G, G' \in \mathscr{G}$ and for every $i \in N$,

  - $k \in F(G) \Leftrightarrow k \in F(G')$ for every $k \in N \setminus \{i\}$,
  - $i \in G_k \Leftrightarrow i \in G'_k$ for every $k \in N$,
    imply
  - $i \in F(G) \Leftrightarrow i \in F(G')$.

- *Liberal principle* (L) if, for every $G \in \mathscr{G}$,

  - $[k \in G_k$ for some $k \in N]$ implies $F(G) \neq \emptyset$, and
  - $[k \notin G_k$ for some $k \in N]$ implies $F(G) \neq N$.

According to the symmetry axiom, a collective identity aggregator should not discriminate between any two individuals (1) who are similar in the eyes of all members of the society and (2) whose opinions on all society members are also symmetric. In other words, if i and j are similar in a particular profile of views $G \in \mathscr{G}$, then these individuals have to be either both socially accepted or both socially unaccepted as having the corresponding identity.

The next axiom, independence, requires that the question whether i is declared as a group member or not depends only on the views about i and the other member's identity as members of the group to be formed.

Finally, the liberal principle states that it is impossible no one to be socially accepted, though there is an individual i who considers oneself to have the corresponding social identity. Similarly, it is impossible that everyone is socially accepted, though there is a society member i who thinks that he does not have the identity considered.

Theorem 1 (Sung and Dimitrov, [21], 2005). *The strong liberal CIF F^L is the only CIF that satisfies axioms (SYM), (I), and (L). Moreover, the three axioms are independent.*

3.2 Separability Axioms

The liberal principle (L) used in the characterization of F^L does not seem to be a very plausible requirement and can hardly be motivated. As shown by Miller ([17], 2008), an alternative and very intuitive characterization of the liberal aggregator can be reached by imposing two types of separability axioms together with standard anonymity and non-degeneracy axioms.

As to explain the separability axioms, let us consider two group identification problems: the first one is to determine the group of Asians in a given society (the corresponding profile of individual opinions is denoted by A), while the second one is to find the group of Whites in the same society (the profile is denoted by W). Additionally, $A \cap W = \left((A_i \cap W_i)_{i \in N} \right)$ and $A \cup W = \left((A_i \cup W_i)_{i \in N} \right)$ denote the individual opinions about who is 'Asian and White', and 'Asian or White', respectively. Consider now the following two axioms a CIF F may satisfy.

- *Meet separability*: $F(A \cap W) = F(A) \cap F(W)$.
- *Join separability*: $F(A \cup W) = F(A) \cup F(W)$.

These requirements display the fact that the result of the aggregation exercise should not depend on which group (Asians or Whites) is under consideration. According to meet separability, the result should be the same no matter if one aggregates opinions about 'Asian and White' people or first generates the groups of socially accepted Asians and socially accepted Whites separately, and then takes the intersection of the two groups. Similarly, join separability requires the aggregation rule to yield the same group no matter if one aggregates opinions about 'Asian or White' people or first generates the groups by aggregating opinions about Asians and about Whites separately, and then takes the union of the two groups.

The next axiom, non-degeneracy, excludes constant aggregation rules where an individual is included or excluded from the set of socially accepted people regardless of the individual opinions.

- *Non-degeneracy*: For every $i \in N$ there exist profiles of individual opinions G and G' such that $i \in F(G)$ and $i \notin F(G')$.

Finally, an anonymity condition requires that the group of socially accepted individuals should not depend on their names. Names are switched through a permutation π of N. Thus, for a given permutation π, i is the new name of the individual formerly known $\pi(i)$). For a given profile G, let πG be the profile in which the names are switched. Finally, let $\pi F(G)$ consists of the socially accepted society members when the profile is G and their names were permuted according to π.

- *Anonymity*: $F(\pi G) = \pi F(G)$.

In other words, the axiom requires that if individual i is qualified in profile πG, then individual $\pi(i)$ is qualified in profile G.

Theorem 2 (Miller, [17], 2008). *The strong liberal CIF F^L is the only CIF that satisfies meet separability, join separability, non-degeneracy, and anonymity. Moreover, all four axioms are independent.*

4 Liberalism versus Consensus in Group Identification Problems

The "liberal" aggregation is an extreme case for determining collective identities in which one's opinions about other society members are not taken into account. We discuss in this section two methods for social aggregation where these opinions also matter. The first method was introduced by Samet and Schmeidler ([20], 2003). These authors axiomatically characterize a class CIFs called consent rules which is parametrized by the weights given to individuals in determining their own identity. The class of consent rules contains liberalism and majoritarianism as extreme cases. The second method views a collective identity function as a recursive procedure with two main ingredients: an initial set of individuals and an extension rule according to which new individuals are added to the initial set. Two such procedures were discussed in Kasher ([15], 1993) and Kasher and Rubinstein ([16], 1997) and axiomatically characterized by Dimitrov et al. ([11], 2007).

4.1 Consent Rules

Samet and Schmeidler ([20], 2003) introduce the class of consent rules as to capture the idea that, provided that one considers oneself as having a corresponding social identity, one needs nevertheless a certain number of supporters in order to be socially accepted; and, on the other hand, if an individual does not consider himself as having the corresponding identity, then there should be again a given number of individuals who also think in this way. The reader is referred to the original work of Samet and Schmeidler for different examples, where the two numbers can be different. Moreover, this class of rules is merely a subset of the one covered in Barberà et al. ([4]1991), where the set of alternatives the individuals vote for does not necessarily coincide with the set of voters.

More precisely, the class of consent rules is parametrized by two positive integers s and t with $s + t \leq n + 2$. A CIF F is a consent rule (with consent quotas s and t) if for any profile G and any individual i, the following two conditions are met:

- $i \in G_i$ implies [$i \in F(G)$ if and only if $\left| \left\{ j \in N \mid i \in G_j \right\} \right| \geq s$];
- $i \notin G_i$ implies [$i \notin F(G)$ if and only if $\left| \left\{ j \in N \mid i \notin G_j \right\} \right| \geq t$].

Thus, i's determination of oneself is adopted by the society if there are at least s other members who support i; correspondingly, if i does not classify oneself as having the corresponding identity, then he needs at least t other members who also think so. Hence, the larger the quota s (t) is, the greater is the society's power to act against one's own opinion.

The first axiom used by Samet and Schmeidler ([20], 2003) is the following:

- *Monotonicity*: For any two profiles G, G' we have that [$G_i \subseteq G'_i$ for every $i \in N$] implies [$F(G) \subseteq F(G')$].

In other words, the CIF is weakly increasing with respect to set inclusion.

- *Independence*: If G and G' are profiles such that for some $j \in N$, $j \in G_i$ if and only if $j \in G'_i$ for all $i \in N$, then $j \in F(G)$ if and only if $j \in F(G')$.

The idea behind independence is that the social identity of an individual j is independent of what individuals think about individuals other than j.

Theorem 3 (Samet and Schmeidler, [20], 2003). *A CIF satisfies monotonicity, independence, and anonymity if and only if it is a consent rule. Moreover, all three axioms are independent.*

4.2 Procedural Group Identification

Another possibility to extend the study of the group identification problem is by adding a *procedural* view in the analysis. By following Dimitrov et al. ([11], 2007) we present two axiomatic characterizations of two recursive procedures for determining "who is a member of a social group" : a consensus-start-respecting procedure which is the one introduced by Kasher ([15], 1993) and a liberal-start-respecting procedure which adds a procedural view to the "liberal" aggregation of Kasher and Rubinstein ([16], 1997).

The structure of both procedures consists of two components: an initial set of individuals and a rule according to which new individuals are added to this initial set. As the names of the procedures suggest, the initial set of the first procedure consists of all the individuals who are defined as group members by everyone in the society, while the initial set of the second procedure collects all the individuals who define themselves as members of the social group. The extension rule for both procedures is the same: only those individuals who are considered to be appropriate group members by someone in the corresponding initial set are added. The application of this rule continues inductively until there is no possibility of expansion any more.[1] Hence, the initial sets can be seen as sets of individuals whose opinions have priority to the opinions of those agents who are finally considered as members of the subgroup.

4.2.1 Definitions

For any profile $G \in \mathcal{G}$, let $K_0(G) = \{i \in N \mid i \in G_k \text{ for all } k \in N\}$. We define a CIF being *consensus-start-respecting*, to be denoted by $K(G)$, as follows: for each positive integer t, let $K_t(G) = K_{t-1}(G) \cup \{i \in N \mid i \in G_k \text{ for some } k \in K_{t-1}(G)\}$; and if for some $t \geq 0$, $K_t(G) = K_{t+1}(G)$, then $K(G) = K_t(G)$. For any $G \in \mathcal{G}$, let $L_0(G) = \{i \in N \mid i \in G_i\}$. With the help of $L_0(G)$, we define a CIF being *liberal-start-respecting*, to be denoted by $L(G)$, as follows: for each positive integer t, let $L_t(G) = L_{t-1}(G) \cup \{i \in N \mid i \in G_k \text{ for some } k \in L_{t-1}(G)\}$; and if for some $t \geq 0$, $L_t(G) = L_{t+1}(G)$, then $L(G) = L_t(G)$.

---

[1] For an axiomatic characterization of the aggregator selecting the agents that are indirectly designated by all individuals in the society the reader is referred to Houy ([13], 2006).

Notice that the extension rule for L and K is the same, but the initial set is different: the liberal-start-respecting procedure starts with $L_0(G)$ which consists of all members of the society who view themselves as Gs. The intuition behind each step of the expansion is in line with Kasher's ([15], 1993) argument: every socially accepted G as being newly added brings a possibly unique new view of being a G collectively with him; since a collective identity function is supposed to aggregate those views, it must pay attention to this new individual's G-concept in order to cover the whole diversity of views in the society about the question "what does it mean to be a G?".

To illustrate the above procedures for defining collectively accepted group members, consider the following example. Let $N = \{1,2,3\}$ and consider the profile $G = (G_1, G_2, G_3)$ with $G_1 = \{1,2\}, G_2 = \{2,3\}$ and $G_3 = \{2\}$. Then, for this profile, $K_0 = \{2\}, K_1 = K_0 \cup \{3\} = \{2,3\}, K_2 = K_1$. Therefore, the collectively accepted group members according to the consensus-start-respecting procedure are collected in the set $K = \{2,3\}$. For the same profile G of individual views, we have $L_0 = \{1,2\}, L_1 = L_0 \cup \{3\} = \{1,2,3\}, L_2 = L_1$. Therefore, for the given profile of views, and as a result of the application of the liberal-start-respecting procedure, we have $L = \{1,2,3\}$.

4.3 Axioms and Characterizations

The first axiom used for the characterization of the two procedures is consensus and it has been initially used by Kasher and Rubinstein ([16], 1997) to reach logically their "liberal" aggregation rule.

- A CIF $F \in \mathscr{F}$ satisfies *consensus* (C) if for all $G \in \mathscr{G}$,

    - $[i \in G_k$ for every $k \in N]$ implies $[i \in F(G)]$, and
    - $[i \notin G_k$ for every $k \in N]$ implies $[i \notin F(G)]$.

This axiom is very plausible when imposed as a requirement on a collective identity function. It says that, if an individual is defined as a group member by everyone in the society, then this individual should be considered as a socially accepted group member; and, correspondingly, if no one views this individual as a group member, then he or she should not deserve the social acceptance as a group member.

As we shall see below, the combination of appropriately defined independence and equal treatment axioms and the consensus axiom results in the desired characterizations of the procedural collective identity functions defined above.

4.3.1 Independence Axioms

Let us first consider the following two independence axioms.

- A CIF $F \in \mathscr{F}$ satisfies *irrelevance of an outsider's view* 1 (IOV1) if for all $G, G' \in \mathscr{G}$ and for all $i, j \in N$,

- $G'_j = G_j \cup \{i\}$, and
- $G^j_l = G_l$ for all $l \in N \setminus \{j\}$,
  imply
- $[j \notin F(G)$ and $i \notin G'_k$ for some $k \in N] \Rightarrow [i \in F(G)$ iff $i \in F(G')]$.

- A CIF $F \in \mathscr{F}$ satisfies *irrelevance of an outsider's view* 2 (IOV2) if for all $G, G' \in \mathscr{G}$ and for all $i, j \in N$ with $i \neq j$,

- $G'_j = G_j \cup \{i\}$, and
- $G^j_l = G_l$ for all $l \in N \setminus \{j\}$,
  imply
- $[j \notin F(G)] \Rightarrow [i \in F(G)$ iff $i \in F(G')]$.

Irrelevance of an outsider's view 1 stipulates that if someone is collectively defined as not having the corresponding identity, then this person's view about any society member is not relevant in deciding his or her collective identity. Note however that, for the consensus-start-respecting procedure, there is one case, in which the view of an outsider cannot be deemed as irrelevant; this case corresponds to the situation in which everyone in the society except the outsider j considers i as having the corresponding identity, so that the change of j's view in favour of i is (via (C)) relevant for the social identification of i. As the reader can see, we exclude this case in (IOV1) by requiring that there is a $k \in N$ such that $i \notin G'_k$. It should also be noted that (IOV1) is weaker than the exclusive self-determination axiom used by Samet and Schmeidler ([20], 2003) and it is satisfied by the consensus-start-respecting procedure but not by the liberal-strat-respecting one. The second independence axiom, (IOV2), is of the same spirit and it is designed for the liberal-start-respecting procedure. The difference here is that, in order to avoid the situation in which an individual becomes crucial for his own social determination (from being outsider to being insider) we require $i \neq j$ in its formulation.

4.3.2 Equal Treatment Axioms

The next two axioms display two slightly different ways in which the opinions of the individuals who are already considered as appropriate group members can be treated symmetrically.

- A CIF $F \in \mathscr{F}$ satisfies *minimal robustness* (MR) if for all $G, G' \in \mathscr{G}$ and for all $i, j, k \in N$,

- $i \in G_j \cap G_k$
- $G'_j = G_j \setminus \{i\}$
- $G^j_l = G_l$ for all $l \in N \setminus \{j\}$,
  imply
- $[j \in F(G)$ and $k \in F(G')] \Rightarrow [i \in F(G)$ iff $i \in F(G')]$.

Minimal robustness requires that if an individual i is considered to be an appropriate group member by two individuals j and k in a given profile, and if in a new profile j does not consider i as an appropriate group member anymore but k still does, and nothing else has changed, then, when j is a collectively accepted group member in the original profile and k is a collectively accepted member in the new profile, it must be true that i is collectively accepted in the original profile if and only if i is collectively accepted in the new profile. This axiom essentially says that the social status of an individual i is robust with respect to changes in the view of an insider j about i, provided that there is another collectively accepted member k who supports i.

- A CIF $F \in \mathscr{F}$ satisfies *equal treatment of insiders' views* (ETIV) if for all $G, G' \in \mathscr{G}$ and for all $i, j, k \in N$,

  - $i \in G_j \setminus G_k$
  - $G'_j = G_j \setminus \{i\}, G'_k = G_k \cup \{i\}$
  - $G'_l = G_l$ for all $l \in N \setminus \{j, k\}$,
    imply
  - $[j \in F(G)$ and $k \in F(G')] \Rightarrow [i \in F(G)$ iff $i \in F(G')]$.

Equal treatment of insiders' views requires that if an individual i is considered to be an appropriate group member by an individual j but not by an individual k in a given profile, and if in a new profile j does not consider i as an appropriate group member anymore but k does, and nothing else has changed, then, when j is a collectively accepted group member in the original profile and k is a collectively accepted member in the new profile, it must be true that i is collectively accepted in the original profile if and only if i is collectively accepted in the new profile. This axiom essentially requires that a CIF should treat the views of all the members who have the corresponding social identity equally.

4.3.3   Characterizations

As shown by Dimitrov et al. ([11], 2007), if a CIF satisfies (C) and (IOV1), then $K(G)$ acts as an upper bound for $F(G)$ at any profile $G \in \mathscr{G}$; adding (MR) to these axioms results in a characterization of the consensus-start-respecting procedure.

Theorem 4 (Dimitrov et al., [11], 2007). *A CIF $F \in \mathscr{F}$ satisfies (C), (MR) and (IOV1) if and only if $F = K$. Moreover, all three axioms are independent.*

As it turns out, the combination of (C) and (IOV2) plays a similar role for a CIF $F \in \mathscr{F}$ as the role of the combination of (C) and (IOV1): it produces an upper bound for F at any profile $G \in \mathscr{G}$. This upper bound is exactly the set of socially accepted group members at G according to the liberal-start-respecting procedure. However, in order to complete the characterization of the latter procedure, the same monotonicity requirement (MON) as in Samet and Schmeidler ([20], 2003) is additionally needed.

Theorem 5 (Dimitrov et al., [11], 2007). *A CIF $F \in \mathscr{F}$ satisfies (C), (MON), (ETIV) and (IOV2) if and only if $F = L$. Moreover, all four axioms are independent.*

5 Other Themes and Contributions

In this concluding section we briefly review several other modelling frameworks that point toward directions for future research.

5.1 Simple Collective Identity Functions

A simple collective identity function as introduced by Cengelci and Sanver ([8], 2010) is an aggregation rule that can be expressed in terms of winning coalitions. More precisely, a coalition S of individuals is winning over a person i if and only if i is qualified as having the corresponding social identity whenever all members of S think so and everyone in $N \setminus S$ has the opposite opinion.

As shown by Cengelci and Sanver ([8], 2010), not all collective identity functions can be expressed in terms of winning coalitions. Basically, the class of such CIFs coincides with the set of CIFs that satisfy the independence condition used by Samet and Schmeidler ([20] 2003). In addition, by imposing certain properties on the family of winning coalitions over each individual, Cengelci and Sanver ([8], 2010) relate simple collective identity function to particular CIFs of the literature.

5.2 Gradual Opinions

In all works referred so far, the opinion of an individual i regarding regarding if another individual j possesses a certain identity or attribute was dichotomous in the sense that j either has the corresponding identity according to i or does not have it. By contrast, Ballester and García-Lapresta ([1], [2], [3], 2008, 2009) assume that each individual gradually assesses all individuals including oneself, i.e., opinions are members of the unit interval or a finite scale. Then one applies an aggregation operator and a threshold as to determine the group of socially accepted society members.

The main research focus in the cited works is then on suitably defined sequential procedures for generating a final subgroup of individuals. More precisely, the above authors formulate properties of aggregators and families of thresholds that guarantee the convergence of the corresponding sequential processes.

5.3 Group Identity Functions

Let us finally consider an environment in which every individual has a view about how the society should be partitioned into classes, where the classes are not ranked. A *group identity function* assigns then to each profile of views a societal decomposition into classes. A first possible setup is to view the number of classes as given. Within this setup, Houy ([14], 2007) provides axiomatic characterizations of the *liberal* group identity function according to which an individual is put in a given class if and only if he classifies oneself in that class. A more general setup is to consider the number of classes as being *endogenously* determined. This is in contrast to the previous framework of collective identity functions where the number of social groups was assumed to be fixed and their names matter.

The most studied rule in this second setup of aggregating partitions is the *conjunctive* aggregator which classifies two individuals in the same social group if and only if everyone in the society thinks so. This function belongs to the class of rules characterized by Fishburn and Rubinstein ([12], 1986) in the context of the aggregation of equivalence relations (see also Mirkin [18], Barthélemy et al. [6], Barthélemy [5]) and it was recently axiomatized by Houy ([14], 2007) in the context of group identification. The central axiom in these characterizations is a binary independence condition requiring the decision of whether or not two individuals belong to the same social class to depend only on the individual classifications with respect to *these* two individuals. We refer the reader to Dimitrov et al. ([9], 2009) and Chambers and Miller ([7], 2010) for characterizations of aggregators of equivalence relations in which separability axioms play a crucial role. Vannucci ([22], 2008) studies the property in a more general model of group identification in which opinions may be in the form of partitions of the society.

In contrast to the works cited above, the central condition in Dimitrov and Puppe ([10], 2010) is a *non-bossiness condition* which requires the group identity function to depend only on *one* cell from the individual partition of each society member - namely on the cell the corresponding individual classifies himself in. Intuitively, non-bossiness thus states that the decision of whether or not two individuals belong to the same class should not depend on the view of unconcerned individuals. Non-bossiness neither implies nor is implied by binary independence.

The non-bossiness condition allows one to describe the social classification in terms of an *opinion graph* on the set of individuals. A directed edge (i, j) in this graph corresponds to the situation in which individual i classifies himself in the same group with individual j. The group identity functions introduced in Dimitrov and Puppe ([10], 2010) correspond to particular ways of decomposing this graph. Specifically, any group identity function satisfying a positive liberalism condition and a simple sovereignty requirement decomposes the graph into particular refinements of its *weakly connected components*.

References

1. Ballester, M.A., García-Lapresta, J.L.: A model of elitist qualification 17, Group Decision and Negotiation pp. 497–513 (2008a)
2. Ballester, M.A., García-Lapresta, J.L.: Sequential consensus for selecting qualified individuals of a group. International Journal of Uncertainty, Fuzziness and Knowledge-Based Systems 16, 57–68 (2008b)
3. Ballester, M.A., García-Lapresta, J.L.: A recursive group decision making procedure for choosing qualified individuals. International Journal of Intelligent Systems 24, 889–901 (2009)
4. Barberà, S., Sonnenschein, H., Zhou, L.: Voting by committees. Econometrica 59, 595–609 (1991)
5. Barthélemy, J.-P.: Comments on Aggregation of equivalence relations. Fishburn, P.C., Rubinstein, A. (eds.) Journal of Classification 5, 85–87 (1988)

6. Barthélemy, J.-P., Leclerc, B., Monjardet, B.: On the use of ordered sets in problems of comparison and consensus of classifications. Journal of Classification 3, 187–224 (1986)
7. Chambers, C., Miller, A.: Rules for information aggregation, Working Paper (2010)
8. Cengelci, M., Sanver, R.: Simple collective identity functions. Theory and Decision 68, 417–443 (2010)
9. Dimitrov, D., Marchant, T., Mishra, D.: Separability and aggregation of equivalence relations, Working Paper (2009)
10. Dimitrov, D., Puppe, C.: Non-bossy social classification, Working Paper (2010)
11. Dimitrov, D., Sung, S.-C., Xu, Y.: Procedural group identification. Mathematical Social Sciences 54, 137–146 (2007)
12. Fishburn, P.C., Rubinstein, A.: Aggregation of equivalence relations. Journal of Classification 3, 61–65 (1986)
13. Houy, N.: He said that he said that I am a J, Economics Bulletin, vol. 4(4), pp. 1–6 (2006)
14. Houy, N.: I want to be a J! quotedblright: Liberalism in group identification problems. Mathematical Social Sciences 54, 59–70 (2007)
15. Kasher, A.: Jewish collective identity. In: Goldberg, D.T., Krausz, M. (eds.) Jewish Identity, pp. 56–78. Temple University Press, Philadelphia (1993)
16. Kasher, A., Rubinstein, A.: On the question 'Who is a J?': A social choice approach. Logique & Analyse, 385–395 (1997)
17. Miller, A.: Group identification. Games and Economic Behavior 63, 188–202 (2008)
18. Mirkin, B.: On the problem of reconciling partitions. In: Blalock, H.M. (ed.) Quantitative Sociology: International Perspectives on Mathematical and Statistical Modeling, pp. 441–449. Academic Press, London (1975)
19. Rubinstein, A., Fishburn, P.C.: Algebraic aggregation theory. Journal of Economic Theory 38, 63–77 (1986)
20. Samet, D., Schmeidler, D.: Between liberalism and democracy. Journal of Economic Theory 110, 213–233 (2003)
21. Sung, S.-C., Dimitrov, D.: On the axiomatic characterization of 'Who is a J?'. Logique & Analyse 192, 101–112 (2005)
22. Vannucci, S.: The liberatarian identification rule in finite atomistic lattices, Working Paper (2008)

Consensus versus Dichotomous Voting[*]

Annick Laruelle and Federico Valenciano

Abstract. Consensus means general agreement among possibly different views, while dichotomus voting rules are a means of making decisions by using votes to settle differences of view. How then can it often be the case that a committee whose only formal mechanism for decision-making is a dichotomus voting rule reaches a consensus? In this paper, based on a game-theoretic model developed in three previous papers, we provide an answer to this question.

Journal of Economic Literature Classification Numbers: C7, D7.

Keywords: Consensus, bargaining, voting, committees.

1 Introduction

The editors have indicated that the purpose of this volume devoted to consensus -the first of a series on collective decision-making- is to cover a wide range of approaches to a diversity of contexts where one form or other of

Annick Laruelle
Departamento de Fundamentos del Análisis Económico I, Universidad del País Vasco, Avenida Lehendakari Aguirre, 83, E-48015 Bilbao, Spain and IKERBASQUE, Basque Foundation for Science, 48011, Bilbao, Spain
e-mail: `a.laruelle@ikerbasque.org`

Federico Valenciano
Departamento de Economía Aplicada IV, Universidad del País Vasco, Avenida Lehendakari Aguirre, 83, E-48015 Bilbao, Spain
e-mail: `federico.valenciano@ehu.es`

[*] This research is supported by the Spanish Ministerio de Ciencia e Innovacion under project ECO2009-11213, co-funded by the ERDF. Both authors also benefit from the Basque Government's funding to Grupo Consolidado GIC07/146-IT-377-07.

E. Herrera-Viedma et al. (Eds.): Consensual Processes, STUDFUZZ 267, pp. 135–143.
springerlink.com   © Springer-Verlag Berlin Heidelberg 2011

consensus is the most important ingredient. As they state explicitly in their invitation to submit contributions, consensus processes may involve human beings, but also animals and computers. In this paper we deal with consensus among rational beings.

According to the *Oxford English Dictionary*, consensus means "general agreement". One could add that usually, in many everyday situations, consensus means general, perhaps even unanimous, agreement *between possibly different views*. Reaching a consensus is often a complex and difficult process involving adjustments, concessions, threats and bluffing, with no general rules. It is largely dependent on the particular context, where social rules, customs, past experience and communication constraints play a role. By contrast, dichotomous voting rules are in principle simple mechanisms for making decisions by using a vote to settle differences of view: the winning side enforces the decision to accept or reject the proposal on the table. Thus, in a sense, dichotomous voting rules are in spirit completely opposed to the idea of consensus. Nevertheless, it is often the case that a committee whose only formal mechanism to make decisions is a specified dichotomous voting rule reaches a consensus about an issue. Moreover, in many such cases the final vote is a purely formal act ratifying the agreement resulting from a consensual process.

Our attention was drawn to this widely ignored fact in the social choice literature by an interview with David Galloway, an experienced officer who has worked for the Council of the European Union for many years. The notes that we took immediately after this interview included some puzzling comments from Galloway that lingered in our mind: "weights matter, the distribution of weights is in fact the most delicate issue to agree upon whenever an enlargement occurs", but "most of the decisions in the Council of the EU are made by unanimity", moreover "most of the time the Council bargains", "after a few rounds, experienced negotiators know rather accurately where the final agreements will lie", "agreements are more easily reached when they do not need unanimity, as negotiators risk being left out of the agreement"[1]. Thus the conclusion is clear: most of the time the Council looks for, and often reaches, a consensus. Nevertheless the binary majority rule plays a role even if the final vote is often merely a formal ratification of it.

How can this conundrum be explained? In this paper, based on a game-theoretic model developed in three previous papers of ours motivated by this issue, we provide an answer to this question.

2   A Two-Person Model of Consensus

Our analysis is based on what can be seen as the first model of consensus in game-theoretic literature. In Nash's (1951) bargaining model two individuals

---

[1] Personal interview with the authors on 23.06.2002. See Galloway (2001) for an excellent account of the Treaty of Nice.

bargain about a set of feasible alternatives that can benefit both differently, so that some agreements would be more favourable for one or the other. They can have any alternative in this set *as long as they agree on which one.* But if they fail to reach an agreement they will remain where they are at the moment when the negotiation starts and will gain nothing. From this summary description, it seems clear that the *Nash bargaining problem can be seen as a model of consensus problem*, in fact, the simplest such model: *two individuals in search of consensus.*

How does he solve it? Nash's model and solution can be summarised like this[2]. Both individuals are assumed to be "highly rational" (in Nash's original words) in several senses. First, both have precise and unambiguous preferences over the set of feasible alternatives. They can even specify their preferences in lotteries over them[3] and in a consistent way as prescribed by the model of rational choice under risk which had then recently been introduced by von Neumann and Morgenstern (1944). Each individual is assumed to know the other's preferences as well as his/her own ("common knowledge" in current terminology). This unrealistic ideal situation can be imagined as that of two characters in a science fiction movie who are mutually transparent and share knowledge of their rationality and preferences, and skip steps and save time to reach an agreement by taking advantage of this. How? Merely by pushing their "high rationality" a bit further, they are assumed to share a rational desideratum about certain conditions that a "reasonable" agreement should satisfy. First, any option to which there is an alternative that is preferred by both should be discarded ("efficiency"). Second, if an option that gives x utility to player 1 and y to 2 were considered a good agreement in a bargaining problem, then in the problem resulting from both players interchanging their preferences a good agreement should give y utility to player 1 and x to 2 ("anonymity"). Finally, if in a given problem a certain option is considered as the best agreement, then in any problem where there are less feasible options but their preferences are the same and the option considered optimal is still reachable that option should be the optimal agreement ("independence of irrelevant alternatives"[4]). Nash proves then that if our science fiction characters share these reasonable views[5] about what a rational agreement should be, they will not waste a minute on bargaining and discussing. In a lightning-consensus they would immediately agree: "That's settled, let's

---

[2] We describe it in informal terms, trying to convey the basic ideas underlying the model but skipping all the technicalities.

[3] A "lottery" is a probabilty distribution over a finite number of alternatives, i.e. a ticket that gives each alternative a given probability.

[4] In fact, Nash did not give names to his rationality conditions (later called "axioms"). In particular, what we have referred to (following part of the subsequent literature) as "independence of irrelevant alternatives" should not be confused with a condition with the same name used in the context of social choice.

[5] There is a forth condition relative to the representation of preferences that we omit here.

go for it!" They simultaneously arrive at what the reasonable agreement is with no need even for words, because this set of conditions characterises a unique good agreement *for every such problem.*

The extreme beauty of the model lies, as do its limitations, in the perfect rationality and reciprocal transparency of the players. Nevertheless, as a term of reference it has played and still plays an important role, and was the main source of inspiration for this work.

3   Binary Dichotomous Rules

As is well known, aggregating preferences is a complicated issue: no general procedure satisfying what can be seen as a set of minimal conditions exists (Arrow's impossibility theorem). Moreover, whatever the procedure based on the preferences of the individuals is, it is subject to the possibility of incentives to declare preferences other than their actual ones (Gibbard-Sattherwaite's theorem). Nevertheless, *none of these problems arises when there are only two options* involved. Perhaps that is why dichotomous rules are the most widely used rules in collective decision-making bodies, even in situations where most of the decisions to be made are not dichotomous. A binary dichotomous rule specifies for each yes/no vote profile whether the proposal is accepted or not. Simple examples are the following: the *unanimity* rule, according to which only a unanimous yes entails acceptance; the *dictatorship*, where the proposal is accepted if and only if a particular voter supports it; the *simple majority*, which establishes as "winning" all vote profiles where more than half the voters support the proposal; and the *weighted majority* rule, in which each voter has a weight and the proposal is accepted if the sum of the weights of the supporters is greater than a specified "quota".

More generally, a *binary dichotomous n-person voting rule* can be specified as follows. Let $N = \{1, 2, .., n\}$ denote the set of seats. If only "yes" and "no" votes are allowed[6] a vote configuration can be summarised by the set of "yes"-voters, so that each subset $S \subseteq N$. An N-voter binary dichotomous voting rule consists of a list of winning vote configurations W, i.e. those which entail the acceptance:

$$W = \{S \subseteq N : S \text{ leads to the acceptance}\}.$$

It is assumed that (i) $N \in W$, (ii) $\varnothing \notin W$, (iii) if $S \subseteq T$ and $S \in W$ then $T \in W$, and (iv) if $S, T \in W$ then $S \cap T \neq \varnothing$.

Thus the way a rule W works is obvious: a proposal is submitted to the committee, its members vote and a voting configuration of "yes"-voters S results, if it is winning (i.e. $S \in W$) the proposal is accepted, otherwise it is rejected.

---

[6] For dichotomus rules sensitive to abstention and absenteeism see Laruelle and Valenciano (2010).

4 Consensus under a Dichotomous Rule

Let us come back now to the problem outlined in the introduction. A set of individuals whose only commonly accepted decision-making rule is a binary dichotomous voting rule looks for consensus among a set of feasible agreements. Unless this set consists of only two options the least that can be said is that they have a problem, but one with which many decision-making bodies must grapple every day. Here we provide an ideal model inspired by Nash's classical two-person bargaining model outlined in Section 2.

Consider first the following issue: Can Nash's bargaining model be extended to more than two individuals? The answer is yes. Under the same model and assumptions, the result holds even if it becomes less and less credible as the number of individuals increases: the greater their number the less credible is the assumption of common knowledge of preferences. If it is doubtful whether two individuals could share all that they are supposed to share according to Nash's model, what can we say about a hundred or a million? Nevertheless we maintain Nash's common knowledge assumption to provide an ideal answer that may serve as a term of reference.

In fact, Nash's model and solution of the two-person consensus problem and its extension to n-person problems can be seen as a particular case of a more general problem. In Nash's model there are two elements. One is explicit -the players preferences-, and the other is implicit -the unanimity rule- for reaching consensus. But *when more than two individuals are involved there are many rules apart from unanimity*, which is implicit in the two-person case. Can Nash's model be extended to this situation?

4.1 A Cooperative Approach

Consider the following minimal two-ingredient model[7] of a committee of n individuals in search of consensus over a set of feasible agreements. As in the n-person extension of the Nash bargaining model, let a set $D \subseteq R^N$ (consisting of all utility vectors feasible by agreement of the n players) summarise their preferences, along with the *status quo* or *disagreement point* $d \in D$ representing the utilities that the players will receive if they fail to reach an agreement. This pair $B = (D,d)$ is the unique ingredient in a classical n-person bargaining model and the first ingredient of our model. The second ingredient is the dichotomous n-person voting rule W under which the committee makes decisions.

Let $\mathcal{B}$ denote the class of all admissible pairs $B = (D,d)$ and let $\mathcal{W}$ denote the set of all binary dichotomous n-person voting rules. Paralleling Nash's approach, we impose some conditions on a map $\Phi : \mathcal{B} \times \mathcal{W} \to R^N$, for vector

---

[7] Here we provide the outline of the model but skip some technical details. These details can be seen in Laruelle and Valenciano (2007, 2008a and 2008b).

$\Phi(B, W) \in R^N$ to be interpretable as providing a reasonable consensus for problem B under rule W.

Now denote $D_d := \{x \in D : x \geq d\}$. As admissible problems we consider those for which $D_d := \{x \in D : x \geq d\}$ is *bounded*, and D is a *closed, convex* and *comprehensive* (i.e., $x \leq y \in D \Rightarrow x \in D$) and *non-level* (i.e., for any $x, y \in D_d$ that belong to the boundary of D, $x \geq y \Rightarrow x = y$) set containing d, such that there exists some $x \in D$ s.t. $x > d$. As obvious prerequisites for a reasonable consensus solution Φ we impose *feasibility* and *individual rationality*: $\Phi(B, W) \in D_d$, if $B = (D, d)$.

In addition to these, the following conditions on Φ, which are adaptations of Nash's (1950, 1953) and/or Shapley's (1953) characterizing properties, can be considered as reasonable:

1. *Efficiency:* A utility vector dominated by another will never be considered as a reasonable consensus. (Formally: for all $(B, W) \in \mathcal{B} \times \mathcal{W}$, there is no $x \in D$, s.t. $x > \Phi(B, W)$.)

2. *Anonymity:* A reasonable consensus will not depend on the distribution of labels or names among the players. (Formally: for all $(B, W) \in \mathcal{B} \times \mathcal{W}$, and any permutation $\pi{:}N \rightarrow N$, and any $i \in N$, $\Phi_{\pi(i)}(\pi(B, W)) = \Phi_i(B, W)$, where $\pi(B, W) := (\pi B, \pi W)$.)

3. *Independence of irrelevant alternatives:* A reasonable consensus in a problem will remain reasonable if the feasible set shrinks as long as the status quo remains unchanged and the first consensus remains feasible. (Formally: if $B, B' \in \mathcal{B}$, with $B = (D, d)$ and $B' = (D', d')$, such that $d' = d$, $D' \subseteq D$ and $\Phi(B, W) \in D'$, then $\Phi(B', W) = \Phi(B, W)$, for any $W \in \mathcal{W}$.)

4. *Invariance w.r.t. positive affine transformations:* The solution should not be affected by the choice of zero and scale of representation of the utilities of the players. (Formally: for all $(B, W) \in \mathcal{B} \times \mathcal{W}$, and all $\alpha \in R_{++}^N$ and $\beta \in R^N$,

$$\Phi(\alpha * B + \beta, W) = \alpha * \Phi(B, W) + \beta,$$

where $\alpha * B + \beta = (\alpha * D + \beta, \alpha * d + \beta)$, denoting $\alpha * x := (\alpha_1 x_1, .., \alpha_n x_n)$, and $\alpha * D + \beta := \{\alpha * x + \beta : x \in D\}$.)

5. *Null player:* Players whose seats are completely irrelevant according to rule W should gain nothing. (Formally: for all $(B, W) \in \mathcal{B} \times \mathcal{W}$, if $i \in N$ is a null player in W, then $\Phi_i(B, W) = d_i$.)

Now let us denote by $Nash^w(B)$ the w-weighted asymmetric Nash bargaining solution (Kalai, 1977) of bargaining problem B for a vector of nonnegative weights $w = (w_i)_{i \in N}$, that is

$$Nash^w(B) := \arg \max_{x \in D_d} \prod_{i \in N} (x_i - d_i)^{w_i}.$$

These solutions basically emerge when anonymity or symmetry is dropped in the conditions that characterise Nash's bargaining solution. This makes sense when the environment is not entirely symmetric and gives some players advantages over others. This is reflected in the possibly different weights w_i for different players. In fact, each w_i is interpreted as the "bargaining power" of player i (see e.g. Binmore, 1998).

The following result shows how conditions 1-5 drastically restrict the possible answers to the question raised.

Theorem 1. *(Laruelle & Valenciano, 2008b) A solution $\Phi : \mathcal{B} \times \mathcal{W} \to R^N$ satisfies efficiency, anonymity, independence of irrelevant alternatives, invariance w.r.t. affine transformations and the null player property, if and only if*

$$\Phi(B, W) = Nash^{\varphi(W)}(B), \tag{1}$$

for some map $\varphi : \mathcal{W} \to R^N$ that satisfies anonymity and and gives weight zero to null players.

In view of the interpretation of asymmetric Nash bargaining solutions, we can provide the following interpretation of (1). According to this formula, a "reasonable" consensus is provided by the asymmetric Nash bargaining solution of problem B for weights (i.e., bargaining powers), which depend on the only source of asymmetry included in the model: the voting rule W. But note that these weights are not fully determined by the conditions assumed. These plausible conditions only constrain somewhat their dependence on W: these weights must be an *anonymous* function of the rule that gives zero weight to null players.

But what is this function φ? Why this indeterminacy? The reason is this: Under the ideal information conditions embodied in the assumption that all the elements in the model are common knowledge, the plausible conditions restrict the answer but fail to provide a precise answer because there are not enough elements in the model for them to do so. When the voting rule is not symmetric the particular protocol according to which bargaining under the voting rule takes place will obviously influence the outcome, but the model itself gives no hint about that protocol. Thus it is not surprising that assumptions that suffice to provide a precise answer in the case of a symmetric rule (note that if the rule is symmetric (1) is the classical Nash bargaining solution!)[8], do not suffice if the rule is not symmetric. This is clarified by a noncooperative approach to the same problem.

4.2 A Non Cooperative Approach

A noncooperative approach to the same problem clarifies the above-mentioned insufficient specification. Starting from the same elements that comprise the model, B and W, a noncooperative model of the bargaining process requires

---

[8] If a condition (standard but not fully plausible) is added then $\varphi(W)$ is the Shapley-Shubik (1954) index of the rule (Laruelle and Valenciano, 2009).

a minimal specification of *how* the bargaining process takes place. In other words, the bargaining protocol must be specified. To quote Laruelle and Valenciano (2008a):

"Consensus requires the acceptance of a proposal by all the players. This involves the specification of a procedure for acceptance/rejection of proposals, which entails two basic issues: (i) How is the proposer chosen?, and (ii) How are disagreements dealt with? (p. 344)"

With respect to the first point, in that paper we consider a variety of proposer-selecting protocols in which the voting rule plays a role and determines function φ. Among them a particularly simple one that yields an interesting result is what we call the Shapley-Shubik protocol, consisting of:

"Starting from the empty coalition, choose one player at random each time from the remaining players until a winning coalition S is formed. Then choose one of the players decisive in S at random."

It can be shown that then the probability of each player being the proposer is given by his/her Shapley-Shubik index for the current voting rule. As to point (ii) (how to deal with disagreement), we assume that if everybody accepts the proposal the game ends. If any player does not accept it, the process recommences with probability r ($0 < r < 1$), and the game ends in failure or "breakdown" and payoffs d with probability $1 - r$.

In this way a rather simple and reasonable protocol is specified in which the rule plays a role and the treatment of the disagreement reflects the search for general consensus (recommencing if it is not achieved) and the risk of failure: the greater the patience (represented by r) of the committee the lower this risk. This potentially infinite (albeit with probability 0) game admits in principle a plethora of strategies for players depending on the history. A subset of this extremely wide set of strategies is that of *stationary strategies*, i.e. those where the player uses the same rule whenever the game recommences regardless of the history of the game so far. It can be then proved that for the Shapley-Shubik protocol there exists a *stationary subgame perfect equilibrium*, that is, a subgame perfect equilibrium strategy profile consisting of stationary strategies. Moreover as $r \to 1$ the associated payoff vector tends to

$$Nash^{Sh(W)}(B),$$

where $Sh(W)$ denotes the Shapley-Shubik of the rule W.

5 Conclusion

We have provided a coherent model of the situation described in the introduction and summarised by the title of this paper from a two-fold game-theoretic point of view: cooperative and non cooperative. As has been pointed out, the assumption of common knowledge of all the elements in the model is perhaps the point where these ideal models are furthest from reality. If this is true for the classical Nash's two-person model of consensus, then it must become still more so as the number of individuals increases. Nevertheless, however unrealistic the model might seem[9], the situation it modelises by incorporating the minimal ingredients, i.e., a group of individuals bargaining for consensus "with the help of a dichotomous voting rule", is an everyday fact insufficiently studied by social theorists.

References

[1] Binmore, K.: Game Theory and the Social Contract II, Just Playing. MIT Press, Cambridge (1998)

[2] Galloway, D.: The Treaty of Nice and Beyond: Realities and Illusions of Power in the EU. Sheffield Academic Press, London (2001)

[3] Kalai, E.: Nonsymmetric Nash Solutions and Replications of 2-person Bargaining. In: International Journal of Game Theory, vol. 6, pp. 129–133 (1977)

[4] Laruelle, A., Valenciano, F.: Bargaining in committees as an extension of Nash's bargaining theory. Journal Economic Theory 132, 291–305 (2007)

[5] Laruelle, A., Valenciano, F.: Non cooperative Foundations of Bargaining Power in Committees. Games and Economic Behavior 63, 341–353 (2008a)

[6] Laruelle, A., Valenciano, F.: Voting and Collective Decision-Making: Bargaining and Power. Cambridge University Press, Cambridge (2008b)

[7] Laruelle, A., Valenciano, F.: Bargaining in Committees of Representatives: the 'Neutral' Voting Rule. Journal of Theoretical Politics 20(1), 93–106 (2008c)

[8] Laruelle, A., Valenciano, F.: Cooperative Bargaining Foundations of the Shapley-Shubik Index. Games and Economic Behavior (Special Issue in Honor of Martin Shubik) 65(1), 242–255 (2009)

[9] Laruelle, A., Valenciano, F.: Quaternary dichotomous voting rules, Discussion Paper 80/2010. In: Departamento de Economía Aplicada IV, Universidad del País Vasco (2010)

[10] Nash, J.F.: The Bargaining Problem. Econometrica 18, 155–162 (1950)

[11] Nash, J.F.: Two-person Cooperative Games. Econometrica 21, 128–140 (1953)

[12] Shapley, L.S.: A Value for N-person Games. Annals of Mathematical Studies 28, 307–317 (1953)

[13] Shapley, L.S., Shubik, M.: A Method for Evaluating the Distribution of Power in a Committee System. American Political Science Review 48, 787–792 (1954)

[14] von Neumann, J., Morgenstern, O. (1947, 1953). In: Theory of Games and Economic Behavior, Princeton University Press, Princeton (1944)

---

[9] This model may also provide arguments in favor of a recommendation for the choice of voting rule in a committeee of representatives different from the most usual ones in the literature (Laruelle and Valenciano, 2008b and 2008c).

On a Priori Evaluation of Power of Veto

Jacek Mercik

Abstract. The main goal of the paper is the evaluation of power connected with veto attribute of the decision maker. A special kind of action attributed to some players is the right to veto, i.e. to stop the action of others permanently or temporarily. In certain cases, it is possible to calculate a value of power of veto attributed to the decision maker and to give the exact value of the power index as well. In other cases, it is only possible to compare the situation with and without veto attribute. In this paper we would like to analyse the power of a player with a right to veto, expecting that the difference between the power of player with veto and his power without veto allows us to evaluate directly or indirectly the power of veto itself.

1 Introduction

Existing constitutional design of any democratic country must favour a consensus among the society it addresses. As it is pointed out by John Rawls (Rawls, 1987):

> "In a constitutional democracy one of its most important aims is presenting a political conception of justice that can not only provide a shared public basis for the justification of political and social institutions but also helps ensure stability from one generation to the next. ... such a basis must be I think, even when moderated by skilful constitutional design, a mere modus vivendi, dependent on a fortuitous conjunction of contingencies. What is needed ... thereby specifying the aims the constitution is to achieve and the limits it must respect. In addition, this political conception needs to be such that there is some hope of its gaining the support of an overlapping consensus, that is a consensus in which it is affirmed by the opposing religious, philosophical and moral doctrines likely to thrive over generation in a more or less just constitutional democracy, where the criterion of justice is that political conception itself."

Therefore, relations between all constitutional and government organs must be moderated and evaluated depending on their way of decision making. Montesquieu's

Jacek Mercik
Wroclaw University of Technology

E. Herrera-Viedma et al. (Eds.): Consensual Processes, STUDFUZZ 267, pp. 145–156.
springerlink.com   © Springer-Verlag Berlin Heidelberg 2011

tripartite of power may be realized via different attributes of involved sides. Among those attributes one may find the right to veto. We think, that priori veto is rather strengthening the position of beholder. So, any considerations about consensus process must include evaluation of veto attribute as well, for to preserve the balance between sides.

The main goal of the paper is the evaluation of power connected with veto attribute of the decision maker. In certain cases, it is possible to calculate a value of power of veto attributed to the decision maker and to give the exact value of the power index as well. In other cases, it is only possible to compare the situation with and without veto attribute. Actually, significant numbers of power indices are in use for evaluation of power of player. The main differences between these indices are the ways in which coalition members share the final outcome of their cooperation and the kind of coalition players choose to form. The latter started from negation of equiprobability (Laplace criterion) of possible coalition transformation into a winning coalition (as it was done in Owen, 1977) and consequently it leads to different assumptions and results. A special kind of action attributed to some players is the right to veto, i.e. to stop the action of others permanently or temporarily. In this paper we would like to analyse the power of a player with a right to veto, expecting that the difference between the power of player with veto and his power without veto allows us to evaluate directly or indirectly the power of veto itself.

2 Veto

The meaning of veto can be explained by the following artificial example: $\{2; 1_a, 1_b, 1_c\}$ where the voting is a majority voting (voting quota equals 2) and weights of all voters a, b, c ($N=3$) are equal and fixed at 1. As it can be seen, there are four winning coalitions: $\{a,b\}$, $\{a,c\}$, $\{b,c\}$, $\{a,b,c\}$. The first three coalitions are vulnerable and the veto (called the veto of the first degree) of any coalition's members transforms it from winning into non-winning one. The classical example of such a veto is a possible veto of permanent members of the Security Council of United Nations.

The last coalition, $\{a,b,c\}$, is different: a single member's veto can be overruled by two other members. This type of veto is called the second degree veto. A very typical example of such a veto is a presidential veto, which under certain circumstances can be overruled.

We also believe, that any real life decision process of legislation can be framed with the above simple games and, for example, associations between "a" and "house of representatives", "b" and "senate" and "c" and "president" are well-founded. In that case, "a" can represent a sub-game, for which weight 1 means that a bill has passed through the House of Representatives, and so on. It is also noticed, that "president" is a single player game, as 1_c in our example. Then, when analysing a bill passing through a legislative way, we are in fact analysing a supra-game where players are games on different levels. This lets us to use standard power indices which must be modified for certain conditions and assumptions.

3 Power Index of Veto

Before we start the evaluation of power of veto, we need to define a standard power concept and a power index consequently. Let $N = \{1, 2, ..., n\}$ be the set of players (individuals, parties) and $\omega_i \, (i = 1, ..., n)$ be the (real, non-negative) weight of the i-th agent and τ be the total sum of weights of all players. Let γ be a real number such as $0 < \gamma < \tau$ (minimal sum of weights necessary to approve a proposal). The $(n+1)$-tuple $[\gamma, \omega] = [\gamma; \omega_1, \omega_2, ..., \omega_n]$ such that

$$\sum_{i=1}^{n} \omega_i = \tau, \omega_i \geq 0, \, 0 < \gamma \leq \tau,$$

we call a weighted voting body of the size $n = card \, \{N\}$ with quota γ, total weight τ and allocation of weights $\omega = (\omega_1, \omega_2, ..., \omega_n)$. We assume that each player i casts all his votes either as "yes" votes, or as "no" votes. Any non-empty subset of players $S \subseteq N$ we shall call a voting coalition. Given an allocation ω and a quota γ, we shall say that $S \subseteq N$ is a winning voting coalition, if $\sum_{i \in S} \omega_i \geq \gamma$ and a losing voting coalition, if $\sum_{i \in S} \omega_i < \gamma$. Let

$$T = \left[(\gamma, \omega) \in R_{n+1} : \sum_{i=1}^{n} \omega_i = \tau, \, \omega_i \geq 0, 0 \leq \gamma \leq \tau \right]$$

be the space of all coalitions of the size n, total weight τ and quota γ.

Most of measures of power are designed to evaluate a priori power of players in a setting that is not structured by any rules except a voting rule. A power index (as it is formulated in Turnovec et al. 2008) is a vector valued function $\Pi : T \rightarrow R_n^+$ that maps the space T of all coalitions of the size n into a non-negative quadrant of R_n. A power index for each of the coalition players represents a "reasonable expectation" that it will be "decisive" in the sense that its vote (YES or NO) will determine the final outcome of voting.

Generally, there are two properties, related to the positions in voting of the coalition players, which are being used as a starting point for quantification of an a priori voting power: swing position and pivotal position of a coalition player.

Let $(i_1, i_2, ..., i_n)$ be a permutation of players of the committee, and let player k be in position r in this permutation, i.e. $k = i_r$. We shall say that a player k of the coalition is in a pivotal situation (has a pivot) with respect to a permutation $(i_1, i_2, ..., i_n)$, if

$$\sum_{j=1}^{r} \omega_{i_j} \geq \gamma \quad and \quad \sum_{j=1}^{r} \omega_{i_j} - \omega_{i_r} < \gamma$$

The most known pivotal position based on the a priori power measure was introduced by Shapley (1953) and Shapley and Shubik (1954), and is called SS-power (SS-index). The Shapley-Shubik index is therefore defined as follows:

$$SS(\gamma, \omega) = \frac{p_i}{n!}$$

where p_i is the number of pivotal positions of the coalition player i and $n!$ is the number of permutations of all coalition players (number of different strict orderings).

The most known swing position based on the a priori power measure is called Penrose-Banzhaf power index (Penrose, 1946; Banzhaf, 1965, 1968). Let S be a winning configuration in a coalition $[\gamma, \omega]$ and $i \in S$. We say that a member i has a swing in configuration S if

$$\sum_{k \in S} \omega_k \geq \gamma \quad and \quad \sum_{k \in S \setminus \{ i \}} \omega_k < \gamma$$

Let s_i denote the total number of swings of the member i in the committee $[\gamma, \omega]$. Then PB-power index of player i is defined as

$$PB_i(\gamma, \omega) = \frac{s_i}{\sum_{k \in N} s_k}$$

This form is usually called a relative PB-index. Original Penrose definition of power of the member i was

$$PB_i^{abs}(\gamma, \omega) = \frac{s_i}{2^{n-1}},$$

which is nothing else but the probability that a given member will be decisive (probability to have a swing). This form is usually called an absolute PB-index.

It is clear that both swing and pivotal attempts are connected with veto. If we assume that coalition is already formed, and one of its member defeats (swings), then this is equivalent to veto (the first or the second degree). So, possible power of veto should be rather measured somehow with participation of power index based on swings. The PB power index is the first one which can be used for it, but as far as we do not know which power index is better fitted for veto measuring, we still keep in minds other power indices.

4 Johnston Power Index

Other a priori indices are also in use depending on a situation when voting is taken. Among them there are: the Coleman coalition prevent index (1971) – $CP(i)$; the Coleman action initiation index (1971) – $CI(i)$; the Coleman group capacity index (power to act) (1971) – $C(A)$; the Rae index (1969) – $R(i)$; the Zipke index (Nevison, 1979) – $Z(i)$; the Brams-Lake index (1978) – $BL(i)$; the Deegan-Packel index (1979)

$- DP(i)$; the Holler index (1982) $- H(i)$; and the Johnston index (1978) $- J(i)$. The last one is suggested by Brams (Brams, 1990) as the best suited (as swing type power index) for veto analysis and it will be presented more detailed below.

Definition: a winning coalition is vulnerable if, among its members, there is at least one in swing position, whose swing would cause the coalition to lose. Such a member is called critical. If only one player is critical, then this player is uniquely powerful in the coalition.

Defining Johnston power index, first we count number of players being in swing position in vulnerable coalition c. Reciprocal of number of swings in coalition c is the share of i-th member swings in critical coalition c in total number of swings in critical coalition c, $f(c)$. For example, if there are only two such players in the coalition c, thus $f(c)=1/2$.

The Johnston power of player i is the sum of the reciprocal of number of swings in vulnerable coalition c in which i is critical, divided by the total number of reciprocal of number of swings in vulnerable coalition c of all players, or i's proportion of reciprocal of number of swings in coalition c.

Formally, if $V \subseteq S$ is a coalition of players, for each coalition $c \in V$, we define the set $f_i(c)$

$$f_i(c) = \begin{cases} f(c) & \textbf{\textit{i is swing in c}} \\ 0 & \textbf{\textit{otherwise}} \end{cases},$$

And Johnston power index:

$$J(i) = \frac{\displaystyle\sum_{c \in V} f_i(c)}{\displaystyle\sum_{j=1}^{n}\sum_{c \in V} f_j(c)}$$

Let us consider the following example (Mercik, 2009): game [4; 3, 2, 1], i.e. voting where there are three voters with 3, 2 and 1 votes each. Majority for the decision is 4 (quota). There are the following vulnerable coalitions in this game: (3, 2), (3, 1) and (3, 2, 1) (vulnerable coalitions must be winning coalitions).

Table 1 The example of Johnston power index for the game [4; 3, 2, 1] (Source: Mercik, 2009).

Vulnerable coalitions	Number of vulnerable coalitions	Critical swings			Reciprocal of number of critical swings		
		3 votes player	2 votes player	1 vote player	3 votes player	2 votes player	1 vote player
(3, 2)	1	1	1	0	½	½	0
(3, 1)	1	1	0	1	½	0	½
(3, 2, 1)	1	1	0	0	1	0	0
Total	3	3	1	1	2	½	½
J(i)					4/6	1/6	1/6

It can be noticed that value (4/6, 1/6, 1/6) of Johnston power index in this example differs from Penrose-Banzhaf power index (3/5, 1/5, 1/5) and it is equal to Shapley-Shubik power index (4/6, 1/6, 1/6). Probabilistic interpretation of Johnston power index (a probability that a certain player is in swing position) shows that Johnston power index belongs to the family of "swing" indices. Shapley-Shubik index describes probability that a certain player is in pivotal position, so SS index belongs to the "pivotal" family of indices.

5 Evaluation of Power of Veto of the First Degree

Let us remind that veto of the first degree means that the veto can not be overruled. It means that any winning coalition must include a veto empowered player or such a player should stop to use his/her veto. Consequently, it means that probabilities that coalitions will occur, are not equal. In the literature, such attempt is commonly recognizable as games with the pre-coalition structure (Owen, 1977) and priori power indices must be modified respectively.

As it was previously mentioned, a good example of decisive body where some (at least) members are empowered with veto attribute is the UN Security Council. Veto of one of the permanent members of the UN Security Council (i.e. China, France, Russia, the United Kingdom and the United States) can not be overruled, so this is a typical example of the veto of the first degree. The rest, ten non-permanent members of the Council have no right to veto resolutions of the Council. In table 2, there are shown the results of calculation of different power indices for the UN Security Council. It can be noticed that veto attributed to a permanent member of the UN Security Council makes them from 10 (based on swing type index) to 103 times more powerful (based on pivotal type index) – this is an indirect estimation of power of veto of permanent members.

Table 2 Values of different power indices for the UN Security Council (P_j stands for permanent members, N_j stands for non-permanent members).

Power index	Value	Power ratio P_j/N_j
SS power index $SS(P_j)$ $SS(N_j)$	 0.1963 0.0019	 103.85
Penrose-Banzhaf power index $PB(P_j)$ $PB(N_j)$	 0.1669 0.0165	 10.09
Coleman Power to prevent action index CP(i) $CP(P_j)$ $CP(N_j)$	 1.0000 0.0990	 10.10
Coleman Power to initiate action index CI(i) $CI(P_j)$ $CI(N_j)$	 0.0266 0.0026	 10.23

Analysing legislative way, it can be noticed that some decisions are made sequentially. Traditional power measures use the notion of a composite voting game introduced in order to deal with inter-body decision-making (see for example Felsenthal and Machover, 1998). At each stage players make their decisions and final result is a consequence of all of them. At fig. 1 and 2 we symbolically illustrate this process.

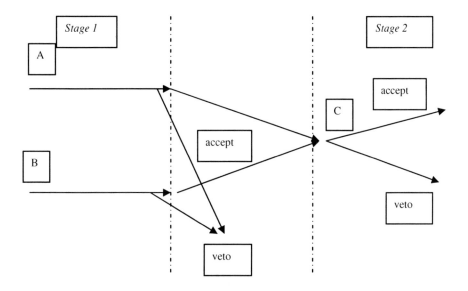

Fig. 1 Two-stage process of decision making with two parallel decision-makers (A and B) and one final decision-maker.

If A, B and C are single players (not bodies) as presented in fig. 1, estimation of a priori power of veto depends on logic of parallel part of the process: when both A and B must accept before C accepts, then value of power of veto is equal for A, B and C each. When only one of A and B must accept before C accepts, all veto power belongs to C. Whichever power index is in use, it must fulfil already mentioned conditions.

It is also clear, that when at least one of the A, B and C players is a decisive body, the veto power has to be distributed among members of the decisive body respectively. In that case, we may use one of the "swing" type indices and evaluate veto part of the power by comparing power of a certain player with and without veto attribute.

If A, B and C are single players as presented in fig. 2, each of the decision-makers has the same veto power. For decisive bodies, marked with A, B and C, the standard "swing" type index can be used and all veto power has to be distributed among members of a decisive body.

Any other veto of the first degree cases is the combination of those two already analysed examples and veto power index can be evaluated in the same way, i.e. by comparing power with and without veto.

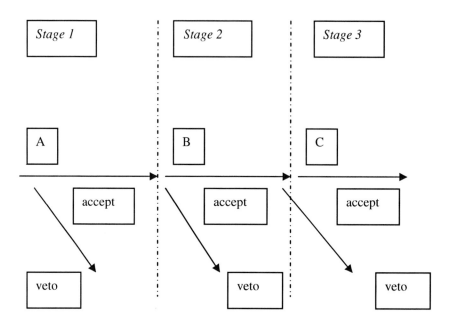

Fig. 2 Three-stages process of decision making with three serial decision-makers (A, B and C).

6 Evaluation of Power of Veto of the Second Degree

Possibility of overruling the veto (veto of the second degree) changes the situation drastically: the same decision maker may accept and/or veto at least more than once. It can be seen that inter-body decision is made and positions of decision-makers are different, depending on their ability to overrule a veto.

Let us have a look at the following example. In Poland, in the process of legislation, any bill accepted by the *Sejm* (lower chamber of the Polish parliament – 460 representatives) is considered by the *Senate* (100 senators), which may accept, amend or reject a bill. If a bill is amended or rejected by the *Senate*, then it comes back to the *Sejm*. The *Sejm* may, by absolute majority, reject the Senate's objection. After that, a bill accepted by the *Sejm* comes to the President of Poland who can accept and sign a bill within 21 days or may declare veto and send a bill back to the *Sejm*.

The presidential veto is considered as a cognizable attribute of the president regarding any bill resolved by the parliament. According to the Constitutional Act, the president signs and declares a bill in the official monitor (gazette). In the case of important state interests or poor quality of constituted law, the president may reject a bill. Presidential rejection of a bill (veto) has a conditional character: the

Sejm may accept a bill once more by a majority of 3/5 of votes in the presence of at least half of the members of the *Sejm* (representatives). In this case, the president has to sign a bill within seven days and publish the bill in the official monitor. The real effectiveness of the president's veto is therefore strongly subordinated to the present structure of parties in the *Sejm*.

Therefore, we have coalitions: $\left(z, p_{j_1}, p_{j_2}, ..., p_{j_{460}}\right)$ where z means the president and p means a representative. Notice, that as far as the *Senate* has no right to veto a bill (any *Senate*'s amendment or veto may be overruled by the *Sejm* by simple majority), the *Senate* does not participate in the game. Assuming that the parliament has no party structure, the winning coalition is as follows:

- $\left(z, p_{j_1}, p_{j_2}, ..., p_{j_n}\right)$, where $n \geq 231$ (we also assume that all representatives participate in every voting).

Among the winning coalitions there are the following vulnerable coalitions:
- for $n = 231$ all players, i.e. the president and 231 representatives, are critical, i.e. each of them can swing,
- for $232 \leq n < 276$ only the president can swing.

Note, that for $n \geq 276$ all coalitions are winning but no member of such a coalition is in a swing position.

Using Johnston's power index one may find (Mercik, 2009) that those values of index of power for the *Sejm* and the president are:

- for the president: 0.92342978817
- for a representative: 0.0001664570
- for the Sejm as a whole: 0.0765702118

It means that the President of Poland is more than 12 times stronger than the House of Representatives, when no party structure is assumed. The position of the president changes radically when the party structure matters. In the tab. 3 one may find adequate calculations. In that case, party structure matters and it changes power of the president, decreasing it deeply, even when the president is attributed with a veto. To some extent, we may say that power of the president in this case consists of power of veto, and in this situation Johnston power index for the president measures only the power of veto directly.

Table 3 Summary of calculations of Johnston power index for different assumptions about the Polish Parliament (Mercik, 2009) compared to the USA (values for the USA are from the book of Brams, 1990).

	Johnston power index		
	No party structure	With party structure	The USA (no party structure)
The President	0.9234	0.0067	0.7700
The House of representatives	0.0766	0.9933	0.0736
The Senate	0	0	0.1560

The immanent characteristics of the veto of the second degree is the inter-body decision making and the necessity of strategic thinking. Any decision-maker has to look ahead and reason back. Strategy of a certain legislator must include evaluation and sometimes it is not possible at all without strategies of other decision bodies (players). In this sense, also veto becomes conditional and depending on other decision-makers. In fig. 3, one may see the legislative way a bill goes through the Polish parliamentary system. The House of Representatives (the *Sejm*) may be in action from 1 to 3 times during the legislation:

- one time when a bill is accepted by the Senate and the president without restrictions,
- two times when the Senate has restrictions and the president has none,
- three times when the Senate and the president have restrictions.

Each of these situations may result in different estimation of power of every player (i.e. the *Sejm*, the Senate and the president) and only the president has the right to veto the bill. Therefore, the estimation of veto power is connected with the president only but depending on the power of other players.

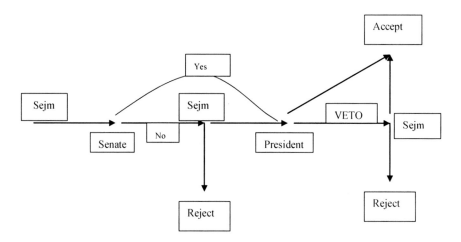

Fig. 3 Legislative way through the Polish parliamentary system (The *Sejm* stands for the House of Representatives).

The system of presidential vetoes may differ from country to country[1], but the general idea is the same: players must look ahead and reason back. It seems that the strategic type power index may be in use for the power evaluation. Among them probably the most adequate could be the index proposed by Steunenberg, Schmidtchen and Koboldt (1999). In their paper one may find also the example of

---

[1] The Polish example is rather untypical because the Senate is not participating in the president's veto overruling. For example, in the USA overruling of presidential veto requires the cooperation of the House of Representatives and the Senate (3/5 of members from both chambers).

evaluation of power for the European Union. In the paper, (Schmidtchen, Steunenberg 2002) they argue that this type of power index can be also used for a priori and a posteriori analysis of power.

7 Final Remarks

In most cases, the power of the right to veto cannot be measured directly, because this right is only the part of the characteristics of players. Therefore, evaluation of chances for consensus can be done indirectly too. However, we can indirectly estimate the influence of the right to veto on the power of a player by comparing its power both with and without this right. Quite intuitively, the right of veto will increase the power of a player in most cases. It is not so obvious how large this increase will be and in some cases power is associated only with the right to veto (as is the case of the President of Poland and the parliament with a party structure). This last example shows measure of veto power in an absolute term. In most cases it is not possible to do it so directly.

Constitutional analyses try to identify the power of the players, including the power of veto, anticipated from behind a veil of ignorance. In fact, strategies of all players need to fix a right of veto as a permanent element of the situation. The veto right of the first degree can be modelled using a standard "swing" type power index. The veto right of the second degree can be estimated by the modified Banzhaf or Johnston power in simple cases, but a new index remains to be shown for complicated games (compound games), with more general structure.

Literature

Banzhaf III, J.F.: Weighted voting doesn't work: a mathematical analysis. Rutgers Law Review 19, 317–343 (1965)

Banzhaf III, J.F.: One man, 3.312 votes: a mathematical analysis of the electoral college. Villanova Law Review 13, 304–332 (1968)

Brams, S.J.: Negotiation games. Routledge, New York (1990)

Brams, S.J., Lake, M.: Power and Satisfaction in a Representative Democracy. In: Ordeshook, P.C. (ed.) Game Theory and Political Science, pp. 529–562. NYU Press (1978)

Coleman, J.S.: Control of collectivities and the power of a collectivity to act. In: Lieberman, B. (ed.) Social Choice, New York, pp. 277–287 (1971)

Deegan Jr., J., Packel, E.W.: A new index of power for simple n-person games. International Journal of Game Theory 7, 113–123 (1979)

Felsenthal, D., Machover, M.: The measurement of voting power. Edward Elgar, Cheltenham (1998)

Holler, M.J.: Forming coalitions and measuring voting power. Political Studies XXX (2), 266–271 (1982)

Johnston, R.J.: On the measurement of power: Some reactions to Laver. Environment and Planning A 10, 907–914 (1978)

Mercik, J.: A priori veto power of the president of Poland. Operational Research and Decisions 4, 61–75 (2009)

Chr Nevison, H.: Structural power and satisfaction in simple games. In: Applied Game Theory, pp. 39–57. Physica-Verlag, Wuerzburg (1979)

Owen, G.: Values of Games with A Priori Unions. Mathematical Economy and Game Theory. Springer, Berlin (1977)

Penrose, L.S.: The Elementary Statistics of Majority Voting. Journal of the Royal Statistical Society 109, 53–57 (1946)

Rae, D.W.: Decision rules and individual values in constitutional choice. American Political Science Review 63, 40–56 (1969)

Rawls, J.: The idea of overlapping consensus. Oxford Journal of Legal Studies 7(1), 1–25 (1987)

Schmidtchen, D., Steunenberg, B.D.: Strategic Power in Policy Games. In: Holler, M., Kliemt, H., Schmidtchen, D., Streit, M. (eds.) Power and Fairness, Jahrbuch fur Neue Politische Okonomie, Mohr Siebeck, Tubingen (2002)

Shapley, L.: A value for n-person games. Annals of Mathematical Studies 28, 307–317 (1953)

Shapley, L.S., Shubik, M.: A Method of Evaluating the Distribution of Power in a Committee System. American Political Science Review 48, 787–792 (1954)

Steunenberg, B.D., Schmidtchen, D., Koboldt, C.: Strategic power in the European Union: Evaluating the Distribution of Power in Policy Games. Journal of Theoretical Politics 11, 339–366 (1999)

Turnovec, F., Mercik, J., Mazurkiewicz, M.: Power indices methodology: decisiveness, pivots and swings. In: Braham, M., Steffen, F. (eds.) Power, freedom, and voting. Springer, Heidelberg (2008)

Part III: Various Aspects of Consensus, Its Measuring and Reaching

Settings of Consensual Processes: Candidates, Verdicts, Policies

Hannu Nurmi*

Abstract. Social choice theory deals with mutual compatibilities of various choice criteria or desiderata. It thus provides a natural angle to look at methods for finding consensus. We distinguish between three types of settings of consensus-reaching. Firstly, we may be looking for the correct decision. This is typically the setting where the participants have different degrees of expertise on an issue to be decided. Also jury decision making falls into this category. Secondly, the setting may involve the selection of one out of a set of candidates for *e.g.* a public office. Thirdly, we may be looking for a policy consensus. This setting is otherwise similar to the candidate choice setting, but usually involves more freedom in constructing new alternatives. Once these settings and relevant results in each one of them have been reviewed, we discuss the implications of some choice paradoxes to consensus-reaching methods.

Keywords: Condorcet's jury theorem, social choice functions, voting procedures, metric representation, voting paradoxes.

1 Introduction

The settings in which collective decision making is deemed appropriate typically involve public goods, *e.g.* collective security, unpolluted environment,

Hannu Nurmi
Public Choice Research Centre and Department of Political Science and Contemporary History, University of Turku, FI-20014 TURKU, Finland
e-mail: hnurmi@utu.fi

* This work has been supported by the Academy of Finland through its support to Public Choice Research Centre and by European Science Foundation through its support to the project Social Software for Elections, Allocation of Tenders and Coalition/Alliance Formation. Comments of Mario Fedrizzi on an earlier version are gratefully acknowledged.

E. Herrera-Viedma et al. (Eds.): Consensual Processes, STUDFUZZ 267, pp. 159–177.
springerlink.com   © Springer-Verlag Berlin Heidelberg 2011

logistic infrastructure, public health. Sometimes these goods can be provided through decentralized mechanisms, such as individual voluntary contributions or market mechanism, but often these lead to (Pareto) sub-optimal supply of the good or to externalities that have to be handled by outside parties. Hence, decoupling the expressed opinions regarding the level of public goods provision from the distribution of costs of those goods is often assumed to be the only way to guarantee a near-optimal level of provision. The standard view is that the decoupling succeeds in voting, i.e. the voters are assumed to reveal their true opinions in voting on public goods. In other words, voting can be a solution to the failure of market mechanism in the case of public –in contradistinction to private – goods.

So, to the extent that we are dealing with public policies, voting seems at least *prima facie* a natural way of aggregating opinions. Voting is also very common method to elect persons to assemblies and other public positions. These are normally construed as opinion aggregating settings as well. One is looking for office-holders that best reflect the opinion distribution prevailing in the electorate. In none of these circumstances one is questioning the validity or correctness of the opinions underlying voting. They are "given" and should not be tampered with in fair elections.

There are, however, also settings in which voting is used to form a collective opinion on issues where one can – usually *a posteriori* – determine whether the opinion is correct, *i.e.* on issues of fact. These may, and often do, also pertain to policies, but whilst in public policies the evaluation criteria are based on values that differ from individual to another, in these settings the evaluation can be done using objective criteria shared by the individuals in the voting body. These settings have been somewhat in the sidelines of the modern social choice theory, but they play a prominent role in the history of group decision making.

This article focuses on three types of settings: those involving candidates for public offices (*e.g.* assemblies or presidencies), those dealing with the choice of public policy and those focusing on making the right decision in some objective sense of the term. We shall begin with the last one, *i.e.* settings where the problem is to find the correct decision on the basis of opinions of people none of whom can be assumed to know the correct decision for sure. This setting is typical in certain types of courts of law, *viz.* those making use of juries. This is the reason we shall be calling this setting the jury decision making, fully cognizant that similar problems are faced with many other decision making bodies consisting of experts, *e.g.* boards of corporations, medical teams *etc.* In these bodies it is natural to assume that the individuals share a common interest, *i.e.* to find the correct decision. The method whereby the consensus decision is reached is valuable only when it improves upon the chances of making correct decisions *vis-à-vis* the prevailing method. There is thus a pretty clear-cut criterion for evaluating group decision methods. Once we have dealt with the jury decision making, we turn to candidate and policy choice procedures. Not all methods used in these settings are *prima facie* consensual. *E.g.* the first-past-the-post system in single member constituencies is apparently not geared towards finding a consensus candidate since it results in the candidate who gets more votes that any other contestant. We will, however, see that nearly all voting systems – first-past-the-post included – can be seen as consensual in the sense of resulting in outcomes that are as close as possible to the opinions expressed by voters.

2   Consensus in Juries

In courts of law the jurors are typically faced with determining whether a person standing the trial is guilty or not guilty for the crime he/she (hereinafter, he) is charged with. Obviously of paramount importance is whether the verdict issued by the jury is correct or not. Whether some juror's opinion weighs less than that of another juror is of secondary importance. One wants the outcome to be correct: "guilty" when the person charged is really guilty and "not guilty" when he is not. Similarly when a panel of medical doctors – cardiologist, dermatologist, neurologist, *etc.* – is making a decision regarding the treatment of a patient suffering from a variety of symptoms, what one is interested in is not whether a specialist's view is deemed equally important than that of another specialist, but whether the treatment decided upon is successful in curing the patient. The setting of expert and jury decision making is aptly characterized by Robert Dahl [8]:

> ... whenever you believe that [expert] 1 is significantly more competent than [expert] 2 or 3 to make a decision that will seriously affect you, you will want the decision to be made by 1. You will not want it to be made by 2 or 3, nor by any majority of 1, 2 and 3.

The main point of the quote is that for the outcome may well be successful even though a majority of expert opinions are completely ignored. As we shall see, this would not in general be considered acceptable in settings where candidates or public policies are being dealt with in the absence of an objective criterion of correct decision.

Jury decision making has a special place in the history of social choice theory. In the 18'th century Marquis de Condorcet suggested a result that later on became known as Condorcet's Jury Theorem (CJT). A proof of the theorem is presented in [5]. Consider a group N consisting of $n(> 2)$ individuals, each with identical probability p of being right on a dichotomous issue. Assume that the individuals are voting independently of each other.

Theorem 1. *The probability P of the majority being right depends on the individuals' probability p of being right in the following way:*

1. If $0.5 < p < 1$, then $P > p$, P increases with n and, when $n \to \infty$, P converges to 1.

2. *If* $0 < p < 0.5$, *then* $P < p$, P *decreases when* n *increases and* $P \to 0$
   when $n \to \infty$.
3. *If* $p = 0.5$, *then* $P = 0.5$, *for all values of* n.

What the result, thus, says is that the consensus reached through the simple
majority rule is, in fact, superior to what any individual alone could achieve
if the individuals are even slightly more often right than wrong. Moreover,
in large groups the majority is almost certain to be right. To counteract this
result, the theorem also says that when the individuals are more often wrong
than right, the majority is less often right than the individuals and is almost
certainly wrong when the groups size increases.

CJT's empirical relevance is, of course, much diminished by the assumption
that every individual has the same probability of being right. Yet, it says a
great deal about the nature of the majority rule. As an aggregation method
it seems to enhance the disposition to be right among the individuals. In a
way it is akin to those game structures that increase the superior player's
probability of turning out victorious [17, 31]. More specifically, if we can
assign probabilities to players being victorious, there are games which increase
the more probable winner's probability of winning from what it is at the
outset. According to CJT the simple majority rule has the same tendency
when it is compared with any individual's probability of being right under
the assumption that the latter is larger than $1/2$.

While simple majority rule can make the group more competent than
any of the individuals if the latter are all equally competent, one is certainly
entitled to ask what happens when the competence varies among individuals.
A very important theorem has been proven by Owen et al.[33] for the case
where individual competence values (probabilities of being right), denoted by
p_i may differ for different individuals $i \in N$. Let now $\bar{p} = \sum_i p_i/n$. Owen *et al.*
show that as long as $1/2 < \bar{p} < 1$ and $n > 2$, $P > \bar{p}$ and $P \to 1$ as $n \to \infty$. So,
in essence, the asymptotic part of CJT remains largely intact if one allows for
some variability in individual competence values. The non-asymptotic part,
on the other hand, no longer necessarily holds for all individuals. In other
words, there may be individuals who are more competent than the majority.
The "may" should be taken seriously, though. It is quite possible that even
in this more general setting the majority is more competent than any of
the individuals, thus contradicting Dahl's position above. An example from
[5] demonstrates this. Suppose that the group consists of three individuals so
that $p_1 = 0.8, p_2 = p_3 = 0.7$. Individual 1 is thus significantly more competent
than 2 or 3. Yet, $P = 0.826$, *i.e.* the majority is more competent than the
most competent individual. So, you might, indeed, be inclined to seek the
advice of the group rather than its most competent member.

Despite this qualification to Dahl's contention, it is clear that adding persons with competence values just barely exceeding $1/2$ does not increase the
competence of the majority. In other words, the asymptotic part of the generalized CJT is conditional on competence distribution. The following theorem
of [29, 39] gives a restriction on the validity of the asymptotic part.

Theorem 2. *Assume that the individuals vote independently of each other and that $p_i > 1/2$ for all $i \in N$. Let the individuals be labeled in a non-increasing order of competence, that is, $p_i \geq p_j$ if $i < j$. The asymptotic part of CJT does not hold if*

$$\frac{p_1}{1 - p_1} > \prod_{i=2}^{n} \frac{p_i}{1 - p_i}.$$

Both Theorem 1 and Theorem 2 deal with individuals casting their votes independently of other individuals. How important is this assumption? It certainly affects the results. Berg [6] has shown that introducing small positive dependence between individual votes decreases majority competence from what it would be under independence assumption. In contrast, a small negative dependence between individuals increases the majority competence from its level under independence.

So, sometimes the majority rule seems to work well in the sense of actually improving the quality of decisions over what they would be if the decisions were made by any individual. Then again, sometimes the most competent individual may be more competent than the majority. But surely the majority rule with equal weight assigned to each individual is not the only way of making decisions. Conceivably, one could improve the quality of decisions (increase the probability of them being correct) by emphasizing the views of the most competent individuals at the cost of individuals with lower competence values. So, we might ask if there is an optimal assignment of voting weights to individuals such that the probability of the correct majority decision is maximized. This question has been answered in [29, 39]. To wit,

Theorem 3. *Assume that $p_i > 1/2$ for all $i \in N$. Then the decision method that maximizes the probability of the group decision being right is the weighted majority rule with weights w_i being assigned as follows:*

$$w_i = log(\frac{p_i}{1 - p_i}). \tag{1}$$

Consider an example borrowed from [31]. The group consists of five individuals with competence distribution: $0.9, 0.8, 0.8, 0.6, 0.6$. The average competence is 0.74 and majority competence 0.897. This is slightly lower than the competence of the most competent individual.

Using Equation 1 we get the following: $w_1 = 0.380, w_2 = w_3 = 0.240, w_4 = w_5 = 0.07$. With these weight the majority rule yields 0.984 as the group competence.

Determining the competence values for individuals is, of course, a major stumbling block in the way of utilizing the above theorems. Most political decisions are not issues of fact, but of values or desiderata. Grofman et al. [14] have suggested the following theorem to solve this empirical problem.

Theorem 4. *The optimal voting weights assigned to individuals can be approximated by giving each individual i the weight $r_i - 1/2$ where r_i is the proportion of collective decisions in which i has voted with the majority in the past.*

While it may be difficult to equate being right with voting with the majority (of weights), it makes the weighted majority voting a more consensus-directed method than it would otherwise be. If voting with the majority increases your voting weight in the future, it creates an additional incentive to vote with the majority.

3   Consensus in Candidate Selection

The jury decision making is in important ways different from political or social decision making situations. Firstly, in the latter situations the values or criteria of performance may, and usually do, differ among the individuals making the decision. This by itself makes "being right" an ambiguous notion. Secondly, typically the number of alternatives exceeds two. This implies *inter alia* that the majority rule has several non-equivalent extensions.

Historically, the systematic study of candidate selection has preceded that of jury decision making and even today the former is far more common subject of study than the jury decision making. Let us consider first the widely-known plurality rule or first-past-the-post system. *Prima facie* it has no built-in tendency for consensus since simply the most widely held opinion wins. Consider a variation of Jean-Charles de Borda's example presented in the French Royal Academy of Sciences in the latter half of the 18'th century [7, 24].

Table 1 Borda's paradox

4 voters	3 voters	2 voters
A	B	C
B	C	B
C	A	A

The table depicts the preference rankings of 9 voters over three candidates A, B and C arranged from best to worst. Given these opinions, A is clearly the plurality winner since 4 voters regard it best, while B and C is first-ranked by 3 and 2 voters, respectively. The usual way to define the plurality rule is say that this rule elects the candidate ranked first by more voters than any other candidate. The same principle can, however, also be stated as follows: the plurality rule elects the candidate for which the number of voters having some other candidate as their first-ranked is the smallest. In Table 1 for A the number of voters ranking some other candidate first is 5. This number for B is 6, while that of C is 7.[1]

---

[1] The rest of this section is based on [25].

Slightly more formally, the plurality rule can be defined in terms of a consensus state and distance metric. Let us define a distance function over a set P (of points or candidates) as any function $d : P \times P \to R^+$, where R^+ is the set of non-negative real numbers. A distance function d_m is called a metric if the the following conditions are met for all elements x, y, z of P:

1. $d_m(x, x) = 0$,
2. if $x \neq y$, then $d_m(x, y) > 0$,
3. $d_m(x, y) = d_m(y, x)$, and
4. $d_m(x, z) \leq d_m(x, y) + d_m(y, z)$.

Substituting preference relations for elements in the above conditions we can extend the concept of distance function to preference relations. But these conditions leave open the way in which the distance between two relations is measured.

The plurality consensus state is one where all voters rank the same candidate first. We define a discrete metric as follows

$$d_d(R, R') = \begin{cases} 0 & \text{if } R = R', \\ 1 & \text{otherwise.} \end{cases}$$

Then the distance between two preference profiles of the same size is

$$d_d(P, P') = \sum_{i=1}^{n} d_d(R_i, R'_i),$$

where the profile P consists of rankings $R_1, \ldots, R_n$ and the profile P' of rankings $R'_1, \ldots, R'_n$. That is, the distance indicates the number of rankings that differ in the two profiles. Note that we consider the profiles always in such a way that the order of the rankings in the profiles is irrelevant. Also, the distance between a profile P and a set of profiles $\mathbf{S}$ is, similarly,

$$d_d(P, \mathbf{S}) = \min_{P' \in \mathbf{S}} d_d(P, P').$$

As just stated, the unanimous consensus state in plurality voting is one where all voters have the same alternative ranked first. With the metric, in turn, we tally for each alternative, how many voters in the observed profile do not have this alternative as their first ranked one. The alternative for which this number is smallest is the plurality winner. The plurality ranking coincides with the order of these numbers.

Using this metric we have for the plurality winner w_p,

$$d_d(P, \mathbf{W}(w_p)) \leq d_d(P, \mathbf{W}(x)) \quad \forall x \in X \setminus w_p.$$

Thus, the apparently non-consensual plurality method can be seen as a way of looking for the closest consensus profile, given the expressed preferences and the way of measuring distance between preference profiles. Let us now turn to Borda's suggestion, the Borda count. To characterize it with the aid

of a consensus state and distance measure, consider the following measure
proposed by Kemeny [16] (see also [1, 2]).

Let R and R' be two rankings over alternative set X. Then their distance is:

$$d_K(R, R') = |\{(x, y) \in X^2 \mid R(x) > R(y), \ R'(y) > R'(x)\}|.$$

Here we denote by $R(x)$ the number of alternatives worse than x in a ranking
R. This is called inversion metric.

The distance between two preference profiles of the same size is, then, the
sum of the distances between the individual rankings. That is,

$$d_K(P, P') = \sum_{i=1}^{n} d_K(R_i, R_i'),$$

where the profile P consists of rankings $R_1, \ldots, R_n$ and the profile P' of
rankings $R_1', \ldots, R_n'$. Similarly, we can measure the distance between a profile
P and a set of profiles $\mathbf{S}$,

$$d_K(P, \mathbf{S}) = \min_{P' \in \mathbf{S}} d_K(P, P').$$

To characterize the Borda count, consider an observed profile P. For a
candidate x we denote by $\mathbf{W}(x)$ the set of all profiles where x is first-ranked
in every voter's ranking. Clearly in all these profiles x gets the maximum
points. As in the case of plurality rule, we consider these as the consensus
states for the Borda count [28, 11].

For a candidate x, the number of alternatives above it in any ranking of R
equals the number of points deducted from the maximum points. This is also
the number of inversions needed to get x in the winning position in every
ranking. Thus, using the inversion metric, w_B is the Borda winner in profile
P if

$$d_K(P, \mathbf{W}(w_B)) \leq d_K(P, \mathbf{W}(x)) \quad \forall x \in X \setminus w_B.$$

Thus, the Borda count differs from the plurality rule in terms of the metric
used in measuring the distance between the observed profile and the consensus state. On the other hand, both of these methods have an identical
consensus state.

The choice rule known as the Kemeny method is perhaps most explicitly
consensus oriented. It is based on the (Kemeny) inversion metric, but differs
from the Borda count in terms of the consensus state: a Kemeny consensus
state is one where a unanimous preference ranking over all alternatives prevails.
Hence, given a profile of expressed preference rankings, the method determines
the ranking that is closest to the profile in terms of the inversion metric.

More formally, let $U(R)$ denote an unanimous profile where every voter's
ranking is R. Kemeny's rule results in the ranking $\bar{R}$ so that

$$d_K(P, U(\bar{R})) \leq d_K(P, U(R)) \quad \forall R \in \mathcal{R} \setminus \bar{R}$$

where P is the observed profile and $\mathcal{R}$ denotes the set of all possible rankings.

Thus far we have considered two consensus states: one with a unanimously first-ranked candidate and the other with consensus regarding all positions in the ranking. Other types of consensus states can be envisioned. To wit, one could look for a consensus concerning which candidate is the Condorcet winner, *i.e.* a candidate that defeats all the others in pairwise comparisons by a majority of votes. Dodgson's system is defined as a method that elects the candidate that is as close as possible to being a Condorcet winner. It is relatively straight-forward to define this system in terms of a consensus state and distance metric.

For any candidate x we denote by $\mathbf{C}(x)$ the set of all profiles where x is the Condorcet winner. Provided that a Condorcet winner exists in the observed profile, the winner is this candidate. Otherwise, one constructs, for each candidate x, a profile in $\mathbf{C}(x)$ which is obtained from the observed profile P by moving x up in one or several voters' preference orders so that x emerges as the Condorcet winner. Obviously, any candidate can thus be rendered a Condorcet winner. It is also clear that for each candidate there is a minimum number of such preference changes involving the improvement of x's position *vis-à-vis* other candidates that are needed to make x the Condorcet winner. The Dodgson winner w_D is then the candidate which is closest, in the sense of Kemeny's metric, of being the Condorcet winner. That is,

$$d_K(P, \mathbf{C}(w_D)) \leq d_K(P, \mathbf{C}(x)) \quad \forall x \in X \setminus w_D.$$

Dodgson's method is thus characterized by Kemeny's inversion metric combined with a goal state where a Condorcet winner exists.

Combining the inversion metric with the $\mathbf{C}(x)$ consensus state thus gives us the Dodgson method. But what happens if we combine this consensus state with the discrete metric? The result is known as Young' method [12]. We can formally define the winner w_Y as the option with property

$$d_d(P, \mathbf{C}(w_Y)) \leq d_d(P, \mathbf{C}(x)) \quad \forall x \in X \setminus w_Y.$$

Thereby we get the largest subset of voters where a Condorcet winner exists.

Combining the discrete metric with the consensus state $U(R)$ results in the most popular ranking being elected elected.

Consensus can be sought in settings where the starting point is not a preference profile. We can start from outranking matrices. These can, of course, be formed from preference profiles, but one could envision a situation where just the outranking matrix – not the underlying preference profile – is given.

We denote by V the outranking matrix where entry V_{xy} indicates the number of voters in profile P preferring candidate x to candidate y. The diagonal entries are left blank.

We define the metric using outranking matrices as follows: if V and V' are the outranking matrices of profiles P and P' then

$$d_V(P, P') = \frac{1}{2} \sum_{x,y \in X} |V_{xy} - V'_{xy}|.$$

In other words, the distance tells us how much pairwise comparisons differ in the corresponding outranking matrices. This metric is very similar to the inversion metric and, indeed, we find the same Borda count winners using this metric instead.

The interesting case is one where the goal is a Condorcet winner. Consider one of the systems introduced by Condorcet, *viz.* the least-reversal system. In this system the winner is the candidate which can be turned into a Condorcet winner with a minimum number of reversals of pairwise comparisons. That is, the Condorcet least-reversal system winner w_{lr} is the candidate with property

$$d_V(P, \mathbf{C}(w_{lr})) \leq d_V(P, \mathbf{C}(x)) \quad \forall x \in X \setminus w_{lr}.$$

Copeland's procedure has also a similar goal state as Dodgson's and Condorcet's least reversal one, namely, one with a Condorcet winner. Given an observed profile P, one considers the corresponding tournament matrix T where entry $T_{xy} = 1$ if majority of the voters in profile P are preferring candidate x to candidate y. Otherwise $T_{xy} = 0$.

Now, the Condorcet winner is seen as a row in T where all $k - 1$ non-diagonal entries are 1s.

We define yet another distance measure between profiles P and P' with tournament matrices T and T' as

$$d_T(P, P') = \frac{1}{2} \sum_{x,y \in X} |T_{xy} - T'_{xy}|.$$

The Copeland winner w_C is the alternative that wins the largest number of comparisons with other candidates, *i.e.* has the smallest number of 0s in its row in the tournament matrix. Thus the winner w_C is the candidate that comes closest to win every other candidate, that is, using the distance above,

$$d_T(P, \mathbf{C}(w_C)) \leq d_T(P, \mathbf{C}(x)) \quad \forall x \in X \setminus w_C.$$

Obviously, the goal states of Condorcet least-reversal system and Copeland's system are the same, but metrics differ. The latter pays no attention to majority margins, while the former depends on them.[2]

A wide variety of existing or imaginable voting systems can thus be seen as consensus-oriented in the sense of looking for the consensus that is as close as possible to the expressed views of the voters (see [25] for a more extensive

---

[2] See, Klamler (2005) for another distance based characterization of the Copeland rule.

account). Their differences pertain to the goal (consensus) states sought for and the metrics used in defining "closeness".

4   Consensus in Policy Choice

Choices of candidates differ from choices of policies in many ways, but from the view-point of choice procedures perhaps the most essential is the greater flexibility associated with the definition of policy sets from which the choice is to be made. Accordingly, the literature often resorts to definitions in terms of property k-tuples. In other words, the policies are assumed to be associated with points in many-dimensional (policy) spaces. The individuals are supposed to be endowed with complete and transitive preference relations $\succeq$ over all point pairs in the space W. These relations are, moreover, assumed to be representable by utility functions in the usual way, that is

$$x \succeq y \Leftrightarrow u(x) \geq u(y), \forall x, y \in W$$

In strong spatial models individual i's evaluations of alternatives are assumed to be related to a distance measure d_i defined over the space. Moreover, each individual i is assumed to have an ideal point x_i in the space so that

$$x \succeq y \Leftrightarrow d_i(x, x_i) \leq d_i(y, x_i), \forall x, y \in W$$

The first spatial results pertaining to consensual processes suggested that even a purely majoritarian system known as the amendment procedure tends to converge into an equilibrium policy in case there is only one dimension and when there is a core policy, *i.e.* a policy that is majority undefeated by any other policy [9]. And the existence of a core policy in one-dimensional policy spaces is guaranteed under a seemingly plausible assumption concerning voter preferences, *viz.* that they are single-peaked [7].

Single-peakedness – albeit a sufficient, not necessary condition for equilibrium – is, however, a very unusual characteristic of voter preferences in multi-dimensional Euclidean policy spaces [19]. Furthermore, a core-related equilibrium concept, the Plott equilibrium, turns out to exist only under very special assumptions concerning voter preference distributions in multi-dimensional spaces [36]. One of the assumptions driving these results is that in pairwise comparisons of policy options, the voters always vote for the option that is closer to their ideal point. If all voters' preferences can be represented by a smooth utility function in the policy space, then the implications of the nonexistence of the core are dramatic, indeed, as shown by the famous theorems of McKelvey [21, 22]: the agenda-setters control over voting outcomes is complete despite the apparent responsiveness of the pairwise voting outcomes. Not even the Pareto set – *i.e.* the set of Pareto-undominated policies – restricts the possible outcomes if the agenda-setter has complete freedom in building the agenda of pairwise majority comparisons, if the voters are rational but myopic and if the core is empty. The conditions under which the core is empty have been studied by many authors [4], [23],[38]. The overall upshot of these studies is that the larger the number of dimensions, the more certain it is that the core is generically empty. [3]

So, the amendment rule does not have the same convergence to (median) equilibrium properties in multi-dimensional spaces as it does in one-dimensional ones. There are, however, mechanisms that counteract this tendency towards arbitrariness of outcomes *vis-à-vis* the ideal point configuration. One of these is pointed out by Kramer [20] who shows that if the policy options enter the agenda so that each one is vote-maximizing with respect to the previous one and if the decision rule is based on the min-max number, the trajectory of vote-maximizing policy options is bound to enter the Pareto set and remain there. The min-max M number is defined as follows. Let $n(x, y)$ be the number of voters preferring x to y. Then

$$v(x) = min_y n(x, y)$$

$$M = max_x v(x)$$

Let now $\succ_i$ denote voter *i*s strict preference relation. Define now M-majority relation $\succ_M$ as follows:

$$x \succ_M y \text{ iff } | \{i | x \succ_i y\} | \geq M.$$

So, the min-max rule and vote-maximization guarantee a modicum of responsiveness of the voting outcomes to voter opinions by securing Pareto optimality. Besides larger than simple majority rules, also voter behavior may work towards converging the outcomes to the Pareto set or some plausible subset thereof. More specifically, if the voters are all strategic – instead of myopically rational as assumed in McKelvey's theorems – the voting outcomes will be found in a specific subset of the Pareto set, *viz.* the uncovered set [26]. This set, however, is in general too large to characterize sophisticated – *i.e.* non-myopic but rational – voting outcomes. For precise characterization, its subset, the Banks set is appropriate [3]. [4] This contains all sophisticated voting outcomes and only them.

One assumption underlying the spatial models of consensual processes is that the voters vote for the policy which is closer to their ideal point than its competitor. This assumption has received relatively scant attention in the literature (see, however, [15]). Yet, one can plausibly question its general validity. In other words, policy alternative a may be closer than b to a voter's ideal point on all dimensions and, yet, the voter may quite reasonably vote for b. Consider a modified setting of what is known as Ostrogorski's paradox [37].

---

[3] See, however, Saari's much more positive results on q-rules [38].
[4] See also [27]. [30] gives a brief overview.

Table 2 Ostrogorski's paradox

issue	dimension 1	dimension 2	dimension 3	majority policy
criterion A	X	X	Y	X
criterion B	X	Y	X	X
criterion C	Y	X	X	X
criterion D	Y	Y	Y	Y
criterion E	Y	Y	Y	Y
winner	Y	Y	Y	?

An individual voter is faced with the choice between two policy options X or Y. There are five criteria A–E used in evaluating the options on three dimensions 1-3. On criterion A, X is closer to the individual's ideal point than Y on 2 dimensions out of 3. The same is true of criteria B and C, while on criteria D and E, Y is closer to his ideal point on all three dimensions. Suppose the individual considers all criteria and dimensions of equal importance. Then criteria A, B and C would dictate him to vote for X, while criteria D and E would suggest Y. It would thus be quite plausible that the individual votes for X since the majority of criteria suggests this. However, Y would outperform X on *all* dimensions, again by a majority of criteria. Hence even when an option is closer to the individual on all dimensions, it makes perfect sense for him to vote for another option.

This modification of Ostrogorski's paradox casts serious doubt on the universal applicability of the "the closer, the better" assumption underlying most spatial models. It is worth pointing out that this conclusion is independent of the specific metric used in measuring closeness of policy options.

What we are dealing with here is a version of a general aggregation problem and the specific culprit or source of problems is the majority rule used in aggregating rows and columns in Table 2.

This suggests a more general question: under which conditions the same outcomes are reached when the entries are real numbers in the unit interval and the outcomes are determined in two different ways:

- by first aggregating over rows and then aggregating the results (rows first method), and
- by first aggregating the entries in each column and thereupon the results (columns first method).

Answer is found in the setting of deprivation measures [10]. Consider two functions

$$g : [0,1]^m \to [0,1] \tag{2}$$

$$h : [0,1]^n \to [0,1] \tag{3}$$

The former is interpreted as the method for aggregating individual deprivation degrees on various attributes (income, housing, education, etc) into an overall degree of deprivation of an individual. The latter is the method for aggregating the individual deprivation degrees into an overall degree of deprivation of the society.

Under this special setting our question can now be phrased as follows. Under which conditions one can arrive at the same result concerning the overall degree of deprivation in a society following two paths?

- Form first an index of deprivation of each individual by aggregating his deprivation values in all m attributes and then compute the social degree of deprivation from those n index values, and
- compute for each attribute a deprivation index by aggregating the n individual deprivation values on that attribute, and aggregate then these indices into a social one.

Let the matrix $A = [a_{ij}]$ be a $n \times m$ matrix of entries $a_{ij} \in [0, 1]$. Each entry is the degree of individual i' deprivation on attribute j. Denote by a_i the row vector representing individual i's deprivation degrees on all m attributes and by a^j the column vector giving each individual's deprivation degree on attribute j.

Now we ask under which conditions the following holds:

$$h(g(a_1), \ldots, g(a_n)) = g(h(a^1), \ldots, h(a^m)) \qquad (4)$$

Dutta et al. (2003) show that only a very restricted set of functions satisfies Equation 4. The conditions are

1. that g be continuous, strictly increasing in all arguments and satisfies non-diminishing increments, and
2. that h be continuous, symmetric and strictly increasing.

Moreover, it is assumed that if all arguments of g are zero (unity, respectively), so is the value of g. The same assumption is made regarding h. Let $g* : [0, 1]^m \to [0, 1]$ so that for any $x \in [0, 1]^m$ there is a set of weights $w_1, \ldots w_m$ each between 0 and 1 with $\sum w_j = 1$ and $g*(x) = w_1 x_1 + \cdots + w_m x_m$. Denote by $G*$ the set of all such functions $g*$. Moreover, let $h*(y) = (y_1 + \cdots + y_n)/n$ and denote by $H*$ the consisting of this function.

Then all functions $g* \in G*$ together with the function $h*$ satisfy Equation 4 and, conversely, any pair of functions (g, h) that satisfies Equation 4 must be such that $g \in G*$ and $h = h*$.

The converse result is perhaps more important for our purposes. What it says is that in order to derive the same result using rows first and columns first methods, the indices attached to rows must be weighted averages of the row entries and the overall index value must be the arithmetic mean of the row indices.

Consensus-searching by majority rule, thus, seems to be riddled by many problems in multi-dimensional policy spaces. Replacing dichotomous variables and majority rule aggregation with membership degrees and linear aggregation rules seems to be a plausible way to proceed in searching for policy consensus.

5   Consensual Processes and Paradoxes

Paradoxes of voting have been studied throughout the history of social choice theory. Sometimes they have provided motivation for new voting systems, sometimes they have suggested new considerations that should be taken into account in the study of voting. Some of the paradoxes are directly relevant to consensual processes. In this section we deal with some – perhaps most obviously relevant – of them.

When a group of individuals gathers together to search for a consensus, then an obvious desideratum is that everyone's opinion counts. After all, the outcome cannot be called a consensus if some individuals have not been "heard". This, in turn, means that every individual should have at least a minimal motivation to participate in the group decision making, minimal in the sense that he is never better-off not participating than by expressing his opinion. This minimal motivation implies that the procedure used in consensus-reaching be invulnerable to the no-show paradox [13]. This paradox occurs whenever a group of voters can by abstaining, *ceteris paribus*, bring about a preferable outcome than by voting according to their preferences. Voting systems in which such a paradox may happen in some profile are vulnerable to the no-show paradox. Table 3 exhibits a strong version of the paradox, *viz.* one where the abstinence brings about the first-ranked option of the abstainers.

Table 3 Strong no-show paradox

2 seats	3 seats	2 seats	2 seats
c	b	a	a
b	a	c	b
a	c	b	c

Let us assume that the consensus is sought through the amendment procedure, *i.e.* by pairwise comparisons of options so that the loser of each comparison is eliminated, while the winner takes on the next option on the agenda. Suppose that b and c are compared first, whereupon the winner confronts a in the final contest. This agenda leads to the victory of b.

Suppose now that the two right-most voters had abstained. Then c would have won the first comparison and a the second. Hence the abstainers' first option would have won. Now obviously a procedure that might lead to a no-show paradox does not encourage participation in the process.

This conclusion also holds for procedures in which it may be beneficial for an individual not to reveal his whole preference ranking (even assuming that such a behavior does not disqualify the submitted ballots). Systems where a partial revelation of one's preferences might yield a better outcome than a wholesale one, are vulnerable to the truncation paradox. Table 4 shows that Nanson's Borda-elimination method is vulnerable to the truncation paradox. The method – it will be recalled – is based on the Borda count so that at the outset one computes the Borda scores of all options. Thereafter all options with at most average score are eliminated and new scores are computed for the options in the reduced profile. Again options with at most average Borda scores are eliminated and the process continues until only one option is left or there is a tie between several options. As Nanson (1882) pointed out, the procedure guarantees that, in case there is a Condorcet winner option, it will end up victorious.

Table 4 Nanson's method and truncation paradox

5 voters	5 voters	6 voters	1 voter	2 voters
A	B	C	C	C
B	C	A	B	B
D	D	D	A	D
C	A	B	D	A

In Table 4 Nanson's method results in B. However, if the 2 voters with preference ranking CBDA reveal only their first-ranked option, C, the outcome is C, obviously a superior option from their point of view.[5]

Abstention can obviously be regarded as an extreme form of preference truncation and, thus, these two paradoxes are closely related. To the same family of paradoxes belongs also the twins' paradox or the " twins not welcome" phenomenon. This paradox occurs whenever adding k copies or "clones" of voter i leads to an outcome which is worse than the original one for i. Table 5 shows that Dodgson's method is vulnerable to this paradox. The method is apparently consensus oriented in looking for the candidate that is closest to being a Condorcet winner. Here closest is interpreted in the sense of inversion metric. Obviously, when there is a Condorcet winner in a profile, it is also the Dodgson winner.

In this profile B is the (strong) Condorcet winner. Adding 20 copies of the one voter with ranking EDBAC leads to A being closest to Condorcet winner. This is worse than B from the point of view of the clones. Hence we have an instance of the twins' paradox.

---

[5] There are various ways of handling truncated preferences in the Borda count. Here we have treated those options over which no preference is indicated as tied ones, *i.e.* assigned each of them the average score. The same result would ensue *a fortiori* if the score 0 is assigned to each of them.

Table 5 Dodgson's method and the twins' paradox

42 voters	26 voters	1 voters	11 voters
B	A	E	E
A	E	D	A
C	C	B	B
D	B	A	D
E	D	C	C

Perhaps the most consensus-oriented of the procedures considered in this article is Kemeny's rule [16] discussed above. It is also vulnerable to the no-show and twins' paradoxes. This is shown in Table 6 which is adapted from [34].

Table 6 Kemeny's rule and non-show paradox

5 voters	4 voters	3 voters	3 voters
D	B	A	A
B	C	D	D
C	A	C	B
A	D	B	C

Here the Kemeny winner is D. Now, add 4 voters with DABC ranking. Then the resulting Kemeny ranking would have had A on top. Hence, we have an instance of the strong no-show paradox. Adding the DABC voters one by one to the 15-voter profile demonstrates the twins' paradox.

Thus, the apparently consensus-oriented system of Kemeny may contain incentives not to participate at all. In fact, all the systems discussed in this section illustrate the general result of Moulin [27]:

Theorem 5. *All systems that necessarily elect the Condorcet winner when one exists are vulnerable to the no-show paradox.*

Pérez [35] has strengthened this result by showing that nearly all Condorcet-consistent voting systems are vulnerable to the strong version of the no-show paradox. This, of course, provides an even stronger incentive not to participate.

6   Concluding Remarks

All commonly used voting systems can seen as consensus-oriented in the sense that each one strives at finding a solution that is as close as possible to the expressed preferences of voters. The systems differ in terms of the type of consensus state they are searching for and the measure used in defining the distance between the observed preferences and the goal state. In the history of social choice, consensus has been sought in jury decision making (verdicts), candidate contests and policy choices. There are obvious similarities between these settings, but also important differences. In jury decision making one is interested in maximizing the probability of a correct decision. In dichotomous decision settings, this problem has received some scholarly attention over centuries, and important results have been achieved. Some of these were re-visited above. Candidate contests, on the other hand, have been studied from the point of view of finding from a given set of alternatives a winner that satisfies some desiderata defined in terms of the voter preferences, *e.g.* Condorcet criteria, Pareto optimality, *etc.* The setting of policy choice differs from the candidate selection in allowing for in principle unlimited number of options. In modeling these settings the spatial models have been extensively utilized. In the preceding we have reviewed some of the principal results pertaining to majority voting – especially the amendment procedure – and pointed out some fundamental problems related to spatial modeling of collective decision making.

References

1. Baigent, N.: Preference proximity and anonymous social choice. The Quarterly Journal of Economics 102, 161–169 (1987)
2. Baigent, N.: Metric rationalization of social choice functions according to principles of social choice. In: Mathematical Social Sciences, vol. 13, pp. 59–65 (1987)
3. Banks, J.: Sophisticated voting outcomes and agenda control. Social Choice and Welfare 1, 295–306 (1985)
4. Banks, J.: Singularity theory and core existence in the spatial model. Journal of Mathematical Economics 24, 523–536 (1995)
5. Ben-Yashar, R., Paroush, J.: A nonasymptotic Condorcet Jury Theorem. Social Choice and Welfare 17, 189–199 (2000)
6. Berg, S.: Condorcet's Jury Theorem, dependency among voters. Social Choice and Welfare 10, 87–96 (1993)
7. Black, D.: The Theory of Committees and Elections. Cambridge University Press, Cambridge (1958)
8. Dahl, R.: After the Revolution. Yale University Press, New Haven (1970)
9. Downs, A.: An Economic Theory of Democracy. Harper & Row, New York (1957)
10. Dutta, I., Pattanaik, P., Xu, Y.: On measuring deprivation and the standard of living in a multidimensional framework on the basis of aggregate data. Economica 70, 197–221 (2003)
11. Farkas, D., Nitzan, S.: The Borda rule and Pareto stability: A comment. Econometrica 47, 1305–1306 (1979)
12. Fishburn, P.: Condorcet social choice functions. SIAM Journal of Applied Mathematics 33, 469–489 (1977)
13. Fishburn, P., Brams, S.: Paradoxes of preferential voting. Mathematics Magazine 56, 201–214 (1983)
14. Grofman, B., Owen, G., Feld, S.: Thirteen theorems in search of the truth. Theory and Decision 15, 261–278 (1983)
15. Humphreys, M., Laver, M.: Spatial models, cognitive metrics, and majority rule equilibria. British Journal of Political Science 40, 11–30 (2010)
16. Kemeny, J.: Mathematics without numbers. Daedalus 88, 571–591 (1959)

17. Kemeny, J., Snell, J.L., Thompson, G.: Introduction to Finite Mathematics. Prentice-Hall, Englewood Cliffs (1956)
18. Klamler, C.: Copeland's rule and Condorcet's principle. Economic Theory 25, 745–749 (2005)
19. Kramer, G.: On a class of equilibrium conditions for majority rule. Econometrica 41, 285–297 (1973)
20. Kramer, G.: A dynamical model of political equilibrium. In: Journal of Economic Theory, vol. 16, pp. 310–334 (1977)
21. McKelvey, R.: Intransitivities in multidimensional voting models and some implications for agenda control. In: Journal of Economic Theory, vol. 12, pp. 472–482 (1976)
22. McKelvey, R.: General conditions for global intransitivities in formal voting models. In: Econometrica, vol. 47, pp. 1085–1112 (1979)
23. McKelvey, R., Schofield, N.: Generalized symmetry conditions at a core point. Econometrica 55, 923–934 (1986)
24. McLean, I., Urken, A.B.: General introduction. In: McLean, I., Urken, A.B. (eds.) Classics of Social Choice, The University of Michigan Press, Ann Arbor (1995)
25. Meskanen, T., Nurmi, H.: Distance from consensus: a theme and variations. In: Simeone, B., Pukelsheim, F. (eds.) Mathematics and Democracy: Recent Advances in Voting Systems and Collective Choice, Springer, Berlin (2006)
26. Miller, N.: Committees, Agendas, and Voting. Harwood Academic Publishers, Chur (1995)
27. Moulin, H.: Choosing from a tournament. Social Choice and Welfare 3, 271–291 (1986)
28. Nitzan, S.: Some measures of closeness to unanimity and their implications. Theory and Decision 13, 129–138 (1981)
29. Nitzan, S., Paroush, J.: Optimal decision rules in uncertain dichotomous choice situation. International Economic Review 23, 289–297 (1982)
30. Nurmi, H.: Voting Paradoxes and how to Deal with Them. Springer, Heidelberg (1999)
31. Nurmi, H.: Voting Proccedures under Uncertainty. Springer, Berlin (2002)
32. Nurmi, H.: A comparison of some distance-based choice rules in ranking environments. Theory and Decision 57, 5–24 (2004)
33. Owen, G., Grofman, B., Feld, S.: Proving distribution- free generalization of the Condorcet Jury Theorem. Mathematical Social Sciences 17, 1–16 (1989)
34. Pérez, J.: Incidence of no show paradoxes in Condorcet choice functions. Investigationes Economicas XIX, 139–154 (1995)
35. Pérez, J.: The strong no show paradoxes are common flaw in Condorcet voting correspondences. Social Choice and Welfare 18, 601–616 (2001)
36. Plott, C.: A notion of equilibrium and its possibility under majority rule. American Economic Review 57, 788–806 (1967)
37. Rae, D., Daudt, H.: The Ostrogorski paradox: A peculiarity of compound majority decision. European Journal of Political Research 4, 391–398 (1976)
38. Saari, D.: The generic existence of a core for g-rules. Economic Theory 9, 219–260 (1997)
39. Shapley, L., Grofman, B.: Optimizing group judgmental accuracy in the presence of uncertainties. Public Choice 43, 329–343 (1984)

Consensus Perspectives: Glimpses into Theoretical Advances and Applications

Miguel Martínez-Panero

Abstract. The polysemic meanings of consensus are surveyed from several points of view, ranging from philosophical aspects and characterizations of several quantification measures within the Social Choice framework, paying also attention to aspects of judgment aggregation as well as fuzzy or linguistic approaches, to practical applications in Decision Making and Biomathematics, among others.

1 Introduction

Consensus [L., fr. consensus, pp. of *consentire*] **1**: group solidarity in sentiment and belief. **2 a**: general agreement, unanimity. **2 b**: the judgement arrived at by most of those concerned.

Consent 1: compliance in or approval of what is done or proposed by another, acquiescence. **2**: agreement as to action or opinion; *specif*: voluntary agreement by a people to organize a civil society and give authority to the government.

Webster's New Collegiate Dictionary

Consensus is a multi-faceted concept. Emerson [40] considers the use of "general agreement" and "majority view" as two widespread opposite senses of the word[1]. And according to Williams [103]:

> Given this actual range, it is now a very difficult word to use, over a range from the positive sense of seeking general agreement, through the sense of a relatively inert or

Miguel Martínez-Panero
PRESAD Research Group – Dep. de Economía Aplicada
Universidad de Valladolid. Av. Valle de Esgueva, 6. 47011 Valladolid, Spain
e-mail: panero@eco.uva.es

---

[1] It is meaningful that Emerson [40] entitled his book after Tocqueville's argument against the "tyranny of the majority" appearing in *Democracy in America*. Perhaps the current controversy on climate change can be understood as the the opposition between the majoritarian belief that global warming is significant and the opinion of those claiming that there is no a general agreement or evidence of this fact.

E. Herrera-Viedma et al. (Eds.): Consensual Processes, STUDFUZZ 267, pp. 179–193.
springerlink.com

even unconscious assent [...], to the implication of a "manipulative" kind of politics seeking to build a silent majority as the power-base from which dissenting movements or ideas can be excluded or repressed. It is remarkable that so apparently mild a word has attracted such strong feelings, but some of the processes of modern electoral and "public opinion" politics go a long way to explain this.

Emerson [39] also asserts that "the general consensus on any one issue is that which is perceived to be the agreed opinion of an overwhelming number, though there may be some who dissent". Thus, this author points out an evident fact, i.e., that this notion is not endowed with an absolute meaning, but a relative one, and nowadays it is usual to deal with levels or measures of consensus, as several of the authors appearing next do.

The chapter, which aims to tackle different approaches to consensus, is organized as follows. In Section 2 we present some philosophical aspects of consensus essentially focused on the doctrine that men are joined together within a society by a contract with explicit or hidden agreements, as Rousseau believed. Then we outline some further developments and connections, such as the link between Rousseau and Condorcet. We also distinguish between the concept of consent and the more technical and recent idea of consensus as appearing in modern Political Science and Sociology. In Section 3 we deal with several formal approaches to consensus mainly from the Social Choice framework, and advances from distance-based, fuzzy or linguistic points of view are presented. In addition, we point out some aspects of an emergent research field focused on *judgment aggregation*. In Section 4 we include some applications as signs of the power of consensus-based methods in practice, and we refer to the way of aggregating different estimates of each candidate through a median-based voting system tested in 2007 (the *majority judgement*), to consensus as a determining condition for publishing in Wikipedia, and about how this idea naturally appears in Biology and Biomathematics. In the last section, we give our concluding remarks.

2   Philosophical Background

There is a long tradition, from Greek and Roman thinkers[2] to Locke, Hobbes, Paine and mainly Rousseau [93], supporting the idea that political authority relies on the consent of the governed through a *social contract*. Thus, the legitimacy of the government should be based on the *general will*[3]. It is interesting to point out how this concept is related to the idea of consensus as appearing in Condorcet's [24] *jury theorem*. This fact has been explained by Grofman and Feld [54], who point out the following three elements of Rousseau's theory then formalized by Condorcet:

1. There is a common good.
2. Citizens are not always accurate in their judgements about what is in the common good.

---

[2] "The doctrine that society is itself a voluntary association was not unheard of in Greek and Roman times" (see Partridge [89, p. 15]).

[3] See Saccamano [96] and Graham Jr. [53].

3. When citizens strive to identify this common good and vote in accordance with their perceptions of it, the vote of the Assembly of the People can be taken to be the most reliable means for ascertaining the common good.

An overview of other recent authors such as Habermas, Rawls, etc. about the subject of consensus-based legitimacy can be found in Knight and Johnson [76]. And it is worth mentioning that Buchanan and Tullock consider their well-known work, *The Calculus of Consent*, as a contractual theory of the State (see Buchanan and Tullock [15, appendix 1]).

We have used the terms "consent" and "consensus" and, in fact, both notions appear jointly in Willams' [103] *Keywords*. According to Partridge [89, p. 71] "within modern Social Theory *consent* has been mainly a term of Political Philosophy, *consensus* mainly a term in Sociology". Indeed, according to this author, it was Comte who introduced in the XIXth century in the vocabulary of Social Sciences the notion of "consensus", in its Latin version. It is well known that Comte founded Sociology from the basis of Biology, and in this particular case the coined term was inspired by Medicine during his age, where consensus meant solidarity among distinct parts of the human body[4]. Thus, in recent years, consensus refers to agreements which provide the conditions of political and social cohesion (on this aspect, see De Dreu and De Vries [33]).

3 Formal Approaches to Consensus

In what follows we present several aspects of consensus and how they have been developed within different formal frameworks.

3.1 Consensus and Social Choice

In Theory of Democracy as understood by van Mill [82] there are two extremely different currents for dealing with conflicts in a legitimate way: "theories of democratic discourse" advocated mainly by Habermas [55], which try to solve them through deliberation and consensus, and "disequilibrium theories of social choice" as proposed by Black [10], Arrow [5] and Riker [92], aggregating individual preferences into a social result.

However, Social Choice is not merely a tool to decide in absence of consensus, and some authors have formalized this notion in the Social Choice context. In this way, Samet and Schmeidler [97] have introduced and then axiomatized "consent rules, that incorporate aspects of majoritarianism and liberalism" and Ballester and García-Lapresta [9] have analyzed a recursive consensus process for selecting qualified individuals of a group[5]. On their part, Bosch [14], Alcalde-Unzu and

---

[4] It seems that this aspect was foreseen by Saint-Simon, although Comte properly developed such idea (see Campillo [22, pp. 69-71]).

[5] The research of these authors is focused on aspects of "group identification". An overview of selected topics on this subject in the context of Social Choice has been outlined by Dimitrov [36] in one of the chapters of this book.

Vorsatz [1, 2] and García-Lapresta and Pérez-Román [50] have proposed several consensus measures and they have also characterized them. On the other hand, García-Lapresta [47] and García-Lapresta and Pérez-Román [49], following Cook and Seiford [29, 30], have devised voting methods penalizing the disagreement, the key idea being the "marginal contribution to consensus" for each agent, and Mata et al. [80, 81] have analyzed strategic aspects in consensus processes. Recently, García-Lapresta et al. [48], also in a Social Choice context, have obtained the set of scoring rules that optimizes consensus among voters by maximizing the collective utility.

It is also worth noting the existence of consensus models for decision making in committees which have nothing to do with voting, for example that appearing in Eklund, Rusinowska and De Swart [38]. These authors have performed a method where a chairman suggests the decision makers adjust or modify their evaluation of some alternatives, attending several criteria in order to improve the agreement, following an idea which recurrently appears in Consensus Theory[6]. In this way, Carlsson et al. [23] have also considered "an advising monitor which tries to contract the decision makers into a mutual decision through soft enforcement", where convergence to consensus is analyzed by means of a topological approach which can measure distances between decision makers and allow these authors to model the trade-off between "degree of consensus" and "strength of majority".

Such metric techniques lead us to the next section, where the notion of distance becomes crucial.

3.2 Distance-Based Approaches to Consensus

The idea of considering distances among preferences (already suggested in Condorcet's writings) has revealed to be very fruitful[7]. In this way, consensus can be somehow understood as closeness to unanimity, and several attempts has been done to minimize (from distance-based assumptions) the aggregate disagreement.

Hornik and Meyer [65] note that such "optimization approach" goes back to Règnier [90]. Nonetheless, it is fair to acknowledge the pioneer analysis of so-called

---

[6] For example, the possibility of changing agents' opinions along the process, in an iterative way, is also essential in the *Delphi method* for expert decision (see Turoff and Linstone [100]). On the Delphi method as a tool for consensus-building see, for example, Hsu and Sanford [63] and the references therein.

[7] This is, for example, the key for the Kemeny [71] rule, inspired by the treatment of statistical data. In fact, as demonstrated by Young [101], this method is just the "maximum likelihood rule" tacitly formulated by Condorcet [24]. See Klamler [74, 75] and Nurmi [87] for Kemeny-type and other distance functions on the set of choice functions, García-Lapresta and Pérez-Román [49] and the references cited there concerning other voting methods using distances for obtaining a collective outcome, and Meskanen and Nurmi [83, 84] who present a comprehensive approach to methods which can be viewed as distance minimizing ones, focusing on the measurement of disagreement. Further insights of this approach can be found in Nurmi [88] and Eckert and Klamler [37] in two chapters of this book.

"compromise solutions" (those minimizing the group regret through appropriate distance functions) by Yu [104] and Freimer and Yu [46]. On the other hand, in order to determine the closest ranking to all decision-makers, Cook and Seiford [29, 30] considered l^1 and l^2 metrics previously introduced by Kemeny and Snell [72] in an important paper and also analyzed by Bogart [11, 12].

It is worth noting that the last of the cited distances is related to the Borda rule[8] and it was independently obtained by Kendall [73] in a statistical estimation context. On the other hand, the l^1 metric has been used by Contreras et al. [27] in order to provide a compromise method for collective decision problems especially suited for situations where the members of the group provide partial or imprecise information about their preferences.

A a general model for distance-based consensus can be found in Cook et al. [28] (see González-Pachón and Romero [51] for a detailed exposition and further developments). The last mentioned authors, González-Pachón and Romero [51, 52], have considered the aforementioned problem of minimization as a Goal Programming model and have obtained consensus weights when the information taken into account by the decision maker is somehow inconsistent, i.e., does not verify some rational assumptions.

In Eckert and Klamler [37] the reader will find a survey on distance-based approaches to aggregation theory dealing with the construction of aggregation rules, the comparison of aggregation procedures and the generalizations of aggregation problems, among other topics. These authors point out that "aggregation problems arise in many other areas than social choice distance-based approaches [and] are not limited to the construction and analysis of voting rules", as happening with the emerging literature on the subject of the next subsection.

3.3 Judgment Aggregation

According to List [77], *judgment aggregation* is concerned with how "a group of individuals aggregate the group members' individual judgments on some interconnected propositions into corresponding collective judgments on these propositions", and such analysis may be done from different collective decision making contexts. This research field runs parallel to Social Choice. In fact there is a *doctrinal paradox* (or *discursive dilemma*) whose homologue in Social Choice would be *Condorcet's*

---

[8] The Borda [13] and Condorcet [24] approaches have been called "*constructive* consensus methods" by Hornik and Meyer [65], but they assert that both rules agree in absence of cycles in the collective outcome, which is not true. There is a wide literature on Borda-Condorcet disagreement, and one of the chapters in this book deals with this topic in an experimental way (see Regenwetter and Popova [91]). At this juncture, Emerson [39, 40] advocates for the Borda rule, which promotes consensus and avoids the "tyranny of the majority". This is the reason why he has devised a modified Borda count system of consensus voting (see Emerson [41]). It is interesting to point out that Emerson [40] also defines a "level of consensus", but specifically intended for his system.

paradox[9]. There is also an *Impossibility Theorem*[10] and further developments to mitigate the resulting impasse, as happening with Arrow's Theorem and its variants. A comprehensive overview on judgment aggregation can be found in List and Puppe [79] (see also List [77] and the references appearing there, all of them available online, as well as Dietrich [34], and Dietrich and List [35]).

Within this framework, Hartmann and Sprenger [57] and Hartmann et al. [56] are developing models of consensus appearance in judgment aggregation situations in order to achieve compromise-based decisions.

3.4 Fuzzy and Linguistic Consensus Models

An updated overview on the use of fuzzy preferences and majorities in obtaining group decision making outcomes and levels of consensus has been presented by Kacprzyk et al. (see [69] and the references appearing there). As for fuzzy and linguistic approaches to consensus models for group decision making it is worth to mention Herrera-Viedma et al. [60, 61, 62], Herrera et al. [58, 59], Alonso [3], Alonso et al. [4] and Cabrerizo et al. [16, 17, 19], among others. In these papers, there usually exists a moderator in the process of obtaining a high level of consensus or agreement among the experts to decide in a group decision making problem (see also Fedrizzi et al. [44], Kacprzyk and Fedrizzi [66], Kacprzyk et al. [67, 68], and the iterative dynamical approach by Fedrizzi et al. [43]). Recently, also guided by a moderator and inspired by a data mining technique, Kacprzyk et al. [70] have provided an interesting tool for consensus reaching: action rules, which take into account different concessions than can be offered to the individuals so that they will change their preferences in order to obtain consensus convergence; and, on their hand, Calvo and Beliakov [20] have introduced penalty functions in order to minimize deviations of individual inputs from the collective output[11]. Advantages, drawbacks and future trends of fuzzy and linguistic approaches to consensus are analyzed in Cabrerizo et al. [18].

---

[9] As exposed by Fishburn [45] in a comprehensive paper concerning the landmarks in Social Choice, "Condorcet felt strongly that a majority candidate ought to be elected when one exists, [but] he realized that a profile need not have a majority candidate, as when m [voters]=n [candidates]=3 and the profile is $(a >_1 b >_1 c, b >_2 c >_2 a, c >_3 a >_3 b)$, in which case a has a 2-to-1 majority over b, b has a 2-to-1 majority over c, and c has a 2-to-1 majority over a. In such case the profile illustrates *Condorcet's effect*, or the *paradox of voting*, or has *cyclical majorities*". The implications of this kind of choice paradoxes to consensus-reaching methods have been analyzed by Nurmi [88] in the pages of this book.

[10] In fact, there are many impossibility theorems in Social Choice, but that seminally due to Arrow [5] has been considered the "Impossibility Theorem" par excellence. Again, according to Fishburn [45] "Arrow's discovery [shows] that a few appealing criteria for social ranking methods are mutually incompatible. Its essential idea is that the problems that arise from Condorcet's paradox of voting [...] cannot be avoided under any reasonable generalization of majority comparisons".

[11] The relationship between these penalty functions and consensus is analyzed by Calvo et al. [21] in one of the chapters of this book.

It is important to note here that Montero [85, 86] has analyzed the impact of fuzziness in Group Decision Making and has also proposed an escape from Arrow's impossibility result to reach a consensus-based solution respecting democratic compelling properties[12]. Even more, when such imposibility result is unavoidable in some rational amalgamation operations, Cutello and Montero [31] have searched for (and then characterized) "good enough" solutions.

4   Applications Dealing with Consensus

This section displays three practical cases where the idea of consensus is taken into account for different purposes in distinct contexts.

4.1   Aggregating Different Assessments through Majority Judgement

According to Balinski and Laraki [7], Galton's only comment related to Social Choice Theory "was discarded as a small contribution by Black". These are Galton's words (cited in Black [10]), which support the idea of mitigate the influence of the extreme opinions:

> Each voter [...] has equal authority with each of his colleagues. How can the right conclusion be reached, considering that there may be as many different estimates as there are members? That conclusion is clearly *not* the *average* of all the estimates, which would give a voting power to cranks in proportion to their crankiness. One absurdly large or small estimate would leave a greater impress on the result than one of reasonable amount, and the more an estimate diverges from the bulk of the rest, the more influence would it exert. I wish to point out that the estimate to which least objection can be raised is the *middlemost* estimate, the number of votes that it is too high being exactly balanced by the number of votes that it is too low. Every other estimate is condemned by a majority of voters as being either too high or too low, the middlemost alone escaping this condemnation.

This argument, advocated by Surowiecki [99] in the first pages of his book, has also been assumed by Balinski and Laraki, the idea of considering the median (middlemost) of the voters' assessments being the key of a new voting system proposal: the *majority judgement*, which has been empirically tested at Orsay during the 2007 French presidential elections (see Balinski and Laraki [7, 8] for details both of the theoretical background and concerning the aforementioned experiment). The method has also been implemented worldwide concerning 2008 United States elections (see Balinski [6]).

---

[12] As pointed out in Montero [86], "the main criticism to Arrow's approach should be the underlying Boolean assumption, present in every concept in his model. This binary view is taken for granted, and therefore hidden to readers. But in Arrow's theorem every condition (modeling liberty or equality) is fuzzy in nature, as consistency is, as consensus and preferences are".

4.2   Consensus for Publishing in Wikipedia

It is commonly known and accepted the role of consensus in prediction markets and the development of online tools as Google or Wikipedia (see Surowiecki [99]). Here we focus on the last one, where consensus is fundamental[13] in its editorial policy (see Figure 1, where the Wikipedia publishing process is shown through a flowchart), and might entail a rejected entry if agreement for acceptance is not present after a reasonable time period. According to Wikipedia rules "consensus need not be fully opposed; if consensus is neutral or unclear on the issue and unlikely to improve, the proposal is likewise rejected" (see [102] for more details). Thus, Wikipedia seems to respect the ancient criterion of truth[14] known as *consensus gentium* (L.: Agreement of the peoples), stating that "which is universal among men carries the weight of truth" (cited in Runes [94]).

4.3   Consensus in Biology and Biomathematics

One tends to think of consensus aspects as something essentially human, but biologists have concluded that "consensus decision making is common in non-human animals, and that cooperation between group members in the decision-making process is likely to be the norm" (see Conradt and Roper [25]; a reply to this insight can be found in List [78], and further advances appear in Conradt and Roper [26]). As examples, these authors cited nest choices in bees and ants, agreement about routes in navigating birds, activity timing in mammals, etc. An in-depth research of consensus decision making by fish has been experimentally carried out in Sumpter et al. [98], where the gregarious behavior of the stickleback[15] has been analyzed.

In this way, according to Day and McMorris [32], who are the authors of a handbook relating Consensus Theory to Biomathematics[16], the archetypal problem of aggregation concerning group choice pioneered by Arrow [5] and then refined and

---

[13] This consensual nature of Wikipedia has been noted, for example, in Essential-Facts.com [42]: "Genuine consensus typically requires more focus on developing the relationships among stakeholders, so that the compromises they achieve are based on willing consent – we want to give this to you, and we want you to give that to us only because you want to. The articles of Wikipedia itself are intended to follow this kind of approach".

[14] In fact, there are theories where the concept of truth crucially relies on consensus as an essential component. And nowadays consensus has become to mean "accepted subject", specially concerning scientific matters. In this way, in the specific context of Medicine, it is common to speak about "consensus statements" on diseases like AIDS, diabetes, etc. There are strict protocols to achieve this kind of agreements. According to Wikipedia, "medical consensus is a public statement on a particular aspect of medical knowledge available at the time it was written, and that is generally agreed upon as the evidence-based, state-of-the-art (or state-of-science) knowledge by a representative group of experts in that area".

[15] Curiously, this very fish is also subject of research by Mr. Pickwick, his paper being parodied at the beginning of Dickens' novel as empty erudition.

[16] This book has been interestingly reviewed by Saari [95] mainly attending its Social Choice background.

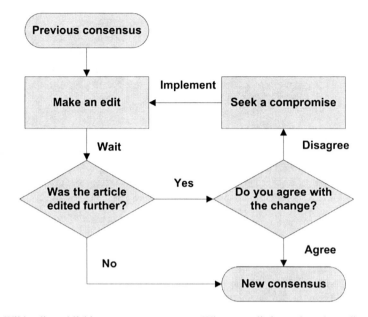

Fig. 1 Wikipedia publishing consensus process: When an edit is made, other editors have these options: accept the edit, change the edit, or revert the edit. These options may be discussed if necessary. *Source: Wikipedia.*

extended by many other authors, can be applied in areas far from their origins in Group Choice and Welfare Economics[17]. As pointed out by Janowitz in the foreword to the mentioned book:

> Since consensus theory has its roots in the theory of elections, many contributions have been (and are being) made by political scientists, sociologists and economists. In the context of human behavior, consensus research is carried out by psychologists. Faced with conflicting evidence on evolutionary history, systematic biologists appeal to concepts of consensus, and molecular biologists attempt to apply consensus theory in areas of DNA research. Market researchers find the discipline relevant since it can be viewed as a theory of how large organizations make decisions based on possibly conflicting lines of evidence. Contemporary applications of consensus theory involve airplane and missile navigational systems, methods to combat bioterrorism, drug development based on DNA research, marketing and manufacturing decisions of large companies, and stock market predictions.

Particularly, Biomathematics and Molecular Biology are concerned with consensus in areas where the objects of interest represent partitions of a set, hierarchical structures, phylogenetic trees and alignment of molecular sequences (for deeper explanations see Day and McMorris [32], especially chapter 1, and Janowitz et al. [64]).

---

[17] Taking into account the common structure or pattern between what they call "bioconsensus" and Social Choice, Day and McMorris [32] advocate for the existence of an Arrow's paradigm in the Kuhn's sense.

5 Conclusion

In this chapter, in a non-exhaustive way, we have outlined the idea of consensus from classic authors to nowadays researchers, from theoretical philosophical points of view to practical applications. Progress has been achieved, but we can still expect new results. The list of chapters of this book well may become a guideline for further advances.

Acknowledgements. The financial support of the Junta de Castilla y León (Consejería de Educación, Projects VA092A08 and GR 99), the Spanish Ministerio de Ciencia e Innovación (Project ECO2009–07332) and ERDF are gratefully acknowledged. Helpful comments have also been provided by José Luis García-Lapresta, Enrique Herrera-Viedma, Christian Klamler, Javier Montero, Carlos Romero, Stefano Vannucci and two anonymous referees.

References

1. Alcalde-Unzu, J., Vorsatz, M.: Measuring the similarity of rankings: An axiomatic approach. Mimeo
2. Alcalde-Unzu, J., Vorsatz, M.: Measuring Consensus: Concepts, Comparisons, and Properties. In: Herrera-Viedma, E., et al. (eds.) Consensual Processes. STUDFUZZ, vol. 267, pp. 195–211. Springer, Heidelberg (2011)
3. Alonso, S.: Group Decision Making with Incomplete Fuzzy Preference Relations, Ph. D. Thesis, Universidad de Granada (2006)
4. Alonso, S., Pérez, I.J., Herrera-Viedma, E., Cabrerizo, F.J.: Consensus with linguistic preferences in web 2.0 communities. In: Proceedings of the 2009 Ninth International Conference on Intelligent Systems Design and Applications, pp. 809–814 (2009)
5. Arrow, K.J.: Social Choice and Individual Values. Cowles Foundation Monographs, vol. 12. Wiley, Chichester (1951); Revised second edition (1963); Reprinted by Yale University Press (1978)
6. Balinski, M.: The majority judgement: A new mechanism for electing and ranking (2009),
   http://www.bhutanstudies.org.bt/conference/Papers/3.demo.pdf
7. Balinski, M., Laraki, R.: A theory of measuring, electing and ranking. Ecole Polytechnique – Centre National de la Rechereche Scientifique. Cahier 2006-11 (2006)
8. Balinski, M., Laraki, R.: Election by majority judgement: Experimental evidence. Ecole Polytechnique – Centre National de la Rechereche Scientifique. Cahier 2007-28 (2007)
9. Ballester, M.A., García-Lapresta, J.L.: Sequential consensus for selecting qualified individuals of a group. International Journal of Uncertainty, Fuzziness and Knowledge-Based Systems 16, 57–68 (2008)
10. Black, D.: The Theory of Committees and Elections. Cambridge University Press, Cambridge (1958)
11. Bogart, K.P.: Preference structure I: distance between transitive preference relations. Journal of Mathematical Sociology 3, 49–63 (1973)
12. Bogart, K.P.: Preference structure II: distances between assymetric relations. SIAM Journal on Applied Mathematics 29, 254–262 (1975)
13. de Borda, J.C.: Mémoire sus les élections au scrutin, Historie de l'Academie Royale des Sciences, 1784

14. Bosch, R.: Characterizations of Voting Rules and Consensus Measures, Ph. D. Dissertation, Tilburg University (2005)
15. Buchahan, J., Tullock, G.: The Calculus of Consent. Logical Foundations of Constitutional Democracy. University of Michigan Press (1962)
16. Cabrerizo, F.J., Alonso, S., Pérez, I.J., Herrera-Viedma, E.: A consensus model for Group Decision Making in unbalanced fuzzy linguistic contexts. In: Proceedings of ESTYLF 2008, pp. 449–456 (2008)
17. Cabrerizo, F.J., Alonso, S., Pérez, I.J., Herrera-Viedma, E.: On consensus measures in fuzzy group decision making. In: Torra, V., Narukawa, Y. (eds.) MDAI 2008. LNCS (LNAI), vol. 5285, pp. 86–97. Springer, Heidelberg (2008)
18. Cabrerizo, F.J., Moreno, J.M., Pérez, I.J., Herrera-Viedma, E.: Analyzing consensus approaches in fuzzy group decision making: Advantages and drawbacks. Soft Computing 14(5), 451–463 (2010)
19. Cabrerizo, F.J., Pérez, I.J., Herrera-Viedma, E.: Managing the consensus in group decision making in an unbalanced fuzzy linguistic context with incomplete information. Knowledge-Based Systems 23(2), 169–181 (2010)
20. Calvo, T., Beliakov, G.: Aggregation functions based on penalties. Fuzzy Sets and Systems 161, 1420–1436 (2010)
21. Calvo, T., Beliakov, G., James, S.: On penalty based aggregation functions and consensus. In: Herrera-Viedma, E., et al. (eds.) Consensual Processes. STUDFUZZ, vol. 267, pp. 23–40. Springer, Heidelberg (2011)
22. Campillo, N.: Razón y Utopía en la Sociedad Industrial: Un Estudio sobre Saint-Simon. Universitat de Valencia (1992)
23. Carlsson, C., Ehrenberg, D., Eklund, P., Fedrizzi, M., Gustafsson, P., Lindholm, P., Merkuryeva, G., Riissanen, T., Ventre, A.G.S.: Consensus in distributed soft environments. European Journal of Operational Research 61, 165–185 (1992)
24. Condorcet, Marquis de: Essai sur l'Application de l'Analyse à la Probabilité des Décisions Rendues à la Pluralité des Voix, L'Imprimierie Royale, 1785
25. Conradt, L., Roper, T.J.: Consensus decision making in animals. TRENDS in Ecology and Evolution 20, 449–456 (2005)
26. Conradt, L., Roper, T.J.: Democracy in animals: The evolution of shared group decisions. Proceedings of the Royal Society B: Biological Sciences 274, 2317–2326 (2007)
27. Contreras, I., Hinojosa, M.A., Mármol, A.M.: Ranking Alternatives in Group Decision-Making with Partial Information. A Stable Approach. In: Herrera-Viedma, E., et al. (eds.) Consensual Processes. STUDFUZZ, vol. 267, pp. 41–52. Springer, Heidelberg (2011)
28. Cook, W.D., Kress, M., Seiford, L.M.: A general framework for distance-based consensus in ordinal ranking models. European Journal of Operational Research 96, 392–397 (1996)
29. Cook, W.D., Seiford, L.M.: Priority ranking and consensus formation. Management Science 24, 1721–1732 (1978)
30. Cook, W.D., Seiford, L.M.: The Borda-Kendall consensus method for priority ranking problems. Management Science 28, 621–637 (1982)
31. Cutello, V., Montero, J.: A characterization of rational amalgamation operations. International Journal of Approximate Reasoning 8, 325–344 (1993)
32. Day, W.H.E., McMorris, F.R.: Axiomatic Consensus Theory in Group Choice and Biomathematics. SIAM, Philadelphia (2003)
33. De Dreu, C.K.W., De Vries, N.K. (eds.): Group Consensus and Minority Influence: Implications for Innovation. Blackwell, Malden (2001)

34. Dietrich, F.: A generalised model of judgment aggregation. Social Choice and Welfare 28, 529–565 (2007)
35. Dietrich, F., List, C.: Arrow's theorem in judgment aggregation. Social Choice and Welfare 29, 19–33 (2007)
36. Dimitrov, D.: The Social Choice Approach to Group Identification. In: Herrera-Viedma, E., et al. (eds.) Consensual Processes. STUDFUZZ, vol. 267, pp. 123–134. Springer, Heidelberg (2011)
37. Eckert, D., Klamler, C.: Distance-Based Aggregation Theory. In: Herrera-Viedma, E., et al. (eds.) Consensual Processes. STUDFUZZ, vol. 267, pp. 3–22. Springer, Heidelberg (2011)
38. Eklund, P., Rusinowska, A., De Swart, H.: Consensus reaching in committees. European Journal of Operational Research 178, 185–193 (2007)
39. Emerson, P.: Consensus Voting Systems. Samizdat (1991)
40. Emerson, P.: Beyond the Tyranny of the Majority. Voting Methodologies in Decision-making and Electoral Systems, The de Borda Institute (1998)
41. Emerson, P.: Consensus voting and conflict resolution,
    http://www.opendemocracy.net/deborda/articles/
    consensusvotingandconflictemresolution
42. Essential-Facts.com: Consensus,
    http://www.essential-facts.com/primary/ethics/Consensus.
    html
43. Fedrizzi, M., Fedrizzi, M., Marques-Pereira, R.A., Brunelli, M.: Consensual dynamics in group decision making with triangular fuzzy numbers. In: Proceedings of the 41st Hawaii International Conference on System Sciences (2008)
44. Fedrizzi, M., Kacprzyk, J., Zadrożny, S.: An interactive multi-user decision support system for consensus reaching processes using Fuzzy Logic with Linguistic Quantifiers. Decision Support Systems 4, 313–327 (1988)
45. Fishburn, P.C.: Multiperson Decision Making: A selective review, in J. Kacprzyk and M. Fedrizzi (eds.). In: Multiperson Decision Making Using Fuzzy Sets and Possibility Theory, pp. 3–27. Kluwer Academic Publishers, Dordrecht (1990)
46. Freimer, M., Yu, P.L.: Some new results on compromise solutions for group decision problems. Management Science 22(6), 688–693 (1976)
47. García-Lapresta, J.L.: Favoring consensus and penalizing disagreement in Group Decision Making. Journal of Advanced Computational Intelligence and Intelligent Informatics 12(5), 416–421 (2008)
48. García-Lapresta, J.L., Llamazares, B., Peña, T.: Sccoring rules and consensus. In: Greco, S., Marques Pereira, R., Squillante, M., Yager, R.R., Kacprzyk, J. (eds.) Preferences and Decisions: Models and Applications. Springer, Heidelberg (forthcoming)
49. García-Lapresta, J.L., Pérez-Román, D.: Some consensus measures and their applications in group decision making. In: Ruan, D., Montero, J., Lu, J., Martínez, L., D'hondt, P., Kerre, E.E. (eds.) Computational Intelligence in Decision and Control, pp. 611–616. World Scientific, Singapore (2008)
50. García-Lapresta, J.L., Pérez-Román, D.: Measuring Consensus in Weak Orders. In: Herrera-Viedma, E., et al. (eds.) Consensual Processes. STUDFUZZ, vol. 267, pp. 213–234. Springer, Heidelberg (2011)
51. González-Pachón, J., Romero, C.: Distance-based consensus methods: A Goal Programming approach. Omega, The International Journal of Management Science 27, 341–347 (1999)
52. González-Pachón, J., Romero, C.: Inferring consensus weights from pairwise comparison matrices without suitable properties. Annals of Operations Research 154, 123–132 (2007)

53. Graham Jr., G.J.: Rousseau's concept of consensus. Political Science Quarterly 85, 80–98 (1970)
54. Grofman, B., Feld, S.L.: Rousseau's general will: A Condorcetian perspective. American Political Science Review 82, 567–576 (1988)
55. Habermas, J.: The Theory of the Communicative Action, vol. 1. Beacon Press (1984), vol. 2 (1987)
56. Hartmann, S., Pigozzi, G., Sprenger, J.: Reliable methods of judgment aggregation. Journal of Logic and Computation 20(2), 603–617 (2010)
57. Hartmann, S., Sprenger, J.: Consensus, compromise and judgment aggregation, http://www.accessecon.com/pubs/SCW2008/SCW2008-08-00296S.pdf
58. Herrera, F., Herrera-Viedma, E., Verdegay, J.L.: A model of consensus in group decision making under linguistic assessments. Fuzzy Sets and Systems 78, 73–87 (1996)
59. Herrera, F., Herrera-Viedma, E., Verdegay, J.L.: A rational consensus model in Group Decision Making using linguistic assessments. Fuzzy Sets and Systems 88, 31–49 (1997)
60. Herrera-Viedma, E., Alonso, S., Chiclana, F., Herrera, F.: A consensus model for Group Decision Making with incomplete fuzzy preference relations. IEEE Transactions on Systems Fuzzy Systems 15, 863–877 (2005)
61. Herrera-Viedma, E., Herrera, F., Chiclana, F.: A consensus model for Multiperson Decision Making with different preference structures. IEEE Transactions on Systems, Man and Cybernetics, Part A: Systems and Humans 32, 394–402 (2002)
62. Herrera-Viedma, E., Martínez, L., Mata, F., Chiclana, F.: A consensus support system model for Group Decision-making problems with multi-granular linguistic preference relations. IEEE Transactions on Fuzzy Systems 13, 644–658 (2005)
63. Hsu, C.C., Sandford, B.A.: The Delphi technique: Making sense of consensus. Practical Assessment, Research & Evaluation 12, 1–8 (2007)
64. Janowitz, M., Lapointe, F.J., McMorris, F.R., Mirkin, B., Roberts, F.S. (eds.): Bioconsensus. DIMACS Series in Mathematics and Theoretical Computer Science. American Mathematical Society, Providence (2003)
65. Hornik, K., Meyer, D.: Deriving consensus rankings from benchmarking experiments. Department of Statistics and Mathematics, Wirtschaftsuniversität Wien, Research Report Series 33 (2006)
66. Kacprzyk, J., Fedrizzi, M.: A "soft" measure of consensus in the setting of partial (fuzzy) preferences. European Journal of Operational Research 34, 316–325 (1988)
67. Kacprzyk, J., Fedrizzi, M., Nurmi, H.: Group decision making and consensus under fuzzy preferences and fuzzy majority. European Journal of Operational Research 49, 21–31 (1992)
68. Kacprzyk, J., Nurmi, H., Fedrizzi, M. (eds.): Consensus under Fuzziness. Kluwer, Dordrecht (1996)
69. Kacprzyk, J., Zadrożny, S., Fedrizzi, M., Nurmi, H.: On group decision making, consensus reaching, voting and voting paradoxes under fuzzy preferences and a fuzzy majority: A survey and some perspectives. In: Bustince, H., Herrera, F., Montero, J. (eds.) Fuzzy Sets and Their Extensions: Representation, Aggregation and Models, pp. 263–295. Springer, Heidelberg (2007)
70. Kacprzyk, J., Zadrożny, S., Ras, Z.W.: Action rules in consensus reaching process support. In: Ninth International Conference on Intelligent Systems Design and Applications (ISDA), pp. 827–832 (2009)
71. Kemeny, J.: Mathematics without numbers. Daedalus 88, 571–591 (1959)

72. Kemeny, J., Snell, J.L.: Preference ranking: an axiomatic approach. In: Mathematical Models in the Social Sciences, Ginn, pp. 9–23 (1962)
73. Kendall, M.: Rank Correlation Methods. Hafner (1962)
74. Klamler, C.: On some distance aspects in Social Choice. In: Simeone, B., Pukelsheim, F. (eds.) Mathematics and Democracy, pp. 97–104. Springer, Heidelberg (2006)
75. Klamler, C.: A distance measure for choice functions. Social Choice and Welfare 30, 419–425 (2008)
76. Knight, J., Johnson, J.: Aggregation and deliberation: On the possibility of Democratic legitimacy. Political Theory 22, 277–296 (1994)
77. List, C.: Judgment aggregation: A bibliography on the discursive dilemma, doctrinal paradox and decisions on multiple propositions,
    http://personal.lse.ac.uk/list/doctrinalparadox.htm
78. List, C.: Democracy in animal gropus: A political science perspective. TRENDS in Ecology and Evolution 19, 168–169
79. List, C., Puppe, C.: Judgment aggregation. In: Anan, P., Pattanaik, P.K., Puppe, C. (eds.) The Handbook of Rational and Social Choice: An Overview of New Foundations and Applications, pp. 457–482. Oxford University Press, Oxford (2009)
80. Mata, F., Martínez, L., Martínez, J.C.: Penalizing manipulation strategies in consensus processes. In: Proceedingis of ESTYLF 2008, pp. 485–491 (2008)
81. Mata, F., Martí, L.: An adaptive consensus support system for group decision making (GDM) problems with heterogeneous information. In: Ruan, D., Montero, J., Lu, J., Martí, L. (eds.) Computational Intelligence in Decision and Control, pp. 835–840. World Scientific, Singapore (2008)
82. van Mill, D.: The possibility of rational outcomes from democratic discourse and procedures. The Journal of Politics 58, 734–752 (1996)
83. Meskanen, T., Nurmi, H.: Distance from consensus: A theme and variations. In: Simeone, B., Pukelsheim, F. (eds.) Mathematics and Democracy, pp. 117–132. Springer, Heidelberg (2006)
84. Meskanen, T., Nurmi, H.: Analyzing political disagreement, JASM Working Papers 2, University of Turku, Finland (2006)
85. Montero, J.: Arrow's theorem under fuzzy rationality. Behavioral Science 32, 267–273 (1987)
86. Montero, J.: The impact of fuzziness in social choice paradoxes. Soft Computing 12, 177–182 (2008)
87. Nurmi, H.: A comparison of some distance-based choice rules in ranking environments. Theory and Decision 57, 5–24 (2004)
88. Nurmi, H.: Settings of Consensual Processes: Candidates, Verdicts, Policies. In: Herrera-Viedma, E., et al. (eds.) Consensual Processes. STUDFUZZ, vol. 267, pp. 159–177. Springer, Heidelberg (2011)
89. Partridge, P.H.: Consent & Consensus. Pall Mall (1971)
90. Règnier, S.: Sur quelques aspects mathématiques des problèmes de classification automatique. ICC Bulletin, pp. 175-191 (1965)
91. Regenwetter, M., Popova, A.: Consensus with Oneself: Within-person Choice Aggregation in the Laboratory. In: Herrera-Viedma, E., et al. (eds.) Consensual Processes. STUDFUZZ, vol. 267, pp. 95–121. Springer, Heidelberg (2011)
92. Riker, W.H.: Liberalism Against Populism. A Confrontation between the Theory of Democracy and the Theory of Social Choice, Waveland Press (1982)
93. Rousseau, J.J.: Du Contrat Social ou Principes de Droit Politique. Marc Michel (1762)
94. Runes, D.D. (ed.): Dictionary of Philosophy, Littlefield, Adams and Company, p. 64 (1962)

95. Saari, D.G.: Review to Day-McMorris,
    http://www.maa.org/reviews/axiomaticconsensus.html
96. Saccamano, N.: Rhetoric, consensus, and the law in Rousseau's. Contrat Social 107,
    730–751 (1992)
97. Samet, D., Schmeidler, D.: Between liberalism and democracy. Journal of Economic
    Theory 110, 213–233 (2003)
98. Sumpter, D.J.T., Krause, J., James, R., Couzin, I.D., Ward, A.J.W.: Consensus decision
    making by fish. Current Biology 18, 1773–1777 (2008)
99. Surowiecki, J.: The Wisdom of Crowds, Doubleday (2005)
100. Linstone, H.A., Turrof, M.: The Delphi Method, Techniques and Applications. Addison
    Wesley Publishing, Reading (1975),
    http://is.njit.edu/pubs/delphibook/
101. Young, H.P.: Optimal ranking and choice from pairwise comparisons. In: Grofman, B.,
    Owen, G. (eds.) Information Pooling and Group Decision Making, pp. 113–122. JAI
    Press (1986)
102. Wikipedia publishing policy: Consensus,
    http://en.wikipedia.org/wiki/Wikipedia:Consensus
103. Williams, R.: Keywords: A Vocabulary of Culture and Society. Fontana (1976)
104. Yu, P.L.: A class of solutions for group decision making. Management Science 19,
    936–946 (1973)

Measuring Consensus: Concepts, Comparisons, and Properties

Jorge Alcalde–Unzu and Marc Vorsatz

1 Introduction

Motivation. Human beings do not live in isolation and they have to take many decisions collectively. Examples include the election of firm representatives, the decision of where to build a new school, and the task of how to share natural resources. Equally, the satisfaction of a single individual usually depends on the performance of the group; just think of the problem of shirking in team production. To obtain a good collective performance and, as a consequence, a high individual satisfaction, it is important that collective decisions are taken with *consensus*. This is because in many instances, it is not beneficial for the society as a whole if the decision is imposed by a subset of the individuals —even if this subset includes more than half of the collective— as there may be other alternatives that are more accepted by the rest of the members and that increase the overall satisfaction.

The process of arriving to a consensus is obviously more difficult to achieve if the preferences of the individuals happen to be rather heterogenous. In an extreme case, if individual preferences are opposed, the probability of conflicts between group members rises significantly and the group may even dissolve. Eisenberg [14] describes the factors affecting group cohesiveness and the similarity of the individuals in terms of their opinions and preferences is considered to be a key dimension. Also, several socio–economic studies have confirmed the intuition that the cohesion has an impact on the society in terms of growth and probability of conflicts between their members (see, among others, Alesina and Rodrik [3], Persson and Tabellini

Jorge Alcalde–Unzu
Department of Economics, Public University of Navarre, Campus Arrosadia.
31006 Pamplona, Spain
e-mail: `jorge.alcalde@unavarra.es`

Marc Vorsatz
Fundación de Estudios de Economía Aplicada (FEDEA), Calle Jorge Juan 46,
28001 Madrid, Spain
e-mail: `mvorsatz@fedea.es`

E. Herrera-Viedma et al. (Eds.): Consensual Processes, STUDFUZZ 267, pp. 195–211.
springerlink.com   © Springer-Verlag Berlin Heidelberg 2011

[24], and Montalvo and Reynol-Querol [23]). Consequently, it is important to assess the alikeness of preferences in a group as this provides relevant information about how difficult it is to take decisions that are accep! ted by its members.

Analysis. In our study, we suppose that each individual in a group has a linear ranking that reflects her/his preferences or opinions on a set of alternatives. A group of individuals is represented by a preference profile, a list of individual preferences. With these primitives at hand, we aim at comparing some measures that assign a value from the unit interval to each preference profile, satisfying the implicit assumption that higher values imply a higher degree of *consensus* (this term will be used from now on to say that individual preferences in the group are more alike). Observe that we abstain from modeling the process used to reach a decision, which itself is likely to affect group cohesion.[1]

The first and probably best–known approach to this question is due to Kendall [19]. The measure he proposes, usually referred to as Kendall's tau (τ), is restricted to the case of two individuals and simply calculates the proportion of pairwise comparisons of alternatives the two individuals agree upon. This measure can be easily reinterpreted in terms of the Kemeny [20] distance that calculates the proportion of pairwise comparisons the two rankings do not coincide upon. So, Kendall's τ is simply one minus the Kemeny distance. An extension to the case of any arbitrary number of individuals has been proposed by Hays [17]. He suggests to calculate the *average tau* ($\bar{\tau}$), which is obtained by averaging Kendall's measure over all possible pairs of individuals. Since the minimum of $\bar{\tau}$ may be different from zero, Hays [17] also hints at the possibility of normalizing the measure to the unit interval, leading to the *normalized average tau* ($\tilde{\tau}$).

Next to the Kendall–Hays proposal, we will also consider a newer approach, axiomatized in Alcalde–Unzu and Vorsatz [1], that calculates for any pair of alternatives the absolute value of the difference between the proportion of individuals who prefer one alternative and the proportion of individuals who prefer the other alternative and takes then the average of these numbers over all possible pairs of alternatives. This measure will be called *average sigma* ($\bar{\sigma}$). The *normalized average sigma* ($\tilde{\sigma}$) is then obtained in the same way as the normalized average tau is constructed from the average tau.

It can be observed that the τ and σ measures have similarities and differences. While both use as an input comparisons between pairs of alternatives and both decompose the problem by looking across alternatives and across individuals, they nevertheless aggregate the information contained in the preference profile in a different order. While the τ measures evaluate first the consensus for each pair of individuals and then aggregate this information by taking the (normalized) mean across all pairs of individuals, the σ measures calculate first the similarity of preferences for each pair of alternatives before aggregating this information by taking the (normalized) mean across all pairs of alternatives. Consequently, we address the question whether the order of how the information in the preference profile is aggregated

---

[1] The interested reader can find more about this topic in other chapters of this book.

matters. Concentrating on the normalized measures, Propositions 1–3 show that $\bar{\tau}$ and $\bar{\sigma}$ are (a) identical to ! each other as long as the number of individuals is at most three, (b) equivalent from an ordinal point of view —that is, they rank all preference profiles according to their similarity in the same way— if there are only two alternatives and (c) different from each other even in ordinal terms if there are at least four individuals and three alternatives.

After this initial comparative study, we consider two pairs of properties a consensus measure *should* naturally satisfy. All properties will refer to the impact on the measure if *only* one individual changes his/her ranking over *only* two alternatives that are *contiguous* in the ranking (x and y are contiguous if there does not exist z situated between them in the ranking). This type of change is very powerful in the normative analysis of consensus measures mainly because of two reasons: it involves a very small change between two preference profiles so that these preference profiles are very suitable to impose normative properties between the values of the similarity measure in them and, second, this change is a sufficiently general tool to be able to construct a (non–unique) path between any two preference profiles.

The first pair of properties consists of axioms that are linked to the classic monotonicity property in social choice theory. The basic axiom studies the impact on the measure if an individual changes his opinion on only one contiguous pair of alternatives and the alternative that is made better off with the change was already preferred by a majority over the other alternative. In these cases, *Monotonicity* requires the consensus to increase. Since all the measures discussed before satisfy this basic axiom, we will strengthen this property in order to distinguish the τ and σ measures. In particular, *Strong Monotonicity* requires that a change in the preference for a contiguous pair of alternatives is an increasing function in the margin by which the alternative that has been made better off wins in a pairwise comparison to the other alternative involved in the change. Proposition 4 shows that only the τ measures but not the σ measures satisf! y Strong Monotonicity.

The second pair of axioms try to adopt Arrow's [5] axiom of Independence of Irrelevant Alternatives to the current context. *Independence* requires that the impact of a change of a contiguous pair of alternatives, $\{x, y\}$ does not depend on how individuals rank other pairs of alternatives different from $\{x, y\}$. It turns out again that both the τ and the σ measures satisfy this axiom. However, we show in Proposition 5 that only the family of σ measures satisfy the more demanding version, *Strong Independence*, which asks also that the change in the similarity is independent from the margin by which the alternative that wins in this pairwise comparison defeats the other alternative. Consequently, the choice between the two types of measures can be reinterpreted as the choice between the stronger version of Monotonicity and the stronger version of Independence.

In the final part of our study, we introduce some measures that try to overcome a problem that appears in some applications of the τ and σ measures. It can be argued that in most decision environments, changes in binary relations between alternatives that are socially more important (*i.e.*, alternatives that are on average ranked higher) should be weighted higher, simply because these alternatives are more likely to be finally chosen. Since neither the τ nor the σ take into account these considerations, we will introduce two additional approaches: First, we will present the choice–based distance measure between two individual rankings suggested by Klamler [21] and its generalization from Baldiga and Green [6]. Afterwards, we will present an alternative approach by Alcalde–Unzu and Vorsatz [2] that considers the σ measure as the basic ingredient, but modifies it adequately to overcome this drawback. Instead of using the! mean as an aggregator, weights are given to each pair of alternatives according to their importance at the social level.

Remainder. We proceed as follows. In Section 2, we introduce and compare the τ and σ measures. Also, we present some normative properties consensus measures should satify. In Section 3, we review measures that take into account how important alternatives are at the social level. In Section 4, we conclude.

2   The τ and σ Measures

2.1   Concepts

Consider a finite set of alternatives X of cardinality k and a finite set of individuals $N = \{1, 2, ..., n\}$. We assume that $n \geq 2$ and $k \geq 2$. Elements of X are usually denoted by x, y and z, generic individuals are indexed by i and j. Also, define $\bar{X}$ as the set of subsets of X with cardinal 2. The set of all distinct and unordered pairs of individuals $\bar{N}$ is defined accordingly.

Let P_i be the linear order of individual i on X. That is, P_i is a complete, transitive, and antisymmetric binary relation. The set of all linear orders on X is denoted by $\mathscr{P}$. A preference profile $P = (P_1, ..., P_n) \in \mathscr{P}^n$, is an n–tuple of linear orders. A *consensus measure* is a function $M : \mathscr{P}^n \to [0, 1]$ that assigns a real number $M(P)$ from the unit interval to every preference profile $P \in \mathscr{P}^n$ with the implicit understanding that higher values imply a higher consensus.

For the restricted case of two individuals, $N = \{i, j\}$, Kendall [19] proposed a measure that has received particular attention. To define it, take any preference profile $P = (P_i, P_j) \in \mathscr{P}^2$. Then, for any pair of alternatives $\{x, y\} \in \bar{X}$, $\tau_{x,y}^{i,j}(P) = 1$ if i and j agree on the binary ordering between x and y ($x P_i y \Leftrightarrow x P_j y$). Otherwise, $\tau_{x,y}^{i,j}(P) = 0$. The consensus, $\tau_{i,j}(P)$, is then defined to be the proportion of pairwise comparisons the two individuals agree upon; that is, $\tau_{i,j}(P) = \frac{2}{k(k-1)} \sum_{\{x,y\} \in \bar{X}} \tau_{x,y}^{i,j}(P)$. One proposal to study the general case of an arbitrary set of individuals put forward by Hays [17] is to naturally extend Kendall's τ by calculating the average of all $\tau_{i,j}(P)$ for all pairs of individuals in the society.

Definition 1. The consensus measure M is called *average tau* ($\bar{\tau}$) if for all preference profiles $P \in \mathscr{P}^n$,

$$M(P) = \frac{2}{n \cdot (n-1)} \sum_{\{i,j\} \in \bar{N}} \tau_{i,j}(P).$$

While the average tau considers pairwise comparisons on the individual level as basic ingredient, an alternative approach consists in calculating for each pair of alternatives the difference in the supports. Formally, given the preference profile $P \in \mathscr{P}^n$ and the pair of alternatives $\{x,y\} \in \bar{X}$, let $\#(xPy)$ be the number of individuals who prefer x to y at P. Then, $n_{x,y}(P) = \#(xPy) - \#(yPx)$ denotes the difference between the number of individuals who prefer x to y and the ones who prefer y to x, always evaluated at $P \in \mathscr{P}^n$.

In line with the construction of Kendall's τ, consider first the case when $X = \{x,y\}$. Given any preference profile $P \in \mathscr{P}^n$, the consensus in the society naturally equals the proportion margin by which one alternative wins against the other in the pairwise comparison; that is, $\sigma_{x,y}(P) = \frac{|n_{x,y}(P)|}{n}$. This measure can be extended in a straightforward way to the general case of an arbitrary number of alternatives by calculating the average of all $\sigma_{x,y}(P)$ for all pairs of alternatives. Formally, we obtain the following definition.

Definition 2. The consensus measure M is called *average sigma* $(\bar{\sigma})$ if for all preference profiles $P \in \mathscr{P}^n$,

$$M(P) = \frac{2}{k \cdot (k-1)} \sum_{\{x,y\} \in \bar{X}} \sigma_{x,y}(P).$$

The measures presented so far are defined on the unit interval, but, nevertheless, the range they take may depend on the parameters n and k. Consequently, we also consider normalized versions that are redefined on the unit interval. To do this, denote by $\min\{M\}$ the minimal value M takes over all possible preference profiles $P \in \mathscr{P}^n$. Define $\max\{M\}$ accordingly.

Definition 3. Given the consensus measure M, the measure $\tilde{M}$ is said to be *normalized with respect to M* if for all preference profiles $P \in \mathscr{P}^n$,

$$\tilde{M}(P) = \frac{M(P) - \min\{M\}}{\max\{M\} - \min\{M\}}.$$

We denote the normalized average tau and the normalized average sigma by $\tilde{\tau}$ and $\tilde{\sigma}$, respectively.

2.2 Comparison

Our first aim is to calculate the minimal and maximal values of $\tilde{\tau}$ and $\bar{\sigma}$ in order to construct the corresponding normalized measures. It is worth noting that the limits are for both measures independent of the size of the set of alternatives X and only depend on the size of the society N. We also find that the maximum of both measures is 1 independently of the size of the society, while the minimums converge to 0 as the size of the society becomes infinite.

Lemma 1. *The minimal and maximal values $\tilde{\tau}$ and $\bar{\sigma}$ attain over all preference profiles $P \in \mathscr{P}^n$ are given by*

$$\min\{\bar{\tau}\} = \begin{cases} \frac{n-2}{2(n-1)} & \textit{if } n \textit{ is even} \\ \frac{n-1}{2n} & \textit{if } n \textit{ is odd,} \end{cases} \qquad \min\{\bar{\sigma}\} = \begin{cases} 0 & \textit{if } n \textit{ is even} \\ \frac{1}{n} & \textit{if } n \textit{ is odd,} \end{cases}$$

and

$$\max\{\bar{\tau}\} = \max\{\bar{\sigma}\} = 1.$$

Proof. We calculate first the minimal values of the two measures. Consider any preference profile $P \in \mathscr{P}^n$. By definition,

$$\bar{\tau}(P) = \frac{2}{n \cdot (n-1)} \sum_{\{i,j\} \in \bar{N}} \tau_{i,j}(P)$$

$$= \frac{2}{n \cdot (n-1)} \sum_{\{i,j\} \in \bar{N}} \left(\frac{2}{k \cdot (k-1)} \sum_{\{x,y\} \in \bar{X}} \tau_{i,j}^{x,y}(P) \right)$$

$$= \frac{2}{n \cdot (n-1)} \cdot \frac{2}{k \cdot (k-1)} \sum_{\{x,y\} \in \bar{X}} \sum_{\{i,j\} \in \bar{N}} \tau_{i,j}^{x,y}(P)$$

and

$$\bar{\sigma}(P) = \frac{2}{k \cdot (k-1)} \sum_{\{x,y\} \in \bar{X}} \sigma_{x,y}(P).$$

Consequently, $\bar{\tau}$ takes its minimal value at any preference profile $P \in \mathscr{P}^n$ which is such that $\sum_{\{i,j\} \in \bar{N}} \tau_{x,y}^{i,j}(P)$ is minimal for each pair of alternatives $\{x,y\} \in \bar{X}$. The minimal consensus under $\bar{\sigma}$ is obtained at any preference profile $P \in \mathscr{P}^n$ which minimizes $\sigma_{x,y}(P)$ for each pair of alternatives $\{x,y\} \in \bar{X}$.

Select a pair of alternatives $\{x,y\} \in \bar{X}$ and suppose without loss of generality, that $\#(xPy) = a$ and $\#(yPx) = n - a$. Then, $\sigma_{x,y}(P) = \frac{|\#(xPy) - \#(yPx)|}{n} = \frac{|a - (n-a)|}{n}$ is minimal when $a = \frac{n}{2}$ in case n is even and when $a \in \{\frac{n-1}{2}, \frac{n+1}{2}\}$ in case n is odd. Now, construct a completely polarized profile $\bar{P} \in \mathscr{P}^n$ such that: (*i*) if n is even, $\frac{n}{2}$ individuals have any linear order $P_i \in \mathscr{P}$ and $\frac{n}{2}$ individuals have the linear order $P_j \in \mathscr{P}$ such that for all pairs of alternatives $\{x,y\} \in \bar{X}$, $xP_iy \Leftrightarrow yP_jx$; and (*ii*) if n is odd, $\frac{n+1}{2}$ and $\frac{n-1}{2}$ individuals hold the linear orders P_i and P_j, respectively. It follows from the calculations above that $\bar{\sigma}$ has a minimum in $\bar{P}$. The result follows from calculating the consensus at this profile.

In order to minimize $\sum_{\{i,j\} \in \bar{N}} \tau_{x,y}^{i,j}(P)$ for a given pair of alternatives $\{x,y\} \in \bar{X}$, we have to minimize $\frac{1}{2} \cdot (a(a-1) + (n-a)(n-a-1))$ with respect to a, where a is an integer. It follows from the first order condition that $a^* = \frac{n}{2}$ if n is even and $a^* \in \{\frac{n-1}{2}, \frac{n+1}{2}\}$ if n is odd. The second order condition shows that these critical points are indeed minimums. Hence, $\bar{\tau}$ takes a minimum in the completely polarized preference profile $\bar{P}$ constructed above and the result follows again from calculating the consensus in that particular preference profile.

Finally, it is easy to show that the maximal value under both measures is obtained at any unanimous preference profile $\tilde{P}$; that is, when $x\tilde{P}_iy \Leftrightarrow x\tilde{P}_jy$ for all $\{x,y\} \in \bar{X}$ and all $i, j \in \bar{N}$. The result follows because the consensus at $\tilde{P}$ is 1 under both measures. □

With this result at hand, we are now able to compare the normalized measures. We find first that the two measures are identical to each other whenever $n \leq 3$.

Proposition 1. *If $n \leq 3$, $\tilde{\tau}(P) = \tilde{\sigma}(P)$ for all preference profiles $P \in \mathscr{P}^n$.*

Proof. If $n = 2$, the result follows because both measures reduce to Kendall's τ. If $n = 3$, observe that $\min\{\tilde{\tau}\} = \min\{\tilde{\sigma}\} = 1/3$. Hence, we are going to show that for all preference profiles $P \in \mathscr{P}^n$, $\tilde{\tau}(P) = \tilde{\sigma}(P)$. Take any preference profile $P \in \mathscr{P}^n$. By definition,

$$\tilde{\tau}(P) = \frac{2}{n \cdot (n-1)} \frac{2}{k \cdot (k-1)} \sum_{\{x,y\} \in \bar{X}} \sum_{\{i,j\} \in \bar{N}} \tau_{x,y}^{i,j}(P).$$

Since $n = 3$, for all pairs of alternatives $\{x,y\} \in \bar{X}$, $\sum_{\{i,j\} \in \bar{N}} \tau_{i,j}^{x,y}(P)$ is equal to 3 if all individuals coincide in how they order x and y, and 1 otherwise. Thus, $\sum_{\{i,j\} \in \bar{N}} \tau_{i,j}^{x,y}(P) = |n_{x,y}(P)|$. Hence, $\tilde{\tau}(P) = \frac{2}{3k(k-1)} \sum_{\{x,y\} \in \bar{X}} |n_{x,y}(P)| = \tilde{\sigma}(P)$.   □

Limiting the number of alternatives does not help to find an exact equivalence between the two consensus measures. However, if $k = 2$, the two measures turn out to be ordinally equivalent meaning that if the consensus in a certain preference profile is higher than in another profile according to one measure, then the same is also true for the other measure. Formally, two consensus measures M and M' are said to be *ordinally equivalent* whenever for all preference profiles $P, P' \in \mathscr{P}^n$, $M(P) \geq M(P')$ if and only if $M'(P) \geq M'(P')$.

Proposition 2. *If $k = 2$, $\tilde{\sigma}$ and $\tilde{\tau}$ are ordinally equivalent.*

Proof. Suppose that $X = \{x,y\}$. It follows from straightforward calculations that for all preference profiles $P \in \mathscr{P}^n$, $\tilde{\tau}(P) = \frac{|n_{x,y}(P)|^2}{n^2}$ and $\tilde{\sigma}(P) = \frac{|n_{x,y}(P)|}{n}$ whenever n is even, while $\tilde{\tau}(P) = \frac{|n_{x,y}(P)|^2 - 1}{(n-1)(n+1)}$ and $\tilde{\sigma}(P) = \frac{|n_{x,y}(P)| - 1}{n-1}$ whenever n is odd. Hence, in order to show that the two measures are ordinally equivalent, we only have to prove that $\tilde{\tau}$ and $\tilde{\sigma}$ are monotonic increasing with respect to $|n_{x,y}(P)|$. If n is even we have that

$$\frac{\Delta \tilde{\tau}}{\Delta |n_{x,y}|} = \frac{(|n_{x,y}(P)| + 2)^2 - |n_{x,y}(P)|^2}{n^2} = \frac{4(|n_{x,y}(P)| + 1)}{n^2} > 0$$

and

$$\frac{\Delta \tilde{\sigma}}{\Delta |n_{x,y}|} = \frac{(|n_{x,y}(P)| + 2) - |n_{x,y}(P)|}{n} = \frac{2}{n} > 0.^2$$

On the other hand, if n is odd we have that

$$\frac{\Delta \tilde{\tau}}{\Delta |n_{x,y}|} = \frac{4(|n_{x,y}(P)| + 1)}{(n-1)(n+1)} > 0$$

---

[2] Observe that the minimal possible change in $|n_{x,y}|$ is 2.

and

$$\frac{\Delta\tilde{\sigma}}{\Delta|n_{x,y}|} = \frac{2}{n-1} > 0.$$

$\square$

Our final result in this section establishes that the measures are different from each other even in ordinal terms as soon as $n \geq 4$ and $k \geq 3$.

Proposition 3. *If $n \geq 4$ and $k \geq 3$, there are two preference profiles $P, P' \in \mathscr{P}^n$ such that $\tilde{\sigma}(P) \leq \tilde{\sigma}(P')$ and $\tilde{\tau}(P) > \tilde{\tau}(P')$.*

Proof. Without loss of generality, let $X = \{x, y, z\} \cup T$. Throughout the proof we are going to use the linear orders P_i, P_j, and P_l, which are as follows: $xP_iyP_izP_iT$, $xP_jzP_jyP_jT$, and $yP_lxP_lzP_lT$.[3] We also assume that all these rankings have the same linear preference ordering on T; i.e., for all $v, w \in T$, $vP_iw \Leftrightarrow vP_jw \Leftrightarrow vP_lw$. The analysis is going to depend on whether n is even or odd.

1. Suppose that n is even. Let P be a profile such that $\frac{n}{2}$ individuals have the linear order P_i and $\frac{n}{2}$ individuals the linear order P_j. Then,

$$\bar{\sigma}(P) = \frac{2}{k \cdot (k-1)} \cdot \frac{(\frac{k(k-1)}{2} - 1) \cdot n + 0}{n} = 1 - \frac{2}{k(k-1)}$$

and

$$\bar{\tau}(P) = \frac{\frac{1}{2} \cdot \frac{n}{2} \cdot (\frac{n}{2} - 1) \cdot 2 + \frac{n}{2} \cdot \frac{n}{2} \cdot \frac{\frac{k(k-1)}{2} - 1}{\frac{k(k-1)}{2}}}{\frac{n(n-1)}{2}} = 1 - \frac{n^2}{nk(k-1)(n-1)}.$$

Next, consider a preference profile P' such that $\frac{n}{2}$ individuals have the linear order P_i, $\frac{n}{2} - 1$ individuals have the linear order P_j, and, finally, one individual has the linear order P_l. Then,

$$\bar{\sigma}(P') = \frac{2}{k \cdot (k-1)} \cdot \frac{(\frac{k(k-1)}{2} - 2) \cdot n + (n-2) + 2}{n} = 1 - \frac{2}{k(k-1)} = \bar{\sigma}(P).$$

Since a normalization does not change how two preference profiles are ordered according to their consensus, it must be the case that $\tilde{\sigma}(P') = \tilde{\sigma}(P)$. Hence, $\tilde{\sigma}(P') \geq \tilde{\sigma}(P)$. To finish this part of the proof, it remains to be shown that $\tilde{\tau}(P') < \tilde{\tau}(P)$. We obtain that

$$\bar{\tau}(P') = \frac{\frac{1}{2} \cdot \frac{n}{2} \cdot \frac{n-2}{2} + \frac{1}{2} \cdot \frac{n-2}{2} \cdot \frac{n-4}{2} + (\frac{n}{2} \cdot \frac{n-2}{2} + \frac{n}{2})(1 - \frac{2}{k(k-1)}) + \frac{n-2}{2}(1 - \frac{4}{k(k-1)})}{\frac{n(n-1)}{2}}$$

$$= 1 - \frac{n^2 + 4n - 8}{nk(n-1)(k-1)}.$$

---

[3] When x is an alternative and $T \subset X$ a set of alternatives, we denote by xP_oT the case in which xP_oy for all $y \in T$.

It is now easy to see that $\bar{\tau}(P') < \bar{\tau}(P)$ whenever $n > 2$. Since a normalization does not change how two preference profiles are ordered according to their consensus, it must be the case that $\tilde{\tau}(P') < \tilde{\tau}(P)$.

2. Suppose that n is odd. Let P be a profile such that $\frac{n+1}{2}$ individuals have the linear order P_i and $\frac{n-1}{2}$ individuals the linear order P_j. Then,

$$\bar{\sigma}(P) = \frac{2}{k \cdot (k-1)} \cdot \frac{(\frac{k(k-1)}{2} - 1) \cdot n + 1}{n} = 1 - \frac{2(n-1)}{nk(k-1)}.$$

Moreover,

$$\bar{\tau}(P) = \frac{\frac{1}{2} \cdot \frac{n+1}{2} \cdot \frac{n-1}{2} + \frac{1}{2} \cdot \frac{n-1}{2} \cdot \frac{n-3}{2} + \frac{n+1}{2} \cdot \frac{n-1}{2}(1 - \frac{2}{k(k-1)})}{\frac{n(n-1)}{2}} = 1 - \frac{n^2 - 1}{nk(n-1)(k-1)}.$$

Next, consider a preference profile P' such that $\frac{n+1}{2}$ individuals have the linear order P_i, $\frac{n-3}{2}$ individuals have the linear order P_j, and, finally, one individual has the linear order P_l. Then,

$$\bar{\sigma}(P') = \frac{2}{k \cdot (k-1)} \cdot \frac{(\frac{k(k-1)}{2} - 2) \cdot n + (n-2) + 3}{n} = 1 - \frac{2(n-1)}{nk(k-1)} = \bar{\sigma}(P).$$

Since a normalization does not change how two preference profiles are ordered according to their consensus, it must be the case that $\tilde{\sigma}(P') = \tilde{\sigma}(P)$. Hence, $\tilde{\sigma}(P') \geq \tilde{\sigma}(P)$. To finish this part of the proof, it remains to be shown that $\bar{\tau}(P') < \bar{\tau}(P)$. We obtain that

$$\bar{\tau}(P') = \frac{\frac{1}{2} \cdot \frac{n+1}{2} \cdot \frac{n-1}{2} + \frac{1}{2} \cdot \frac{n-3}{2} \cdot \frac{n-5}{2} + (\frac{n+1}{2} + \frac{n+1}{2} \cdot \frac{n-3}{2})(1 - \frac{2}{k(k-1)}) + \frac{n-3}{2}(1 - \frac{4}{k(k-1)})}{\frac{n(n-1)}{2}}$$

$$= 1 - \frac{n^2 + 4n - 13}{n(n-1)k(k-1)}.$$

It is now easy to see that $\bar{\tau}(P') < \bar{\tau}(P)$ whenever $n > 3$. Since a normalization does not change how two preference profiles are ordered according to their consensus, it must be the case that $\tilde{\tau}(P') < \tilde{\tau}(P)$.   □

We conclude this part with an example that applies the four measures to a particular preference profile.

Example 1. Let $X = \{x, y, z\}$ and $n = 4$. Moreover, suppose that the preference profile $P \in \mathscr{P}^n$ is such that xP_1yP_1z, xP_2zP_2y, yP_3xP_3z, and zP_4xP_4y. We have then that $\sigma_{x,y}(P) = \sigma_{x,z}(P) = \frac{1}{2}$ and $\sigma_{y,z}(P) = 0$. Consequently, $\bar{\sigma}(P) = \frac{1}{3}$. Since $\min\{\bar{\sigma}\} = 0$ and $\max\{\bar{\sigma}\} = 1$, we obtain that $\tilde{\sigma}(P) = \frac{1}{3}$ as well. With respect to the τ measures, we observe first that $\tau_{1,2}(P) = \tau_{1,3}(P) = \tau_{2,4}(P) = \frac{2}{3}$, $\tau_{1,4}(P) = \tau_{2,3}(P) = \frac{1}{3}$, and $\tau_{3,4}(P) = 0$. Hence, $\bar{\tau}(P) = \frac{4}{9}$. Since $\min\{\bar{\tau}\} = \frac{1}{3}$ and $\max\{\bar{\tau}\} = 1$, we obtain that $\tilde{\tau}(P) = \frac{\frac{4}{9} - \frac{1}{3}}{\frac{2}{3}} = \frac{1}{6}$.   □

2.3 Properties

Our comparison of the measures as a function of the size of the set of alternatives k and the size of the set of individuals n has above all revealed that apart from some restrictive cases, the σ and the τ measures are rather distinct from each other even in ordinal terms. Our next objective is therefore to gain more insight into the structure of the measures by distinguishing them according to their underlying properties. In particular, we will concentrate on very basic monotonicity and independence properties. We start this analysis by introducing the necessary notation.

   Given the linear order $P_i \in \mathscr{P}$ for individual i and the pair of alternatives $\{x,y\} \in \bar{X}$, we will say that the ordered pair (x,y) is a *contiguous pair at P_i* if xP_iy and there is no other alternative $z \in X$ such that xP_izP_iy. For any two linear orders $P_i, P_i' \in \mathscr{P}$ and any pair of alternatives $\{x,y\} \in \bar{X}$, P_i' is said to be (x,y)–*different from P_i* if (y,x) is a contiguous pair at P_i, (x,y) is a contiguous pair at P_i', and $zP_iw \Leftrightarrow zP_i'w$ for all pairs of alternatives $\{z,w\} \neq \{x,y\}$. We will also speak of a yP_ix–*change* in order to express the movement from P_i to P_i' by means of interchanging the contiguous pair of alternatives (y,x). Finally, given the preference profiles P,P', an individual $i \in N$, and the pair of alternatives $\{x,y\} \in \bar{X}$, P' is said to be (x,y)–*different from P for individual i !* if P_i' is (x,y)–different from P_i and $P_j' = P_j$ for all $j \neq i$. Intuitively, P' is (x,y)–different from P for individual i if P' can be derived from P by only reversing the comparison between y and x in the preference ranking of individual i.

   The first group of properties we are going to discuss considers changes in the measures when one individual performs a yP_ix–change. Monotonicity states that if, before the flip, there were at least as many individuals who prefer the alternative that is made better off with the change, then the consensus should increase. Formally, this property is defined as follows.

MONOTONICITY: The consensus measure M is *monotone* if for all pairs of alternatives $\{x,y\} \in \bar{X}$, all individuals $i \in N$, and all preference profiles $P,P' \in \mathscr{P}^n$ such that $n_{x,y}(P) \geq 0$ and P' is (x,y)–different from P for i,

$$M(P') > M(P).$$

   Monotonicity is a rather basic and weak requirement and, in this sense, it is not surprising that all measures introduced in the former section satisfy it (as we will see in the next proposition). But this also means that we have to consider properties with more structure in order to be able to distinguish the measures according to how they respond to changes in preferences. The next property, Strong Monotonicity, is derived along this line and suggests that an increment in the consensus thanks to such a change in preferences should be the greater the more individuals prefer the alternative that is made better off.

STRONG MONOTONICITY: The consensus measure M is *strongly monotone* if it is monotone and for all pairs of alternatives $\{x,y\}, \{w,z\} \in \bar{X}$, all individuals $i \in N$, and all preference profiles $P,P',P'' \in \mathscr{P}^n$ such that $0 \leq n_{x,y}(P) < n_{z,w}(P)$, P' is (x,y)–different from P for i, and P'' is (z,w)–different from P for i,

$$M(P'') > M(P').$$

The next proposition shows that Strong Monotonicity helps to separate the measures because this condition, among the measures considered here, is only met by $\bar{\tau}$ and $\tilde{\tau}$.

Proposition 4. *If $n \geq 4$, then*

1. *the consensus measures $\bar{\sigma}$ and $\tilde{\sigma}$ are monotone but not strongly monotone.*
2. *the consensus measures $\bar{\tau}$ and $\tilde{\tau}$ are strongly monotone.*

Proof. Since the proofs for non–normalized and normalized measures are almost identical, we concentrate on the non–normalized measures throughout.

1. We first show that $\bar{\sigma}$ is monotone. Take any two preference profiles $P, P' \in \mathscr{P}^n$ such that for some pair of alternatives $\{x, y\} \in \bar{X}$, $n_{x,y}(P) \geq 0$ and P' is (x, y)–different from P for some individual i. We have to show that $\bar{\sigma}(P') > \bar{\sigma}(P)$. Due to the change in preferences, $n_{x,y}(P') = n_{x,y}(P) + 2$ and $n_{w,z}(P') = n_{w,z}(P)$ for all pairs of alternatives $\{w, z\} \neq \{x, y\}$. Hence, $\sigma_{x,y}(P') > \sigma_{x,y}(P)$ and $\sigma_{w,z}(P') > \sigma_{w,z}(P)$ for all pairs of alternatives $\{w, z\} \neq \{x, y\}$. Consequently, $\bar{\sigma}(P') > \bar{\sigma}(P)$ by definition of the measure.

   Next, we show that $\bar{\sigma}$ is not strongly monotone. Consider any three preference profiles $P, P', P'' \in \mathscr{P}^n$ such that $0 \leq n_{x,y}(P) < n_{w,z}(P)$ for some pairs of alternatives $\{x, y\}, \{w, z\} \in \bar{X}$, while P' and P'', respectively, are (x, y) and (w, z)-different from P for some individual $i \in N$. Then, $n_{x,y}(P') - 2 = n_{x,y}(P) = n_{x,y}(P'')$, $n_{w,z}(P'') - 2 = n_{w,z}(P) = n_{w,z}(P')$, and $n_{a,b}(P'') = n_{a,b}(P') = n_{a,b}(P)$ for all pairs of alternatives $\{a, b\} \notin \{\{x, y\}, \{w, z\}\}$. By definition, we obtain then that $\bar{\sigma}(P') = \bar{\sigma}(P'') = \bar{\sigma}(P)$. Consequently, $\bar{\sigma}$ is not strongly monotone.

2. We show that $\bar{\tau}$ is strongly monotone. Consider any three preference profiles $P, P', P'' \in \mathscr{P}^n$ such that $0 \leq n_{x,y}(P) < n_{w,z}(P)$ for some pairs of alternatives $\{x, y\}, \{w, z\} \in \bar{X}$, while P' and P'', respectively, are (x, y) and (w, z)-different from P for some individual $i \in N$. Due to the change in preferences, $\tau_{i,j}(P') - \frac{2}{k(k-1)} = \tau_{i,j}(P) = \tau_{i,j}(P'')$ for all $j \in N$ with xP_jy and $\tau_{i,l}(P') + \frac{2}{k(k-1)} = \tau_{i,l}(P) = \tau_{i,l}(P'')$ for all $l \in N$ with yP_lx. Also, $\tau_{i,j}(P'') - \frac{2}{k(k-1)} = \tau_{i,j}(P) = \tau_{i,j}(P')$ for all $j \in N$ with wP_jz and $\tau_{i,l}(P'') + \frac{2}{k(k-1)} = \tau_{i,l}(P) = \tau_{i,l}(P')$ for all $l \in N$ with zP_lw. Since $n_{w,z}(P) > n_{x,y}(P) \geq 0$ by assumption, it must be that $\bar{\tau}(P'') > \bar{\tau}(P') > \bar{\tau}(P)$. Consequently, $\bar{\tau}$ is strongly monotone. $\square$

Next, we present two independence properties. The first states that the effect on the consensus measure of a yP_ix–change is independent of the individual preferences over all other pairwise comparisons. To express this idea more formally, consider any two preference profiles P and $\bar{P}$ with the property that no individual changes her/his preferences on the pair $\{x, y\}$ across the two profiles. Suppose also that (y, x) is a contiguous pair of alternatives for individual i in both preference profiles. Consider now the two preference profiles P' and $\bar{P}'$ that are obtained from P and $\bar{P}$, respectively, by changing the ordering between x and y for individual i. The condition states that the measure changes in both situations by the same amount.

INDEPENDENCE: The consensus measure M is *independent* if for all pairs of alternatives $\{x,y\} \in \bar{X}$, all individuals $i \in N$, and all preference profiles $P,P',\bar{P},\bar{P}' \in \mathscr{P}^n$ such that $n_{x,y}(P) \geq 0$, $xP_jy \Leftrightarrow x\bar{P}_jy$ for all $j \in N$, P' is (x,y)–different from P for individual i, and $\bar{P}'$ is (x,y)–different from $\bar{P}$ for individual i,

$$M(\bar{P}') - M(\bar{P}) = M(P') - M(P).$$

From a theoretical point of view, one could consider this property an adoption of Arrow's [5] Independence of Irrelevant Alternatives axiom to our setting. The next proposition will show that all the measures considered here satisfy it. The stronger property we consider next reflects the idea that the effect of interchanging a contiguous pair of alternatives is also independent of the margin by which x wins the pairwise comparison against y.

STRONG INDEPENDENCE: The consensus measure M is *strongly independent* if for all pairs of alternatives $\{x,y\} \in \bar{X}$, all individuals $i \in N$, and all preference profiles $P,P',\bar{P},\bar{P}' \in \mathscr{P}^n$ such that $n_{x,y}(P) \geq 0$, $n_{x,y}(\bar{P}) \geq 0$, P' is (x,y)–different from P for individual i, and $\bar{P}'$ is (x,y)–different from $\bar{P}$ for individual i,

$$M(\bar{P}') - M(\bar{P}) = M(P') - M(P).$$

The final proposition in this section shows that the independence properties also serve to separate the measures. However, here it is the family of τ measures that does not satisfy the stronger property.

Proposition 5. *If $n \geq 4$, then*

1. the consensus measures $\bar{\sigma}$ and $\tilde{\sigma}$ are strongly independent.
2. the consensus measures $\bar{\tau}$ and $\tilde{\tau}$ are independent but not strongly independent.

Proof. Since the proofs for non–normalized and normalized measures are almost identical, we concentrate on the non–normalized measures throughout.

1. We show that $\bar{\sigma}$ is strongly independent. Let $P,P',\bar{P},\bar{P}' \in \mathscr{P}^n$ be profiles such that for some pair of alternatives $\{x,y\} \in \bar{X}$, $n_{x,y}(P) \geq 0$, $n_{x,y}(\bar{P}) \geq 0$, and P' and $\bar{P}'$ are (x,y)–different from P and $\bar{P}$, respectively, for some individual $i \in N$. We have to show that $\bar{\sigma}(\bar{P}') - \bar{\sigma}(\bar{P}) = \bar{\sigma}(P') - \bar{\sigma}(P)$.
   Due to the change in preferences, $n_{x,y}(P') = n_{x,y}(P) + 2$ and $n_{w,z}(P') = n_{w,z}(P)$ for all pairs of alternatives $\{w,z\} \neq \{x,y\}$. Hence, $\sigma_{x,y}(P') = \sigma_{x,y}(P) + \frac{2}{n}$ and $\sigma_{w,z}(P') = \sigma_{w,z}(P)$ for all pairs of alternatives $\{w,z\} \neq \{x,y\}$. Consequently, $\bar{\sigma}(P') = \bar{\sigma}(P) + \frac{4}{nk(k-1)}$ by definition of the measure. Since it is also the case that $n_{x,y}(\bar{P}') = n_{x,y}(\bar{P}) + 2$ and $n_{w,z}(\bar{P}') = n_{w,z}(\bar{P})$ for all pairs of alternatives $\{w,z\} \neq \{x,y\}$, it follows from the same kind of calculations that $\bar{\sigma}(\bar{P}') = \bar{\sigma}(\bar{P}) + \frac{4}{nk(k-1)}$. So, $\bar{\sigma}(\bar{P}') - \bar{\sigma}(\bar{P}) = \bar{\sigma}(P') - \bar{\sigma}(P)$.

2. We show that $\bar{\tau}$ is independent. Let the profiles $P, P', \bar{P}, \bar{P}' \in \mathscr{P}^n$ be such that for some pair of alternatives $\{x,y\} \in \bar{X}$, $n_{x,y}(P) \geq 0$, $xP_jy \Leftrightarrow x\bar{P}_jy$ for all $j \in N$, and P' and $\bar{P}'$ are (x,y)–different from P and $\bar{P}$, respectively, for some individual $i \in N$. We have to show that $\bar{\tau}(P') - \bar{\tau}(P) = \bar{\tau}(\bar{P}') - \bar{\tau}(\bar{P})$. Due to the change in preferences, the consensus $\tau_{i,l}(P)$ increases by $\frac{2}{k(k-1)}$ whenever xP_ly and decreases by the same amount whenever yP_lx. Since $n_{x,y}(P) \geq 0$ by assumption, the total incremental due to the change of preferences amounts to $\bar{\tau}(P') - \bar{\tau}(P) = \frac{2n_{x,y}(P)}{k(k-1)}$. Applying the same kind of argument, one can easily show that $\bar{\tau}(\bar{P}') - \bar{\tau}(\bar{P}) = \frac{2n_{x,y}(\bar{P})}{k(k-1)}$. Finally, $\bar{\tau}(P') - \bar{\tau}(P) = \bar{\tau}(\bar{P}') - \bar{\tau}(\bar{P})$ because $n_{x,y}(P) = n_{x,y}(\bar{P})$ by assumption.

To show that $\bar{\tau}$ is not strongly independent consider the preference profiles $P, P', \bar{P}, \bar{P}' \in \mathscr{P}^n$ such that for some pair of alternatives $\{x,y\} \in \bar{X}$, $n_{x,y}(P) > n_{x,y}(\bar{P}) \geq 0$ and P' and $\bar{P}'$ are (x,y)–different from P and $\bar{P}$, respectively, for some individual $i \in N$. It follows then from the same calculations as above that $\bar{\tau}(P') - \bar{\tau}(P) = \frac{2n_{x,y}(P)}{k(k-1)} > \frac{2n_{x,y}(\bar{P})}{k(k-1)} = \bar{\tau}(\bar{P}') - \bar{\tau}(\bar{P})$. Consequently, $\bar{\tau}$ is not strongly independent. □

3 Social Importance

The measures presented in the last section share the characteristic that they are based on pairwise comparisons, however none of them takes into account how important the alternatives are for the society. To see this, consider the following example.

Example 2. Let $X = \{x,y,z,w\}$ and $n = 4$. Table 1 presents two preference profiles P (to the left) and P' (to the right) and we would like to know how they should be ordered in terms of their consensus.

Observe that the profile P' can be obtained from P by inverting xP_3y, xP_4y, zP_3w, and zP_4w. So the question of in which society there is a higher consensus can be rephrased as follows: Is the consensus higher if all individuals prefer x over y and there is a split opinion between z and w or if all individuals prefer z over w and there is a split opinion between x and y? Given that all individuals in the society prefer both x and y over both z and w, it is reasonable to assume that the consensus in P' should be strictly higher than in P. Since the consensus is the same for all four measures presented in the previous section ($\frac{5}{6}$ under $\bar{\sigma}$, $\tilde{\sigma}$ and $\tilde{\tau}$, and $\frac{8}{9}$ under $\bar{\tau}$), there is the need to come up with new measures that take the social importance of the alternatives explicitly into account. □

Table 1 The preference profiles P and P'.

P_1	P_2	P_3	P_4		P'_1	P'_2	P'_3	P'_4
x	x	y	y		x	x	x	x
y	y	x	x		y	y	y	y
z	z	z	z		z	z	w	w
w	w	w	w		w	w	z	z

This section will be devoted to present two different approaches that have been developed to differentiate between preference profiles as those presented in Example 2. The first approach, which we are going to call choice–based consensus measures, has been put forward by Klamler [21] and Baldiga and Green [6]; the second approach, based on an adaptation of the σ measures, has been developed in Alcalde-Unzu and Vorsatz [2].

3.1 Choice–Based Consensus Measures

A linear order can be interpreted also as a choice function over all possible subsets of alternatives. That is, a preference ranking P over a set X can be redefined in terms of a choice function $c_P : 2^X \setminus \{\emptyset\} \to X$ that selects for any possible subset of alternatives $S \subseteq X$ the maximal element according to the binary relation P; i.e., $c_P(S) = \{x \in S \,|\, xPy \text{ for all } y \in (S \setminus \{x\})\}$. For example, suppose that $X = \{x, y, z\}$ and that $xPyPz$. There is an equivalence between this preference ranking and the choice function c_P such that $c_P(S) = \{x\}$ for all $S \subseteq X$ whenever $x \in S$, $c_P(S) = \{y\}$ for all $S \subseteq X$ whenever $S \cap \{x, y\} = \{y\}$, and $c_P(\{z\}) = \{z\}$.

Based on this interpretation, Klamler [21] proposed an alternative distance measure between two linear orders. This measure evaluates the proportion of subsets of alternatives with at least two alternatives such that the choice functions generated by the two linear orders differ in the selected alternative. This measure can be reinterpreted in terms of a consensus measure (that we will call *beta*) for the restrictive case of only two individuals similarly as the Kemeny measure can be translated to the Kendall's τ. In this case, the consensus between two individuals $\beta_{i,j}$ would be the proportion of subsets of alternatives with at least two alternatives such that the choice functions generated by the linear order agree in the selected alternative:

$$\beta_{i,j}(P) = \frac{|\{S \subseteq X \text{ such that } |S| \geq 2 \text{ and } c_{P_i}(S) = c_{P_j}(S)\}|}{2^k - k - 1}.$$

For the general case of any arbitrary number of individuals, one can then again take the average over all pairs of individuals.

Definition 4. The consensus measure M is called *average beta* $(\bar{\beta})$ if for all preference profiles $P \in \mathscr{P}^n$,

$$M(P) = \frac{2}{n \cdot (n-1)} \sum_{\{i,j\} \in \tilde{N}} \beta_{i,j}(P).$$

Applying $\bar{\beta}$ to the two preference profiles above, we obtain that $\bar{\beta}(P') = \frac{31}{33} > \frac{25}{33} = \bar{\beta}(P)$.

The study by Klamler [21] implicitly supposes that all subsets occur with the same probability, an assumption that has been recently relaxed by Baldiga and Green [6]. In their study, an exogenous probability of occurrence is given to each possible choice set and the consensus measure is then constructed by using these probabilities as weights. [4]

3.2 Majority–Rule Based Consensus Measures

While the studies of Klamler [21] and Baldiga and Green [6] are based on the same idea of Kendall [19] of comparing two rankings and define the general measures by aggregating this information over all pairs of individuals, Alcalde–Unzu and Vorsatz [2] take comparisons between pairs of alternatives —by means of the σ measure— as the basic ingredient but modify the aggregation procedure. To formally introduce this family, some additional notation is needed.

Given a preference profile P and the alternatives $x, y \in X$, x *wins–or–ties* against y (denoted by $x \succsim_P y$) whenever $n_{x,y}(P) \geq 0$. The asymmetric part (the *win* relation) and the symmetric part (the *tie* relation) of $\succsim_P$ —referred to as $\succ_P$ and $\sim_P$, respectively— are defined in the natural way. Let $p_x(P) = \#\{y \in X \setminus \{x\} : x \succ_P y\}$ be the number of alternatives x wins against. Similarly, $i_x(P) = \#\{y \in X \setminus \{x\} : x \sim_P y\}$ refers to the number of alternatives x ties against. Finally, $r_{x,y}(P) = p_x(P) + p_y(P) + \frac{1}{2}(i_x(P) + i_y(P))$ is the number of alternatives x and y jointly win–or–tie against when ties count only half a win.

Definition 5. The consensus measure M is said to belong to the class Φ if there exists $t \geq 0$ such that for preference profiles $P \in \mathscr{P}^n$,

$$M(P) = \frac{\sum\limits_{\{x,y\} \in \bar{X}} \omega(r_{x,y}(P)) \cdot \sigma_{x,y}(P)}{\sum\limits_{\{x,y\} \in \bar{X}} \omega(r_{x,y}(P))},$$

where for all $v \in \{1, \frac{3}{2}, 2, \frac{5}{2}, \ldots, 2k-3\}$, $\omega(v) = 1 + 2t(v-1)$.

The generic measure of the family Φ will be denoted by Φ_t. Any measure of this class can be described as follows: Calculate the values $\sigma_{x,y}$ for all pairs of alternatives and aggregate them by a weighted mean. The key is that the weights indicate how important objects are at the social level. In particular, the weights are a function that depends on the number of alternatives each of the two alternatives under consideration wins and ties against (with a tie counting half a win) according to pairwise comparisons. The measures of the family differ in the degree of this positive dependence. To be more exact, each measure is associated with a parameter $t \in \mathbb{R}_+$ and higher values of t imply that more weight is given to the consensus between alternatives that are socially more important. As a consequence, the measures of the family can differentiate between preference profiles as those of Example 2, with the exception of the case $t = 0$ that reduces to the $\bar{\sigma}$ measure. To be more exact, $\phi_t(P) = \frac{5+24t}{6+24t}$ and $\phi_t(P') = \frac{5+16t}{6+24t}$ so that $\Phi_t(P') > \Phi_t(P)$ for all $t > 0$.

---

[4] The paper of Baldiga and Green [6] constructs measures that evaluate the polarization of the preferences based in this family of distance functions.

4 Concluding Discussion

We have presented some possible ways to measure the consensus and we have some characterizations of these measures.[5] The measures in which we have focused in this paper take pairwise comparisons between alternatives as basic ingredient, as in the Condorcet [9] tradition of social choice. As a consequence, our measures have some relation with the Kemeny distance between two preference orderings. Other possibilities should be to start with other information of the individual orderings different from the pairwise comparisons. For example, if we construct a consensus measure taking the information of the positional comparisons as in the Borda [8] tradition of social choice, it is probably that the measures at which we arrive should have some relation with the distance function proposed by Cook and Seiford [10, 11].

There is a vast literature that studies how to measure the similarity of a group of individuals in some particular dimensions. Some examples include Gini [16] and Atkinson [4] who propose measures of income inequality (see, also Cowell [12] for an overview of this literature); Hutchens [18] who proposes a measure for quantifying how similar are the individuals in terms of their occupations; Kranich [22] and Roemer [25] who evaluate the similarity of the economical and social opportunities that each individual has; and Esteban and Ray [15] and Duclos et al. [13] propose measures of polarization. In this chapter, our objective is to present some alternative approaches to construct measures of the similarity of individual preferences, to compare them and to present some results related to the measures.

Acknowledgements. Jorge Alcalde–Unzu acknowledges financial support from the Spanish Ministry of Education and Science, through the projects ECO2009-11213 and ECO2009-12836. Marc Vorsatz acknowledges financial support from the Spanish Ministry of Education and Science, through the Ramón y Cajal program and project ECO2009-07530.

References

1. Alcalde Unzu, J., Vorsatz, M.: The measurement of consensus: an axiomatic analysis. FEDEA Working Paper 28 (2008)
2. Alcalde–Unzu, J., Vorsatz, M.: Do we agree? measuring the cohesiveness of preferences. FEDEA Working Paper 23 (2010)
3. Alesina, A., Rodrik, D.: Distributive politics and economic growth. Quart. J. Econ. 109, 465–490 (1994)
4. Atkinson, A.: On the measurement of inequality. J. Econ. Theory 2, 244–263 (1970)
5. Arrow, K.: Social choice and individual values, 2nd edn. John Wiley, New York (1963)
6. Baldiga, K., Green, J.: Choice-based measures of conflict in preferences. Harvard University (2009)
7. Bosch, R.: Characterizations of voting rules and consensus measures. Ph. D. Dissertation, Tilburg University, The Netherlands (2006)

---

[5] To our best knowledge, Bosch [7] was the first to analyze some simple measures from an axiomatic point of view.

8. Borda, J.:Mémoire sur les élections au scrutin. Histoire de l'Academie Royale des Sciences, Paris (1781)
9. Condorcet, M.: An essay on the application of probability decision making: An election between three candidates. In: Sommerlad, F., McLean, I. (eds.) The political theory of Condorcet, University of Oxford, Oxford (1989)
10. Cook, W., Seiford, L.: Priority ranking and consensus formation. Management Sci. 24, 1721–1732 (1978)
11. Cook, W., Seiford, L.: On the Borda–Kendall consensus method for priority ranking problems. Management Sci. 28, 621–637 (1982)
12. Cowell, F.: Measuring inequality. In: LSE Perspectives in Economic Analysis. Oxford University Press, Oxford (2008)
13. Duclos, J., Esteban, J., Ray, D.: Polarization: concepts, measurement, estimation. Econometrica 74, 1737–1772 (2006)
14. Eisenberg, J.: Group cohesiveness. In: Baumeister, F., Vohs, K. (eds.) Encyclopedia of social psychology, Thousands Oaks, Sage (2007)
15. Esteban, J., Ray, D.: On the measurement of polarization. Econometrica 62, 819–852 (1994)
16. Gini, C.: Measurement of inequality and incomes. Econ. J. 31, 124–126 (1921)
17. Hays, W.: A note on average tau as a measure of concordance. J. Amer. Stat. Ass. 55, 331–341 (1960)
18. Hutchens, R.: One measure of segregation. Inter. Econ. Rev. 45, 555–578 (2004)
19. Kendall, M.: Rank correlation methods, 3rd edn. Hafner Publishing Company, New York (1962)
20. Kemeny, J.: Mathematics without numbers. Daedalus 88, 577–591 (1959)
21. Klamler, C.: A distance measure for choice functions. Soc. Choice Welfare 30, 419–425 (2008)
22. Kranich, L.: Equitable opportunities: an axiomatic approach. J. Econ. Theory 71, 131–147 (1996)
23. Montalvo, D., Reynal–Querol, M.: Ethnic polarization, political conflict, and civil war. Amer. Econ. Rev. 95, 796–816 (2005)
24. Persson, P., Tabellini, G.: Is inequality harmful for growth? Amer. Econ. Rev. 90, 869–887 (1994)
25. Roemer, J.: Equality of opportunity. Harvard University Press, Boston (1998)

Measuring Consensus in Weak Orders

José Luis García-Lapresta and David Pérez-Román

Abstract. In this chapter we focus our attention in how to measure consensus in groups of agents when they show their preferences over a fixed set of alternatives or candidates by means of weak orders (complete preorders). We have introduced a new class of consensus measures on weak orders based on distances, and we have analyzed some of their properties paying special attention to seven well-known distances.

1 Introduction

Consensus has different meanings. One of them is related to iterative procedures where agents must change their preferences to improve agreement. Usually, a moderator advises agents to modify some opinions (see, for instance, Eklund, Rusinowska and de Swart [12]). However, in this chapter consensus is related to the degree of agreement in a committee, and agents do not need to change their preferences. For an overview about consensus, see Martínez-Panero [22].

From a technical point of view, it is interesting to note that the problem of measuring the concordance or discordance between two linear orders has been widely explored in the literature. In this way, different *rank correlation indices* have been considered for assigning grades of agreement between two rankings (see Kendall and Gibbons [21]). Some of the most important indices in this context are Spearman's rho [31], Kendall's tau [20], and Gini's cograduation index [16]. On the other hand, some natural extensions of the above mentioned indices have been considered

José Luis García-Lapresta
PRESAD Research Group, Dep. de Economía Aplicada, Universidad de Valladolid, Spain
e-mail: lapresta@eco.uva.es

David Pérez-Román
PRESAD Research Group, Dep. de Organización de Empresas y Comercialización
e Investigación de Mercados, Universidad de Valladolid, Spain
e-mail: david@emp.uva.es

E. Herrera-Viedma et al. (Eds.): Consensual Processes, STUDFUZZ 267, pp. 213–234.
springerlink.com   © Springer-Verlag Berlin Heidelberg 2011

for measuring the concordance or discordance among more than two linear orders (see Hays [17] and Alcalde-Unzu and Vorsatz [1, 2]). For details and references, see for instance Borroni and Zenga [6] and Alcalde-Unzu and Vorsatz [1, 2].

In the field of Social Choice, Bosch [7] introduced the notion of *consensus measure* as a mapping that assigns a number between 0 and 1 to every profile of linear orders, satisfying three properties: *unanimity* (in every subgroup of agents, the highest degree of consensus is only reached whenever all individuals have the same ranking), *anonymity* (the degree of consensus is not affected by any permutation of agents) and *neutrality* (the degree of consensus is not affected by any permutation of alternatives).

Recently, Alcalde-Unzu and Vorsatz [1] have introduced some consensus measures in the context of linear orders –related to some of the above mentioned rank correlation indices– and they provide some axiomatic characterizations (see also Alcalde-Unzu and Vorsatz [2]).

In this chapter[1] we extend Bosch's notion of consensus measure to the context of weak orders (indifference among different alternatives is allowed)[2], and we consider some additional properties that such measures could fulfill: *maximum dissension* (in each subset of two agents, the minimum consensus is only reached whenever preferences of agents are linear orders and each one is the inverse of the other), *reciprocity* (if all individual weak orders are reversed, then the consensus does not change) and *homogeneity* (if we replicate a subset of agents, then the consensus in that group does not change). After that, we introduce a class of consensus measures based on the distances among individual weak orders. We pay special attention to seven specific metrics: discrete, Manhattan, Euclidean, Chebyshev, cosine, Hellinger, and Kemeny.

The chapter is organized as follows. Section 2 is devoted to introduce basic terminology and distances used along the chapter. In Section 3 we introduce consensus measures and we analyze their properties. An Appendix contains the most technical proofs.

2 Preliminaries

Consider a set of agents $V = \{v_1, \ldots, v_m\}$ ($m \geq 2$) who show their preferences on a set of alternatives $X = \{x_1, \ldots, x_n\}$ ($n \geq 2$). With $L(X)$ we denote the set of *linear orders* on X, and with $W(X)$ the set of *weak orders* (or *complete preorders*) on X. Given $R \in W(X)$, the *inverse* of R is the weak order R^{-1} defined by $x_i R^{-1} x_j \Leftrightarrow x_j R x_i$, for all $x_i, x_j \in X$.

---

[1] A preliminary study can be found in García-Lapresta and Pérez-Román [15].

[2] Recently, García-Lapresta [14] has introduced a class of agreement measures in the context of weak orders when agents classify alternatives within a finite scale defined by linguistic categories with associated scores. These measures are based on distances among individual and collective scores generated by an aggregation operator.

A *profile* is a vector $\mathbf{R} = (R_1,\ldots,R_m)$ of weak or linear orders, where R_i contains the preferences of the agent v_i, with $i = 1,\ldots,m$. Given a profile $\mathbf{R} = (R_1,\ldots,R_m)$, we denote $\mathbf{R}^{-1} = (R_1^{-1},\ldots,R_m^{-1})$.

Given a permutation π on $\{1,\ldots,m\}$ and $\emptyset \neq I \subseteq V$, we denote $\mathbf{R}_\pi = (R_{\pi(1)},\ldots,R_{\pi(m)})$ and $I_\pi = \{v_{\pi^{-1}(i)} \mid v_i \in I\}$, i.e., $v_j \in I_\pi \Leftrightarrow v_{\pi(j)} \in I$.

Given a permutation σ on $\{1,\ldots,n\}$, we denote by $\mathbf{R}^\sigma = (R_1^\sigma,\ldots,R_m^\sigma)$ the profile obtained from $\mathbf{R}$ by relabeling the alternatives according to σ, i.e., $x_i R_k x_j \Leftrightarrow x_{\sigma(i)} R_k^\sigma x_{\sigma(j)}$ for all $i,j \in \{1,\ldots,n\}$ and $k \in \{1,\ldots,m\}$.

Cardinality of any subset I is denoted by $|I|$. With $\mathscr{P}(V)$ we denote the power set of V, i.e., $I \in \mathscr{P}(V) \Leftrightarrow I \subseteq V$; and we also use $\mathscr{P}_2(V) = \{I \in \mathscr{P}(V) \mid |I| \geq 2\}$.

2.1 Codification of Weak Orders

We now introduce a system for codifying linear and weak orders by means of vectors that represent the relative position of each alternative in the corresponding order.

Given $R \in L(X)$, we codify R with a vector $(a_1,\ldots,a_n) \in \{1,\ldots,n\}^n$, where a_j is the position of alternative x_j in R. Notice that R is completely determined by $(a_1,\ldots,a_n)$.

There does not exist a unique system for codifying weak orders. We propose one based on linearizing the weak order and to assign each alternative the average of the positions of the alternatives within the same equivalence class[3].

Given $R \in W(X)$, let $\succ$ and $\sim$ the asymmetric and the symmetric parts of R, respectively. We assign the position of each alternative in R through the mapping $o_R : X \longrightarrow \mathbb{R}$ defined as

$$o_R(x_j) = n - \left|\{x_i \in X \mid x_j \succ x_i\}\right| - \frac{1}{2}\left|\{x_i \in X \setminus \{x_j\} \mid x_j \sim x_i\}\right|.$$

We denote $\mathbf{o}_R = (o_R(x_1),\ldots,o_R(x_n))$ and, depending on the context, $R \equiv \mathbf{o}_R$ or $\mathbf{o}_R \equiv R$.

Example 1. Consider $R \in W(\{x_1,\ldots,x_7\})$ given in the following table:

R
$x_2\ x_3\ x_5$
x_1
$x_4\ x_7$
x_6

Then,

---

[3] Similar procedures have been considered in the generalization of scoring rules from linear orders to weak orders (see Smith [30], Black [5] and Cook and Seiford [9], among others).

$$o_R(x_2) = o_R(x_3) = o_R(x_5) = \frac{1+2+3}{3} = 7 - 4 - \frac{1}{2}2 = 2$$

$$o_R(x_1) = 7 - 3 - \frac{1}{2}0 = 4$$

$$o_R(x_4) = o_R(x_7) = \frac{5+6}{2} = 7 - 1 - \frac{1}{2}1 = 5.5$$

$$o_R(x_6) = 7 - 0 - \frac{1}{2}0 = 7.$$

Consequently, R is codified by $(4, 2, 2, 5.5, 2, 7, 5.5)$.

Remark 1. If $(a_1, \ldots, a_n) \equiv R \in W(X)$, then $R^{-1} \equiv (n+1-a_1, \ldots, n+1-a_n)$.

Remark 2. For every $R \in W(X)$, it holds

1. $o_R(x_j) \in \{1, 1.5, 2, 2.5, \ldots, n - 0.5, n\}$ for every $j \in \{1, \ldots, n\}$.

2. $\sum_{j=1}^{n} o_R(x_j) = 1 + 2 + \cdots + n = \frac{n(n+1)}{2}$.

Definition 1. We denote by A_W the set of vectors that codify weak orders, i.e.,

$$A_W = \{(a_1, \ldots, a_n) \in \mathbb{R}^n \mid (a_1, \ldots, a_n) \equiv R \text{ for some } R \in W(X)\}.$$

Remark 3. The mapping $o : W(X) \longrightarrow A_W$ that assigns o_R to each $R \in W(X)$ is a bijection. Thus, we can identify $W(X)$ and A_W.

In Proposition 1, we provide a complete characterization of the vectors that codify weak orders.

Given $(a_1, \ldots, a_n) \equiv R \in W(X)$, we denote $M_i(R) = \{m \in \{1, \ldots, n\} \mid a_m = a_i\}$, for $i = 1, \ldots, n$. Given a permutation σ on $\{1, \ldots, n\}$, we denote $R^\sigma \equiv (a_1^\sigma, \ldots, a_n^\sigma)$, with $a_i^\sigma = a_{\sigma(i)}$.

Proposition 1. *Given* $(a_1, \ldots, a_n) \in \mathbb{R}^n$, $(a_1, \ldots, a_n) \equiv R \in W(X)$ *if and only if there exists a permutation* σ *on* $\{1, \ldots, n\}$ *such that* R^σ *satisfies the following conditions:*

1. $a_1^\sigma \leq \cdots \leq a_n^\sigma$.

2. $a_1^\sigma + \cdots + a_n^\sigma = \frac{n(n+1)}{2}$.

3. *For all* $i \in \{1, \ldots, n\}$ *and* $m \in M_i(R^\sigma)$ *it holds*

$$a_m^\sigma = \sum_{l=0}^{k-1} \frac{j+l}{k} = j + \frac{k-1}{2},$$

where $j = \min M_i(R^\sigma)$ *and* $k = |M_i(R^\sigma)|$.

Proof. See the Appendix. □

Remark 4. Notice that if $|M_i(R)| = 1$ for every $i \in \{1, \ldots, n\}$, then $R \in L(X)$. Moreover, if $R \in L(X)$, then there exists a unique permutation σ such that $a_i^\sigma = i$ for every $i \in \{1, \ldots, n\}$.

Remark 5. A_W is stable under permutations, i.e., for every permutation σ on $\{1,\dots,n\}$, if $(a_1,\dots,a_n) \in A_W$, then $(a_1^\sigma,\dots,a_n^\sigma) \in A_W$.

Example 2. Consider $R \in W(\{x_1,\dots,x_8\})$ given in the following table:

$$
\begin{array}{c}
\hline
R \\
\hline
x_3 \\
x_1 \; x_6 \\
x_4 \\
x_5 \; x_7 \; x_8 \\
x_2 \\
\end{array}
$$

Then, $R \equiv (2.5, 8, 1, 4, 6, 2.5, 6, 6)$.

Let σ be the permutation on $\{1,\dots,8\}$ represented by $\begin{pmatrix} 1\,2\,3\,4\,5\,6\,7\,8 \\ 3\,1\,6\,4\,5\,7\,8\,2 \end{pmatrix}$, i.e., $\sigma(1) = 3$, $\sigma(2) = 1$, ..., $\sigma(8) = 2$. Then,

$$
\begin{array}{c}
\hline
R^\sigma \\
\hline
x_1^\sigma \\
x_2^\sigma \; x_3^\sigma \\
x_4^\sigma \\
x_5^\sigma \; x_6^\sigma \; x_7^\sigma \\
x_8^\sigma \\
\end{array}
$$

and, consequently, $R^\sigma \equiv (1, 2.5, 2.5, 4, 6, 6, 6, 8)$.

1. $a_i^\sigma \le a_j^\sigma$, for $1 \le i < j \le 8$.
2. $a_1^\sigma + \cdots + a_8^\sigma = \frac{8(8+1)}{2} = 36$.
3. For $i = 7$, we have $M_7(R^\sigma) = \{m \in \{1,\dots,8\} \mid a_m^\sigma = a_7^\sigma\} = \{5,6,7\}$, $j = \min\{M_7(R^\sigma)\} = 5$, $k = |M_7(R^\sigma)| = 3$ and

$$
a_5^\sigma = a_6^\sigma = a_7^\sigma = \sum_{l=0}^{3-1} \frac{5+l}{2} = 6 = 5 + \frac{3-1}{2} = j + \frac{k-1}{2}.
$$

Definition 2. $W_\le(X) = \{R \in W(X) \mid R \equiv (a_1,\dots,a_n) \text{ and } a_1 \le \cdots \le a_n\}$.

Remark 6. By Proposition 1, for every $R \in W(X)$ there exists some permutation σ on $\{1,\dots,n\}$ such that $R^\sigma \in W_\le(X)$. Notice that if $R \in L(X)$, then σ is unique, but if $R \in W(X) \setminus L(X)$, then there exist more than one σ satisfying $R^\sigma \in W_\le(X)$.

Lemma 1. *For all* $(a_1,\dots,a_n) \equiv R_1 \in W(X)$ *and* $(b_1,\dots,b_n) \equiv R_2 \in W(X)$, *there exists a permutation* σ *on* $\{1,\dots,n\}$ *such that:*

1. $a_1^\sigma \le \cdots \le a_n^\sigma$.
2. *For every* $i \in \{1,\dots,n\}$ *such that* $|M_i(R_1^\sigma)| > 1$, *if* $j = \min M_i(R_1^\sigma)$ *and* $k = |M_i(R_1^\sigma)|$, *then* $b_j^\sigma \ge \cdots \ge b_{j+k-1}^\sigma$.

Proof. See the Appendix.   □

Example 3. In order to illustrate Lemma 1, consider $R_1, R_2 \in W(\{x_1, \ldots, x_8\})$ given in the following tables:

R_1	R_2
x_3	x_4
$x_1 \, x_4 \, x_6$	$x_5 \, x_8$
x_8	x_7
$x_5 \, x_7$	x_1
x_2	x_3
	$x_2 \, x_6$

Then, $R_1 \equiv (3, 8, 1, 3, 6.5, 3, 6.5, 5)$ and $R_2 \equiv (5, 7.5, 6, 1, 2.5, 7.5, 4, 2.5)$. Let σ' be the permutation on $\{1, \ldots, 8\}$ represented by $\begin{pmatrix} 1 \ 2 \ 3 \ 4 \ 5 \ 6 \ 7 \ 8 \\ 3 \ 1 \ 4 \ 6 \ 8 \ 5 \ 7 \ 2 \end{pmatrix}$. Then,

$$R_1^{\sigma'} \equiv (1, \ 3, \ 3, \ 3, \ \ 5, \ 6.5, \ 6.5, \ 8),$$
$$R_2^{\sigma'} \equiv (6, \ 5, \ 1, \ 7.5, \ 2.5, \ 2.5, \ 4, \ 7.5).$$

Let σ'' be the permutation on $\{1, \ldots, 8\}$ represented by $\begin{pmatrix} 1 \ 2 \ 3 \ 4 \ 5 \ 6 \ 7 \ 8 \\ 1 \ 4 \ 2 \ 3 \ 5 \ 7 \ 6 \ 8 \end{pmatrix}$. It is clear that $R_1^{\sigma'} = (R_1^{\sigma'})^{\sigma''}$. Thus, if $\sigma = \sigma' \cdot \sigma''$, i.e., $\sigma(i) = \sigma'(\sigma''(i))$, then σ is represented by $\begin{pmatrix} 1 \ 2 \ 3 \ 4 \ 5 \ 6 \ 7 \ 8 \\ 3 \ 6 \ 1 \ 4 \ 8 \ 7 \ 5 \ 2 \end{pmatrix}$. Therefore,

$$R_1^{\sigma} \equiv (1, \ \ 3, \ \ 3, 3, \ 5, \ 6.5, 6.5, \ 8),$$
$$R_2^{\sigma} \equiv (6, \ 7.5, 5, 1, \ 2.5, \ 4, \ 2.5, 7.5).$$

2.2 Distances

The use of distances for designing and analyzing voting system has been widely considered in the literature. On this, see Kemeny [18], Slater [29], Nitzan [26], Baigent [3, 4], Nurmi [27, 28], Meskanen and Nurmi [23, 24], Monjardet [25], Gaertner [13, 6.3] and Eckert and Klamler [11], among others. A general and complete survey on distances can be found in Deza and Deza [10].

The consensus measures introduced in this chapter are based on distances on weak orders. After presenting the general notion, we show the distances on $\mathbb{R}^n$ used for inducing the distances on weak orders. We pay special attention to the Kemeny distance.

Definition 3. A *distance* (or *metric*) on a set $A \neq \emptyset$ is a mapping $d : A \times A \longrightarrow \mathbb{R}$ satisfying the following conditions for all $a, b, c \in A$:

1. $d(a, b) \geq 0$.
2. $d(a, b) = 0 \Leftrightarrow a = b$.

3. $d(a,b) = d(b,a)$.
4. $d(a,b) \leq d(a,c) + d(c,b)$.

2.2.1 Distances on $\mathbb{R}^n$

Example 4. Typical examples of distances on $\mathbb{R}^n$ or $[0,\infty)^n$ are the following:

1. The *discrete* distance $d' : \mathbb{R}^n \times \mathbb{R}^n \longrightarrow \mathbb{R}$,

$$d'((a_1,\ldots,a_n),(b_1,\ldots,b_n)) = \begin{cases} 1, & \text{if } (a_1,\ldots,a_n) \neq (b_1,\ldots,b_n), \\ 0, & \text{if } (a_1,\ldots,a_n) = (b_1,\ldots,b_n). \end{cases}$$

2. The *Minkowski* distance $d_p : \mathbb{R}^n \times \mathbb{R}^n \longrightarrow \mathbb{R}$, with $p \geq 1$,

$$d_p((a_1,\ldots,a_n),(b_1,\ldots,b_n)) = \left(\sum_{i=1}^{n} |a_i - b_i|^p \right)^{\frac{1}{p}}.$$

For $p = 1$ and $p = 2$ we obtain the *Manhattan* and *Euclidean* distances, respectively.

3. The *Chebyshev* distance $d_\infty : \mathbb{R}^n \times \mathbb{R}^n \longrightarrow \mathbb{R}$,

$$d_\infty((a_1,\ldots,a_n),(b_1,\ldots,b_n)) = \max \left\{ |a_1 - b_1|,\ldots,|a_n - b_n| \right\}.$$

4. The *cosine* distance $d_c : \mathbb{R}^n \times \mathbb{R}^n \longrightarrow \mathbb{R}$,

$$d_c((a_1,\ldots,a_n),(b_1,\ldots,b_n)) = 1 - \frac{\sum\limits_{i=1}^{n} a_i b_i}{\sqrt{\sum\limits_{i=1}^{n} a_i^2} \sqrt{\sum\limits_{i=1}^{n} b_i^2}}.$$

5. The *Hellinger* distance $d_H : [0,\infty)^n \times [0,\infty)^n \longrightarrow \mathbb{R}$,

$$d_H((a_1,\ldots,a_n),(b_1,\ldots,b_n)) = \left(\sum_{i=1}^{n} \left(\sqrt{a_i} - \sqrt{b_i} \right)^2 \right)^{\frac{1}{2}}.$$

Definition 4. Given $D \subseteq \mathbb{R}^n$ stable under permutations, a distance $d : D \times D \longrightarrow \mathbb{R}$ is *neutral* if for every permutation σ on $\{1,\ldots,n\}$, it holds

$$d\left((a_1^\sigma,\ldots,a_n^\sigma),(b_1^\sigma,\ldots,b_n^\sigma)\right) = d\left((a_1,\ldots,a_n),(b_1,\ldots,b_n)\right),$$

for all $(a_1,\ldots,a_n),(b_1,\ldots,b_n) \in D$.

Remark 7. All the distances introduced in Example 4 are neutral.

2.2.2 Distances on Weak Orders

We now introduce a direct way of defining distances on weak orders. They are induced by distances on $\mathbb{R}^n$ by considering the position vectors. It is important to note that for our proposal we only need distances on subsets $D \subseteq \mathbb{R}^n$ such that $A_W \subseteq D$.

Definition 5. Given $D \subseteq \mathbb{R}^n$ such that $A_W \subseteq D$ and a distance $d : D \times D \longrightarrow \mathbb{R}$, the *distance on $W(X)$ induced by d* is the mapping $\bar{d} : W(X) \times W(X) \longrightarrow \mathbb{R}$ defined by

$$\bar{d}(R_1, R_2) = d\big((o_{R_1}(x_1), \ldots, o_{R_1}(x_n)), (o_{R_2}(x_1), \ldots, o_{R_2}(x_n))\big) ,$$

for all $R_1, R_2 \in W(X)$.

Given a distance d_- on $D \subseteq \mathbb{R}^n$ such that $A_W \subseteq D$, we use $\bar{d}_-$ to denote the distance on $W(X)$ induced by d_-.

2.2.3 The Kemeny Distance

The Kemeny distance was initially defined on linear orders by Kemeny [18], as the sum of pairs where the orders' preferences disagree. However, it has been generalized to the framework of weak orders (see Cook, Kress and Seiford [8] and Eckert and Klamler [11], among others).

The *Kemeny* distance on weak orders $d^K : W(X) \times W(X) \longrightarrow \mathbb{R}$ is usually defined as the cardinality of the symmetric difference between the weak orders, i.e.,

$$d^K(R_1, R_2) = |(R_1 \cup R_2) \setminus (R_1 \cap R_2)| .$$

We now consider $d_K : A_W \times A_W \longrightarrow \mathbb{R}$, given by

$$d_K\big((a_1, \ldots, a_n), (b_1, \ldots, b_n)\big) = \sum_{\substack{i,j=1 \\ i<j}}^{n} |\operatorname{sgn}(a_i - a_j) - \operatorname{sgn}(b_i - b_j)| ,$$

where sgn is the *sign function*:

$$\operatorname{sgn}(a) = \begin{cases} 1, & \text{if } a > 0, \\ 0, & \text{if } a = 0, \\ -1, & \text{if } a < 0. \end{cases}$$

Notice that d_K is a neutral distance on A_W.

Taking into account Kemeny and Snell [19, p. 18], it is easy to see that d^K coincides with the distance on $W(X)$ induced by d_K, i.e.,

$$d^K(R_1, R_2) = \bar{d}_K(R_1, R_2) = d_K\big((a_1, \ldots, a_n), (b_1, \ldots, b_n)\big) =$$

$$= \sum_{\substack{i,j=1 \\ i<j}}^{n} |\operatorname{sgn}(a_i - a_j) - \operatorname{sgn}(b_i - b_j)| ,$$

where $(a_1, \ldots, a_n) \equiv R_1 \in W(X)$ and $(b_1, \ldots, b_n) \equiv R_2 \in W(X)$.

3   Consensus Measures

Consensus measures have been introduced and analyzed by Bosch [7] in the context of linear orders. We now extend this concept to the framework of weak orders.

Definition 6. A *consensus measure* on $W(X)^m$ is a mapping

$$\mathcal{M} : W(X)^m \times \mathcal{P}_2(V) \longrightarrow [0,1]$$

that satisfies the following conditions:

1. *Unanimity.* For all $\boldsymbol{R} \in W(X)^m$ and $I \in \mathcal{P}_2(V)$, it holds

$$\mathcal{M}(\boldsymbol{R},I) = 1 \Leftrightarrow R_i = R_j \text{ for all } v_i, v_j \in I.$$

2. *Anonymity.* For all permutation π on $\{1,\dots,m\}$, $\boldsymbol{R} \in W(X)^m$ and $I \in \mathcal{P}_2(V)$, it holds

$$\mathcal{M}(\boldsymbol{R}_\pi, I_\pi) = \mathcal{M}(\boldsymbol{R},I) .$$

3. *Neutrality.* For all permutation σ on $\{1,\dots,n\}$, $\boldsymbol{R} \in W(X)^m$ and $I \in \mathcal{P}_2(V)$, it holds

$$\mathcal{M}(\boldsymbol{R}^\sigma, I) = \mathcal{M}(\boldsymbol{R},I) .$$

Unanimity means that the maximum consensus in every subset of decision makers is only achieved when all opinions are the same. Anonymity requires symmetry with respect to decision makers, and neutrality means symmetry with respect to alternatives.

We now introduce other properties that a consensus measure may satisfy.

Definition 7. Let $\mathcal{M} : W(X)^m \times \mathcal{P}_2(V) \longrightarrow [0,1]$ be a consensus measure.

1. $\mathcal{M}$ satisfies *maximum dissension* if for all $\boldsymbol{R} \in W(X)^m$ and $v_i, v_j \in V$ such that $i \neq j$, it holds

$$\mathcal{M}(\boldsymbol{R}, \{v_i, v_j\}) = 0 \Leftrightarrow R_i, R_j \in L(X) \text{ and } R_j = R_i^{-1} .$$

2. $\mathcal{M}$ is *reciprocal* if for all $\boldsymbol{R} \in W(X)^m$ and $I \in \mathcal{P}_2(V)$, it holds

$$\mathcal{M}(\boldsymbol{R}^{-1}, I) = \mathcal{M}(\boldsymbol{R}, I) .$$

3. $\mathcal{M}$ is *homogeneous* if for all $\boldsymbol{R} \in W(X)^m$, $I \in \mathcal{P}_2(V)$ and $t \in \mathbb{N}$, it holds

$$\mathcal{M}^t(t\boldsymbol{R}, tI) = \mathcal{M}(\boldsymbol{R}, I) ,$$

where $\mathcal{M}^t : W(X)^{tm} \times \mathcal{P}_2(tV) \longrightarrow [0,1]$, $t\boldsymbol{R} = (\boldsymbol{R}, \overset{t}{.\,.\,.}, \boldsymbol{R}) \in W(X)^{tm}$ is the profile defined by t copies of $\boldsymbol{R}$ and $tI = I \uplus \overset{t}{\cdots} \uplus I$ is the multiset of agents[4] defined by t copies of I.

---

[4] List of agents where each agent occurs as many times as the multiplicity. For instance, $2\{v_1, v_2\} = \{v_1, v_2\} \uplus \{v_1, v_2\} = \{v_1, v_2, v_1, v_2\}$.

Maximum dissension means that in each subset of two agents[5], the minimum consensus is only reached whenever preferences of agents are linear orders and each one is the inverse of the other. Reciprocity means that if all individual weak orders are reversed, then the consensus does not change. And homogeneity means that if we replicate a subset of agents, then the consensus in that group does not change.

It is important to note that reciprocity and homogeneity are not always desirable properties. Under reciprocity, disagreements in the top alternatives have the same effect than in the bottom ones. With homogeneity, a society polarized in two groups of the same size with opposite preferences has no consensus at all (see footnote 5).

We now introduce our proposal for measuring consensus in sets of weak orders.

Definition 8. Given a distance $\bar{d} : W(X) \times W(X) \longrightarrow \mathbb{R}$, the mapping

$$\mathscr{M}_{\bar{d}} : W(X)^m \times \mathscr{P}_2(V) \longrightarrow [0,1]$$

is defined by

$$\mathscr{M}_{\bar{d}}(\boldsymbol{R},I) = 1 - \frac{\displaystyle\sum_{\substack{v_i,v_j \in I \\ i<j}} \bar{d}(R_i,R_j)}{\dbinom{|I|}{2} \cdot \Delta_n},$$

where

$$\Delta_n = \max\left\{\bar{d}(R_i,R_j) \mid R_i,R_j \in W(X)\right\}.$$

Notice that the numerator of the quotient appearing in the above expression is the sum of all the distances between the weak orders of the profile, and the denominator is the number of terms in the numerator's sum multiplied by the maximum distance between weak orders. Consequently, that quotient belongs to the unit interval and it measures the disagreement in the profile.

Proposition 2. *For every distance* $\bar{d} : W(X) \times W(X) \longrightarrow \mathbb{R}$, $\mathscr{M}_{\bar{d}}$ *satisfies unanimity and anonymity.*

Proof. Let $\boldsymbol{R} \in W(X)^m$ and $I \in \mathscr{P}_2(V)$.

1. Unanimity.

$$\mathscr{M}_{\bar{d}}(\boldsymbol{R},I) = 1 \quad \Leftrightarrow \quad \sum_{\substack{v_i,v_j \in I \\ i<j}} \bar{d}(R_i,R_j) = 0 \quad \Leftrightarrow$$

$$\forall v_i,v_j \in I \quad \bar{d}(R_i,R_j) = 0 \quad \Leftrightarrow \quad \forall v_i,v_j \in I \quad R_i = R_j.$$

---

[5] It is clear that a society reach maximum consensus when all the opinions are the same. However, in a society with more than two members it is not an obvious issue to determine when there is minimum consensus (maximum disagreement).

2. Anonymity. Let π be a permutation on $\{1,\dots,m\}$.

$$\sum_{\substack{v_i,v_j\in I_\pi \\ i<j}} \bar{d}(R_{\pi(i)},R_{\pi(j)}) = \sum_{\substack{v_{\pi(i)},v_{\pi(j)}\in I \\ \pi(i)<\pi(j)}} \bar{d}(R_{\pi(i)},R_{\pi(j)}) = \sum_{\substack{v_i,v_j\in I \\ i<j}} \bar{d}(R_i,R_j) \,.$$

Thus, $\mathcal{M}_{\bar{d}}(\boldsymbol{R}_\pi,I_\pi) = \mathcal{M}_{\bar{d}}(\boldsymbol{R},I)$. $\square$

If $\mathcal{M}_{\bar{d}}$ is neutral, then we say that $\mathcal{M}_{\bar{d}}$ is the *consensus measure associated with* $\bar{d}$.

Proposition 3. *If* $d : A_W \times A_W \longrightarrow \mathbb{R}$ *is a neutral distance, then* $\mathcal{M}_{\bar{d}}$ *is a consensus measure.*

Proof. By Proposition 2, $\mathcal{M}_{\bar{d}}$ satisfies unanimity and anonymity. Obviously, if $\bar{d}$ is neutral, then $\mathcal{M}_{\bar{d}}$ is neutral and thus $\mathcal{M}_{\bar{d}}$ is a consensus measure. $\square$

Proposition 4. *If* $\bar{d}$ *is the distance on* $W(X)$ *induced by* d', d_1, d_2, d_∞, d_c *or* d_K, *then* $\mathcal{M}_{\bar{d}}$ *is a reciprocal consensus measure. However, for the Hellinger distance,* $\mathcal{M}_{\bar{d}_H}$ *is not a reciprocal consensus measure.*

Proof. See the Appendix. $\square$

In order to prove that the maximum dissension property is satisfied for some of the consensus measures introduced above, we need two lemmas (see the Appendix). With Lemma 2 we establish that, for the distances induced by d_2, d_c, d_H or d_K, the maximum distance between weak orders is not reached when one of the weak orders is not linear. Lemma 3 ensures that, for the same distances, the maximum distance between linear orders is not reached when they are not each one inverse of the other.

Proposition 5. *If* $\bar{d}_-$ *is the distance induced by* d_2, d_c, d_H *or* d_K, *then* $\mathcal{M}_{\bar{d}_-}$ *satisfies the maximum dissension property. However, if* $\bar{d}_-$ *is the distance induced by* d', d_1 *or* d_∞, *then* $\mathcal{M}_{\bar{d}_-}$ *does not satisfy the maximum dissension property.*

Proof. See the Appendix. $\square$

As presented in the following result, none of the introduced consensus measures is homogeneous. In García-Lapresta and Pérez-Román [15] we introduced the Borda consensus measures, based on the Euclidean distance. We note that they are homogeneous and reciprocal, but they do not satisfy the maximum dissension property.

Proposition 6. *The consensus measure* $\mathcal{M}_{\bar{d}}$ *is not homogeneous for any distance* $\bar{d}$ *on* $W(X)$.

Proof. Let $\boldsymbol{R} = (R_1,R_2) \in W(X)^2$ such that $R_1 \neq R_2$ and $I = \{v_1,v_2\} \in \mathscr{P}_2(V)$. We now consider $2\boldsymbol{R} = (R'_1,R'_2,R'_3,R'_4)$, where $R'_1 = R'_3 = R_1$ and $R'_2 = R'_4 = R_2$, and $2I = \{v'_1,v'_2,v'_3,v'_4\}$, with $v'_1 = v'_3 = v_1$ and $v'_2 = v'_4 = v_2$. Then, we have

$$\mathcal{M}_{\bar{d}}^2(2\boldsymbol{R}, 2I) =$$

$$= 1 - \frac{\bar{d}(R_1', R_2') + d(R_1', R_3') + d(R_1', R_4') + d(R_2', R_3') + d(R_2', R_4') + d(R_3', R_4')}{\binom{|2I|}{2} \cdot \Delta_n} =$$

$$= 1 - \frac{4 \cdot \bar{d}(R_1, R_2)}{6 \cdot \Delta_n} = 1 - \frac{2 \cdot \bar{d}(R_1, R_2)}{3 \cdot \Delta_n} \neq 1 - \frac{\bar{d}(R_1, R_2)}{\Delta_n} = \mathcal{M}_{\bar{d}}(\boldsymbol{R}, I) . \qquad \square$$

Proposition 7. *For any distance $\bar{d}$ on $W(X)$, if $\boldsymbol{R} = (R_1, R_2) \in W(X)^2$ is such that $\bar{d}(R_1, R_2) = \max\{\bar{d}(R_i, R_j) \mid R_i, R_j \in W(X)\}$, then it holds:*

$$\lim_{t \to \infty} \mathcal{M}_{\bar{d}}^t(t\boldsymbol{R}, tI) = \frac{1}{2}.$$

Proof. See the Appendix. $\square$

Remark 8. Homogeneity ensures that a society has no consensus at all when it is divided into two groups of the same size each one ranks order the alternatives just in the opposite way to the other group. According to Proposition 6, our consensus measures are not homogeneous. Thus, they perceive some consensus in polarized societies and, by Proposition 7, consensus tends to 0.5 when the number of agents tends to infinity, regardless of the distance used. Notice that this result holds even when the distance used does not verify the maximum dissension property, which guaranteed that the consensus between two profiles is zero if and only if they are opposites.

We summarize the properties of the analyzed consensus measures in Table 1.

Table 1 Summary

	Max. diss.	Reciproc.	Homogen.
$\mathcal{M}_{\bar{d}'}$	No	Yes	No
$\mathcal{M}_{\bar{d}_1}$	No	Yes	No
$\mathcal{M}_{\bar{d}_2}$	Yes	Yes	No
$\mathcal{M}_{\bar{d}_\infty}$	No	Yes	No
$\mathcal{M}_{\bar{d}_c}$	Yes	Yes	No
$\mathcal{M}_{\bar{d}_H}$	Yes	No	No
$\mathcal{M}_{\bar{d}_K}$	Yes	Yes	No

Acknowledgements

This research has been partially supported by the Spanish Ministerio de Ciencia e Innovación (Project ECO2009–07332), ERDF and Junta de Castilla y León (Consejería de Educación, Projects VA092A08 and GR99). The authors are grateful to Jorge Alcalde-Unzu, Miguel Ángel Ballester, Christian Klamler and Miguel Martínez-Panero for their suggestions and comments.

References

1. Alcalde-Unzu, J., Vorsatz, M.: Do we agree? Measuring the cohesiveness of preferences. University of the Basque Country Working Paper 2010/28 (2010)
2. Alcalde-Unzu, J., Vorsatz, M.: Measuring Consensus: Concepts, Comparisons, and Properties. In: Herrera-Viedma, E., et al. (eds.) Consensual Processes. STUDFUZZ, vol. 267, pp. 195–211. Springer, Heidelberg (2011)
3. Baigent, N.: Preference proximity and anonymous social choice. The Quarterly Journal of Economics 102, 161–169 (1987)
4. Baigent, N.: Metric rationalisation of social choice functions according to principles of social choice. Mathematical Social Sciences 13, 59–65 (1987)
5. Black, D.: Partial justification of the Borda count. Public Choice 28, 1–15 (1976)
6. Borroni, C.G., Zenga, M.: A test of concordance based on Gini's mean difference. Statistical Methods and Applications 16, 289–308 (2007)
7. Bosch, R.: Characterizations of Voting Rules and Consensus Measures. Ph. D. Dissertation, Tilburg University (2005)
8. Cook, W.D., Kress, M., Seiford, L.M.: A general framework for distance-based consensus in ordinal ranking models. European Journal of Operational Research 96, 392–397 (1996)
9. Cook, W.D., Seiford, L.M.: On the Borda-Kendall consensus method for priority ranking problems. Management Science 28, 621–637 (1982)
10. Deza, M.M., Deza, E.: Encyclopedia of Distances. Springer, Berlin (2009)
11. Eckert, D., Klamler, C.: Distance-Based Aggregation Theory. In: Herrera-Viedma, E., et al. (eds.) Consensual Processes. STUDFUZZ, vol. 267, pp. 3–22. Springer, Heidelberg (2011)
12. Eklund, P., Rusinowska, A., de Swart, H.: Consensus reaching in committees. European Journal of Operational Research 178, 185–193 (2007)
13. Gaertner, W.: A Primer in Social Choice Theory, Revised Edition. Oxford University Press, Oxford (2009)
14. García-Lapresta, J.L.: Favoring consensus and penalizing disagreement in group decision making. Journal of Advanced Computational Intelligence and Intelligent Informatics 12(5), 416–421 (2008)
15. García-Lapresta, J.L., Pérez-Román, D.: Some consensus measures and their applications in group decision making. In: Ruan, D., Montero, J., Lu, J., Martínez, L., D'hondt, P., Kerre, E.E. (eds.) Computational Intelligence in Decision and Control, pp. 611–616. World Scientific, Singapore (2008)
16. Gini, C.: Corso di Statistica. Veschi, Rome (1954)
17. Hays, W.L.: A note on average tau as a measure of concordance. Journal of the American Statistical Association 55, 331–341 (1960)
18. Kemeny, J.G.: Mathematics without numbers. Daedalus 88, 571–591 (1959)
19. Kemeny, J.G., Snell, J.C.: Mathematical Models in the Social Sciences. Ginand Company, New York (1961)
20. Kendall, M.G.: Rank Correlation Methods. Griffin, London (1962)
21. Kendall, M., Gibbons, J.D.: Rank Correlation Methods. Oxford University Press, New York (1990)
22. Martínez-Panero, M.: Consensus Perspectives: Glimpses into Theoretical Advances and Applications. In: Herrera-Viedma, E., et al. (eds.) Consensual Processes. STUDFUZZ, vol. 267, pp. 179–193. Springer, Heidelberg (2011)
23. Meskanen, T., Nurmi, H.: Analyzing political disagreement. In: JASM Working Papers, University of Turku, Finland (2006)

24. Meskanen, T., Nurmi, H.: Distance from consensus: A theme and variations. In: Simeone, B., Pukelsheim, F. (eds.) Mathematics and Democracy. Recent Advances in Voting Systems and Collective Choice, pp. 117–132. Springer, Heidelberg (2007)
25. Monjardet, B.: "Mathématique Sociale" and Mathematics. "Mathématique Sociale" and Mathematics. A case study: Condorcet's effect and medians 4 (2008)
26. Nitzan, S.: Some measures of closeness to unanimity and their implications. Theory and Decision 13, 129–138 (1981)
27. Nurmi, H.: A comparison of some distance-based choice rules in ranking environments. Theory and Decision 57, 5–24 (2004)
28. Nurmi, H.: Settings of Consensual Processes: Candidates, Verdicts, Policies. In: Herrera-Viedma, E., et al. (eds.) Consensual Processes. STUDFUZZ, vol. 267, pp. 159–177. Springer, Heidelberg (2011)
29. Slater, P.: Inconsistencies in a schedule of paired comparisons. Biometrika 48, 303–312 (1961)
30. Smith, J.: Aggregation of preferences with variable electorate. Econometrica 41, 1027–1041 (1973)
31. Spearman, C.: The proof and measurement of association between two things. American Journal of Psychology 15, 72–101 (1904)

Appendix

Proof of Proposition 1. Consider $(a_1, \ldots a_n) \equiv R \in W(X)$.

1. Obvious.
2. By Remark 2.
3. Consider $i \in \{1, \ldots, n\}$ with $|M_i(R^\sigma)| = k > 1$, and $j = \min M_i(R^\sigma)$. Then, $M_i(R^\sigma) = \{j, j+1, \ldots, j+(k-1)\}$ and $a_j^\sigma = a_{j+1}^\sigma = \cdots = a_{j+(k-1)}^\sigma = a_i^\sigma$.
Since we assign each alternative the average of the positions of the alternatives within the same equivalence class, then we have

$$a_i^\sigma = \frac{\sum\limits_{m \in M_i(R^\sigma)} m}{k} = \frac{j+(j+1)+\cdots+(j+(k-1))}{k} = j + \frac{k-1}{2}.$$

Reciprocally, given $(a_1, \ldots, a_n) \in \mathbb{R}^n$ and a permutation σ verifying conditions 1, 2 and 3, then we can consider that a_i^σ is the relative position of the alternative x_i in $R \in W(X)$. Then, $R \equiv (a_1, \ldots, a_n)$. □

Proof of Lemma 1. By Proposition 1, there exists a permutation σ' on $\{1, \ldots, n\}$ such that $R_1^{\sigma'}$ satisfies 1. We consider $R_2^{\sigma'}$. Let σ'' be a permutation on $\{1, \ldots, n\}$ such that:

1. If $\left|M_i\left(R_1^{\sigma'}\right)\right| = 1$, then $\sigma''(i) = i$.

2. If $\left|M_i\left(R_1^{\sigma'}\right)\right| = k > 1$ and $j = \min M_i\left(R_1^{\sigma'}\right)$, then let σ'' be a permutation on $\{1, \ldots, n\}$ such that $b_{\sigma''(j)}^{\sigma'} \geq b_{\sigma''(j+1)}^{\sigma'} \geq \cdots \geq b_{\sigma''(j+k-1)}^{\sigma'}$. Obviously, $a_{\sigma''(j)}^{\sigma'} = \cdots = a_{\sigma''(j+k-1)}^{\sigma'} = a_i^{\sigma'}$.

Therefore, $\sigma = \sigma' \cdot \sigma''$.   □

Proof of Proposition 4. By Remark 7 and Proposition 3, we only need to prove that the corresponding distances are reciprocal.

1. Case $\bar{d}'$.

   Since $R_i = R_j \Leftrightarrow R_i^{-1} = R_j^{-1}$, we have $\bar{d}'(R_i, R_j) = \bar{d}'(R_i^{-1}, R_j^{-1})$. Consequently, $\mathscr{M}_{\bar{d}'}(\mathbf{R}^{-1}, I) = \mathscr{M}_{\bar{d}'}(\mathbf{R}, I)$.

2. Cases $\bar{d}_p$ $(p \in \{1,2\})$.

   By Remark 1, we have

   $$\bar{d}_p(R_i^{-1}, R_j^{-1}) = \left(\sum_{k=1}^{n} |(n+1 - o_{R_i}(x_k)) - (n+1 - o_{R_j}(x_k))|^p \right)^{\frac{1}{p}} =$$

   $$= \left(\sum_{k=1}^{n} |o_{R_i}(x_k) - o_{R_j}(x_k)|^p \right)^{\frac{1}{p}} = \bar{d}_p(R_i, R_j) .$$

   Thus, $\mathscr{M}_{\bar{d}_p}(\mathbf{R}^{-1}, I) = \mathscr{M}_{\bar{d}_p}(\mathbf{R}, I)$.

3. Case $\bar{d}_\infty$.

   By Remark 1, we have

   $$\bar{d}_\infty(R_i^{-1}, R_j^{-1}) =$$
   $$= \max \left\{ |(n+1 - o_{R_i}(x_k)) - (n+1 - o_{R_j}(x_k))| \mid k \in \{1,\ldots,n\} \right\} =$$
   $$= \max \left\{ |o_{R_i}(x_k) - o_{R_j}(x_k)| \mid k \in \{1,\ldots,n\} \right\} = \bar{d}_\infty(R_i, R_j) .$$

   Thus, $\mathscr{M}_{\bar{d}_\infty}(\mathbf{R}^{-1}, I) = \mathscr{M}_{\bar{d}_\infty}(\mathbf{R}, I)$.

4. Case $\bar{d}_c$.

$$\bar{d}_c(R_i^{-1}, R_j^{-1}) = 1 - \frac{\sum_{k=1}^{n} (n+1 - o_{R_i}(x_k))(n+1 - o_{R_j}(x_k))}{\sqrt{\sum_{k=1}^{n} (n+1 - o_{R_i}(x_k))^2} \sqrt{\sum_{k=1}^{n} (n+1 - o_{R_j}(x_k))^2}} .$$

By Remark 2, we have $\sum_{k=1}^{n} \left((n+1) - (o_{R_i}(x_k) + o_{R_j}(x_k)) \right) = 0$. Thus,

$$\sum_{k=1}^{n} (n+1 - o_{R_i}(x_k))(n+1 - o_{R_j}(x_k)) =$$

$$= (n+1)\left[\sum_{k=1}^{n} \left((n+1) - (o_{R_i}(x_k) + o_{R_j}(x_k))\right)\right] + \sum_{k=1}^{n} o_{R_i}(x_k) o_{R_j}(x_k) =$$

$$= \sum_{k=1}^{n} o_{R_i}(x_k) o_{R_j}(x_k) .$$

By Remark 2, we also have $\displaystyle\sum_{k=1}^{n} ((n+1) - 2o_{R_i}(x_k)) = \sum_{k=1}^{n} ((n+1) - 2o_{R_j}(x_k)) = 0.$

Thus,

$$\sum_{k=1}^{n} (n+1 - o_{R_i}(x_k))^2 = (n+1)\sum_{k=1}^{n} ((n+1) - 2o_{R_i}(x_k)) + \sum_{k=1}^{n} o_{R_i}(x_k)^2 = \sum_{k=1}^{n} o_{R_i}(x_k)^2 ,$$

and

$$\sum_{k=1}^{n} (n+1 - o_{R_j}(x_k))^2 = (n+1)\sum_{k=1}^{n} ((n+1) - 2o_{R_j}(x_k)) + \sum_{k=1}^{n} o_{R_j}(x_k)^2 = \sum_{k=1}^{n} o_{R_j}(x_k)^2 .$$

Consequently,

$$\bar{d}_c(R_i^{-1}, R_j^{-1}) = 1 - \frac{\displaystyle\sum_{k=1}^{n} o_{R_i}(x_k) o_{R_j}(x_k)}{\sqrt{\displaystyle\sum_{k=1}^{n} (o_{R_i}(x_k))^2}\sqrt{\displaystyle\sum_{k=1}^{n} (o_{R_j}(x_k))^2}} = \bar{d}_c(R_i, R_j) .$$

Thus, $\mathscr{M}_{\bar{d}_c}(\mathbf{R}^{-1}, I) = \mathscr{M}_{\bar{d}_c}(\mathbf{R}, I).$

5. Case $\bar{d}_K$.

$$\bar{d}_K(R_1^{-1}, R_2^{-1}) =$$

$$= \sum_{\substack{i,j=1 \\ i<j}}^{n} \left| \mathrm{sgn}\left(n+1 - o_{R_1}(x_i) - (n+1 - o_{R_1}(x_j))\right) - \right.$$

$$\left. - \mathrm{sgn}\left(n+1 - o_{R_2}(x_i) - (n+1 - o_{R_2}(x_j))\right) \right| =$$

$$= \sum_{\substack{i,j=1 \\ i<j}}^{n} \left| \mathrm{sgn}\left(o_{R_1}(x_j) - o_{R_1}(x_i)\right) - \mathrm{sgn}\left(o_{R_2}(x_j) - o_{R_2}(x_i)\right) \right| =$$

$$= \sum_{\substack{i,j=1 \\ i<j}}^{n} \left| \mathrm{sgn}\left(o_{R_1}(x_i) - o_{R_1}(x_j)\right) - \mathrm{sgn}\left(o_{R_2}(x_i) - o_{R_2}(x_j)\right) \right| = \bar{d}_K(R_1, R_2) .$$

Thus, $\mathscr{M}_{\bar{d}_K}(\mathbf{R}^{-1}, I) = \mathscr{M}_{\bar{d}_K}(\mathbf{R}, I).$

6. Case $\bar{d}_H$.

Let us consider $R_1, R_2 \in W(\{x_1, x_2, x_3\})$:

R_1	R_2	R_1^{-1}	R_2^{-1}
x_1	$x_1\ x_2\ x_3$	$x_2\ x_3$	$x_1\ x_2\ x_3$
$x_2\ x_3$		x_1	

The above weak orders are codified by $R_1 \equiv (1, 2.5, 2.5)$, $R_1^{-1} \equiv (3, 1.5, 1.5)$ and $R_2 = R_2^{-1} \equiv (2, 2, 2)$. We have

$$\bar{d}_H(R_1, R_2) =$$

$$= \left(\left(\sqrt{1} - \sqrt{2} \right)^2 + \left(\sqrt{2.5} - \sqrt{2} \right)^2 + \left(\sqrt{2.5} - \sqrt{2} \right)^2 \right)^{\frac{1}{2}} = 0.476761 \neq$$

$$\neq 0.415713 = \left(\left(\sqrt{3} - \sqrt{2} \right)^2 + \left(\sqrt{1.5} - \sqrt{2} \right)^2 + \left(\sqrt{1.5} - \sqrt{2} \right)^2 \right)^{\frac{1}{2}} =$$

$$= \bar{d}_H(R_1^{-1}, R_2^{-1}).$$

Thus, $\mathcal{M}_{\bar{d}_H}(\mathbf{R}^{-1}, I) \neq \mathcal{M}_{\bar{d}_H}(\mathbf{R}, I)$. $\qquad\qquad\square$

Lemma 2. *Let $R_1 \in W(X) \setminus L(X)$ and $R_2 \in W(X)$. If $\bar{d}_-$ is the distance induced by d_2, d_c, d_H or d_K, then there exists $R_3 \in W(X)$ such that $\bar{d}_-(R_1, R_2) < \bar{d}_-(R_3, R_2)$.*

Proof. Consider $(a_1, \ldots, a_n) \equiv R_1 \in W(X) \setminus L(X)$, $(b_1, \ldots, b_n) \equiv R_2 \in W(X)$ and $(a'_1, \ldots, a'_n) \equiv R' \in W(X)$. By Proposition 1 and Lemma 1, and taking into account that all the considered distances are neutral (Remark 7), we can assume without loss of generality:

- $(a_1, \ldots, a_j, \ldots, a_{j+k-1}, \ldots, a_n) \equiv R_1 \in W_\leq(X)$ with $j = \min\{i \mid |M_i(R_1)| > 0\}$, $|M_j(R_1)| = k$ and $a_j = \cdots = a_{j+k-1} = j + \frac{k-1}{2}$.
- $(b_1, \ldots, b_j, \ldots, b_{j+k-1}, \ldots, b_n) \equiv R_2 \in W(X)$ with $b_j \geq \cdots \geq b_{j+k-1}$.

Let now $(a'_1, \ldots, a'_n) = (a_1, \ldots, a_{j-1}, j, \ldots, j+k-1, a_{j+k}, \ldots, a_n) \equiv R_3 \in W(X)$ and

$$m = \begin{cases} \frac{k-1}{2} - 1, & \text{if } k \text{ is odd}, \\ \frac{k-1}{2} - \frac{1}{2}, & \text{if } k \text{ is even}. \end{cases}$$

1. <u>Case $\bar{d}_2$.</u>

$$\bar{d}_2(R_1,R_2) < \bar{d}_2(R_3,R_2) \Leftrightarrow 0 < \left(\bar{d}_2(R_3,R_2)\right)^2 - \left(\bar{d}_2(R_1,R_2)\right)^2 \Leftrightarrow$$

$$\Leftrightarrow 0 < \sum_{i=1}^{n}(a_i'-b_i)^2 - (a_i-b_i)^2 = \sum_{l=0}^{k-1}(j+l-b_{j+l})^2 - \left(j+\frac{k-1}{2}-b_{j+l}\right)^2 =$$

$$= \sum_{l=0}^{m}(j+l-b_{j+l})^2 - \left(j+\frac{k-1}{2}-b_{j+l}\right)^2 + ((j+k-1)-l-b_{j+k-1-l})^2 -$$

$$- \left(j+\frac{k-1}{2}-b_{j+k-1-l}\right)^2 =$$

$$= \sum_{l=0}^{m} 2\left(l-\frac{k-1}{2}\right)^2 + \left((j-b_{j+k-1-l})-(j-b_{j+l})\right)\left((k-1)-2l\right).$$

Since $0 < l < \frac{k-1}{2}$ and $0 < b_{j+k-1-l} \le b_{j+l}$, we have $\bar{d}_2(R_1,R_2) < \bar{d}_2(R_3,R_2)$.

2. <u>Case $\bar{d}_c$.</u> Consider $\|R\| = \sqrt{a_1^2 + \cdots + a_n^2}$ whenever $R \equiv (a_1,\ldots,a_n)$.

$$\|R_3\|^2 - \|R_1\|^2 = \sum_{i=1}^{n}\left((a_i')^2 - (a_i)^2\right) =$$

$$= \sum_{l=0}^{m}(j+l)^2 - \left(j+\frac{k-1}{2}\right)^2 + (j+(k-1)-l)^2 - \left(j-\frac{k-1}{2}\right)^2 =$$

$$= 2\left(\frac{k-1}{2}-l\right)^2 > 0. \tag{1}$$

Thus, $\|R_3\| > \|R_1\|$.

$$\bar{d}_c(R_3,R_2) - \bar{d}_c(R_1,R_2) = \frac{\sum_{i=1}^{n}(a_i b_i)}{\|R_1\| \|R_2\|} - \frac{\sum_{i=1}^{n}(a_i' b_i)}{\|R_3\| \|R_2\|} \overset{\text{by (1)}}{>}$$

$$> \frac{1}{\|R_1\| \|R_2\|} \left(\sum_{i=0}^{n}(a_i b_i - a_i' b_i) \right) =$$

$$= \frac{1}{\|R_1\| \|R_2\|} \left(\sum_{l=0}^{m}\left(\left(j+\frac{k-1}{2}\right)b_{j+l} - (j+l)b_{j+l} + \right. \right.$$

$$\left. \left. + \left(j+\frac{k-1}{2}\right)b_{j+k-1-l} - (j+k-1+l)b_{j+k-1-l} \right) \right) =$$

$$= \frac{1}{\|R_1\| \|R_2\|} \left(\sum_{l=0}^{m}\left(b_{j+l}\left(\frac{k-1}{2}+l\right) + b_{j+k-1-l}\left(l-\frac{k-1}{2}\right) \right) \right) \geq$$

$$\geq \frac{1}{\|R_1\| \|R_2\|} \left(\sum_{l=0}^{m}b_{j+k-1-l}\,2l \right) \geq 0.$$

Thus, $\bar{d}_c(R_1,R_2) < \bar{d}_c(R_3,R_2)$.

3. Case $\bar{d}_H$.

$$\bar{d}_H(R_1,R_2) < \bar{d}_H(R_3,R_2) \Leftrightarrow 0 < \left(\bar{d}_H(R_3,R_2)\right)^2 - \left(\bar{d}_H(R_1,R_2)\right)^2 \Leftrightarrow$$

$$\Leftrightarrow 0 < \sum_{i=1}^{n}\left(\sqrt{a_i'} - \sqrt{b_i}\right)^2 - \left(\sqrt{a_i} - \sqrt{b_i}\right)^2 = \sum_{l=0}^{m}\left(\sqrt{j+l} - \sqrt{b_{j+l}}\right)^2 -$$

$$- \left(\sqrt{j+\frac{k-1}{2}} - \sqrt{b_{j+l}}\right)^2 + \left(\sqrt{(j+k-1)-l} - \sqrt{b_{j+k-1-l}}\right)^2 -$$

$$- \left(\sqrt{j+\frac{k-1}{2}} - \sqrt{b_{j+k-1-l}}\right)^2 =$$

$$= 2\sum_{l=0}^{m}\sqrt{b_{j+l}}\left(\sqrt{j+\frac{k-1}{2}} - \sqrt{j+l}\right) +$$

$$+ \sqrt{b_{j+k-1-l}}\left(\sqrt{j+k-1} - \sqrt{j+k-1-l}\right).$$

Since $0 < l < \frac{k-1}{2}$, we have $\bar{d}_H(R_1,R_2) < \bar{d}_H(R_3,R_2)$.

4. Case $\bar{d}_K$.

$$\bar{d}_K(R_1,R_2) = \sum_{\substack{i,h=1\\i<h}}^{n} |\operatorname{sgn}(a_i - a_h) - \operatorname{sgn}(b_i - b_h)|.$$

$$\bar{d}_K(R_3,R_2) = \sum_{\substack{i,h=1\\i<h}}^{n} |\operatorname{sgn}(a'_i - a'_h) - \operatorname{sgn}(b_i - b_h)|.$$

$$|\operatorname{sgn}(a_i - a_h) - \operatorname{sgn}(b_i - b_h)| =$$

$$= \begin{cases} |\operatorname{sgn}(a'_i - a'_h) - \operatorname{sgn}(b_i - b_h)|, & \text{if } \{i,h\} \nsubseteq M_j(R), \\ |-\operatorname{sgn}(b_i - b_h)| < |\operatorname{sgn}(a'_i - a'_h) - \operatorname{sgn}(b_i - b_h)|, & \text{if } \{i,h\} \subseteq M_j(R). \end{cases}$$

Thus, $\bar{d}_K(R_1,R_2) < \bar{d}_K(R_3,R_2)$.   □

Lemma 3. *Let* $R_1, R_2 \in L(X)$ *such that* $R_2 \neq R_1^{-1}$. *If* $\bar{d}_-$ *is the distance induced by* d_2, d_c, d_H *or* d_K, *then there exists* $R_3 \in W(X)$ *such that* $\bar{d}_-(R_1,R_2) < \bar{d}_-(R_1,R_3)$.

Proof. Consider $R_1, R_2, R_3 \in L(X)$ such that $R_1 \equiv (a_1,\dots a_n)$, $R_2 \equiv (b_1,\dots b_n)$ and $R' \equiv (a'_1,\dots a'_n)$. By Proposition 1 and Lemma 1, and taking into account that all the considered distances are neutral (Remark 7), we can assume without loss of generality that $R_1 \equiv (1,2,\dots,n)$.

If $R_2 \neq R_1^{-1}$, then we consider $j = \min\{i \mid b_i \neq n-i+1\}$ and $k = j+l$ such that $b_k = n-j+1$. Let now $R_3 \equiv (b'_1,\dots,b'_n)$ such that $b'_i = b_j$ for every $i \notin \{j,k\}$, $b'_j = b_k = n-j+1$ and $b'_k = b_j < n-j+1$.

1. Case $\bar{d}_2$.

$$\bar{d}_2(R_1,R_3)^2 - \bar{d}_2(R_1,R_2)^2 = |j - b'_j|^2 + |k - b'_k|^2 - \left(|j - b_j|^2 + |k - b_k|^2\right).$$

$$|j - b'_j|^2 + |k - b'_k|^2 = (j+l-b_k-l)^2 + (j-b_j+l)^2 =$$
$$= (k-b_k)^2 + (j-b_j)^2 + 2l(l + (j-b_j) - (k-b_k)) =$$
$$= (j-b_j)^2 + (k-b_k)^2 + 2l(b_k-b_j) > |j-b_j|^2 + |k-b_k|^2.$$

Thus, $\bar{d}_2(R_1,R_2) < \bar{d}_2(R_1,R_3)$.

2. Case $\bar{d}_c$.

It is clear that $\|R_2\| = \|R_3\|$.

$$\bar{d}_c(R_1,R_3) - \bar{d}_c(R_1,R_2) = \frac{\sum_{i=1}^n i b_i}{\|R_1\|\,\|R_2\|} - \frac{\sum_{i=1}^n i b_i'}{\|R_1\|\,\|R_3\|} =$$

$$= \frac{jb_j + kb_k - (jb_j' + kb_k')}{\|R_1\|\,\|R_2\|} = \frac{jb_j + kb_k - (jb_k + kb_j)}{\|R_1\|\,\|R_2\|} =$$

$$= \frac{(b_k - b_j)(k - j)}{\|R_1\|\,\|R_2\|} > 0.$$

Thus, $\bar{d}_c(R_1,R_2) < \bar{d}_c(R_1,R_3)$.

3. Case $\bar{d}_H$.

$$\bar{d}_H(R_1,R_3)^2 - \bar{d}_H(R_1,R_2)^2 =$$

$$= \left(\sqrt{j} - \sqrt{b_j'}\right)^2 + \left(\sqrt{k} - \sqrt{b_k'}\right)^2 - \left(\sqrt{j} - \sqrt{b_j}\right)^2 + \left(\sqrt{k} - \sqrt{b_k}\right)^2 =$$

$$= \left(\sqrt{j} - \sqrt{b_k}\right)^2 + \left(\sqrt{k} - \sqrt{b_j}\right)^2 - \left(\sqrt{j} - \sqrt{b_j}\right)^2 + \left(\sqrt{k} - \sqrt{b_k}\right)^2 =$$

$$= 2\left(\sqrt{jb_j} + \sqrt{kb_k} - \sqrt{jb_k} - \sqrt{kb_j}\right) =$$

$$= 2\left(\left(\sqrt{k} - \sqrt{j}\right) - \left(\sqrt{b_k} - \sqrt{b_j}\right)\right) > 0.$$

Thus, $\bar{d}_H(R_1,R_2) < \bar{d}_H(R_1,R_3)$.

4. Case $\bar{d}_K$.

$$\bar{d}_K(R_1,R_3) - \bar{d}_K(R_1,R_2) =$$
$$= |\mathrm{sgn}\,(j-k) - \mathrm{sgn}\,(b_j' - b_k')| - |\mathrm{sgn}\,(j-k) - \mathrm{sgn}\,(b_j - b_k)| =$$
$$= |\mathrm{sgn}\,(j-k) - \mathrm{sgn}\,(b_k - b_j)| - |\mathrm{sgn}\,(j-k) - \mathrm{sgn}\,(b_j - b_k)| =$$
$$= |-1 - 1| - |-1 - (-1)| = 2 > 0.$$

Thus, $\bar{d}_K(R_1,R_2) < \bar{d}_K(R_1,R_3)$. $\qquad\qquad\square$

Proof of Proposition 5.

1. Cases $\bar{d}_2$, $\bar{d}_c$, $\bar{d}_H$ and $\bar{d}_K$.
   First of all, notice that $\mathcal{M}_{\bar{d}_-}(\boldsymbol{R}, \{v_i, v_j\}) = 0$ if and only if $\bar{d}_-(R_i, R_j) = \Delta_n$. By Lemma 2 and Lemma 3, $d_-(R_1, R_2) = \Delta_n$ if and only if $R_2 = R_1^{-1}$.

2. Cases $\bar{d}'$, $\bar{d}_1$, and $\bar{d}_\infty$.
   Let us consider the following profile $\boldsymbol{R} = (R_1, R_2) \in L(X)^2$:

R_1	R_2	R_1^{-1}
x_1	x_3	x_3
x_2	x_1	x_2
x_3	x_2	x_1

Notice that $R_2 \neq R_1^{-1}$. Since the above linear orders are codified by $R_1 \equiv (1,2,3)$, $R_2 \equiv (2,3,1)$ and $R_1^{-1} \equiv (3,2,1)$, we have

a. $\bar{d}'(R_1,R_2) = \bar{d}'(R_1,R_1^{-1}) = 1$ and $\mathcal{M}_{\bar{d}'}(\boldsymbol{R},\{v_1,v_2\}) = 0$.
b. $\bar{d}_1(R_1,R_2) = \bar{d}_1(R_1,R_1^{-1}) = 4$ and $\mathcal{M}_{\bar{d}_1}(\boldsymbol{R},\{v_1,v_2\}) = 0$.
c. $\bar{d}_\infty(R_1,R_2) = \bar{d}_\infty(R_1,R_1^{-1}) = 2$ and $\mathcal{M}_{\bar{d}_\infty}(\boldsymbol{R},\{v_1,v_2\}) = 0$. $\square$

Proof of Proposition 7. Let $\boldsymbol{R} = (R_1,R_2) \in W(X)^2$ with $\bar{d}(R_1,R_2) = \max\{\bar{d}(R_i,R_j) \mid R_i,R_j \in W(X)\} = \Delta_n$ and $I = \{v_1,v_2\}$.

For every $t \in \mathbb{N}$, we have $t\boldsymbol{R} = (R_1,R_2,\ldots,R_{2t})$, where $R_{2k-1} = R_1$ and $R_{2k} = R_2$ for every $k \in \{1,2,\ldots,t\}$.

We should calculate the limit of the following expression:

$$\mathcal{M}_{\bar{d}}^t(t\boldsymbol{R},tI) = 1 - \frac{\displaystyle\sum_{\substack{v_i,v_j \in tI \\ i<j}} \bar{d}(R_i,R_j)}{\dbinom{|tI|}{2} \cdot \Delta_n}.$$

Since

$$\bar{d}(R_i,R_j) = \begin{cases} 0, & \text{if } i,j \text{ are both even}, \\ 0, & \text{if } i,j \text{ are both odd}, \\ \Delta_n, & \text{otherwise}, \end{cases}$$

we obtain

$$\sum_{\substack{v_i,v_j \in tI \\ i<j}} \bar{d}(R_i,R_j) = \sum_{i=1}^{2t-1}\sum_{j=i+1}^{2t} \bar{d}(R_i,R_j) = \left(\sum_{i=1}^{t} i + \sum_{j=1}^{t-1} j\right) \cdot \Delta_n = t^2 \cdot \Delta_n.$$

On the other hand, we have

$$\binom{|tI|}{2} = \binom{2t}{2} = 2t^2 - t.$$

Consequently,

$$\lim_{t \to \infty} \mathcal{M}_{\bar{d}}^t(t\boldsymbol{R},tI) = 1 - \lim_{t \to \infty} \frac{t^2 \cdot \Delta_n}{(2t^2 - t) \cdot \Delta_n} = \frac{1}{2}.$$ $\square$

A Qualitative Reasoning Approach to Measure Consensus

Llorenç Roselló, Francesc Prats, Núria Agell, and Mónica Sánchez

Abstract. This chapter introduces a mathematical framework on the basis of the absolute order-of-magnitude qualitative model. This framework allows to develop a methodology to assess the consensus found among different evaluators who use ordinal scales in group decision-making and evaluation processes. The concept of entropy is introduced in this context and the algebraic structure induced in the set of qualitative descriptions given by evaluators is studied. We prove that it is a weak partial semilattice structure that in some conditions takes the form of a distributive lattice. The definition of the entropy of a qualitatively-described system enables us, on one hand, to measure the amount of information provided by each evaluator and, on the other hand, to consider a degree of consensus among the evaluation committee. The methodology presented is able of managing situations where the assessment given by experts involves different levels of precision. In addition, when there is no consensus within the group decision, an automatic process measures the effort necessary to reach said consensus.

1 Introduction

In many situations decision-making processes entail qualitative data. Group decision-making systems need new efficient algorithms to measure consensus among evaluators. Existing approaches to group decision-making (GDM) use different ways to deal with the difficulty in managing a lack of complete information and the problem that not all the participants in the group have equal expertise. In [35] the variable precision rough set is used as a tool to support group decision-making. In [3] a model inspired by the ELECTRE I method is proposed to address these kinds of problems,

Llorenç Roselló · Francesc Prats · Mónica Sánchez
Department of Applied Mathematics 2, Universitat Politècnica de Catalunya,
Barcelona, Spain

Núria Agell
ESADE, Universitat Ramon Llull, Barcelona, Spain

E. Herrera-Viedma et al. (Eds.): Consensual Processes, STUDFUZZ 267, pp. 235–261.
springerlink.com   © Springer-Verlag Berlin Heidelberg 2011

using evidence theory to represent imprecise and uncertain data. The model presented in [19] expresses uncertainty in the preference values using interval values.

Many different approaches have been developed to measure the degree of consensus. In [11, 12] a consensus model in group decision-making is presented in a linguistic framework, along with the study of a consensus-reaching process and linguistic consistency measures. [10] provides another dynamic consensus model for decision-making within committees based on a degree of consensus. In [5] the different consensus approaches in fuzzy group decision-making problems are analyzed and their advantages and drawbacks are discussed. Nevertheless, no specific work has been devoted to measuring consensus in the context of order-of-magnitude Qualitative Reasoning.

Qualitative Reasoning (QR) is a subarea of Artificial Intelligence that seeks to understand and explain human beings' ability to reason without having precise information [14, 24]. The main goal of QR is to develop systems that permit operating in conditions of insufficient or no numerical data in such a way that the principle of relevance is preserved, that is, each variable is valued with the level of precision required [13]. In group decision evaluation processes, different levels of precision have to be often handled simultaneously depending on the information available to each evaluator.

On the basis of on Absolute Order-of-Magnitude Qualitative Models (OM) [33] of Qualitative Reasoning (QR) [14], this chapter proposes a methodology to integrate evaluators' opinions and measure their accuracy and degree of consensus. The model of consensus proposed will allow the implementation of an automatic system to measure the consensus within a committee without the figure of a moderator [11, 20]. On the other hand, the use of the presented measure within an automatic system for group decision-making will help to detect and avoid any potential subjectivity arising from conflicts of interests among the evaluators in the group.

The methodology presented allows integrating the representation of existing uncertainty within the group. The concept of entropy in a qualitative evaluation permits us to calculate each evaluator's precision and the degree of consensus within the decision group. When there is no consensus within the group, an automatic process to achieve this consensus and compute the degree of consensus is activated. Although the proposed methodology does not deal with the decision-making process itself, its main advantage is its ability to evaluate this process, managing situations where expert assessment involves different levels of precision.

This chapter is structured as follows. Section 2 presents the proposal's theoretical framework. It concludes with the definitions of entropy for a qualitatively-described system and an index to measure the precision of the alternatives' qualitative descriptions induced by each evaluator. Section 3 defines the degree of consensus among the evaluation committee and gives a simple example to demonstrate the application of this measure. Section 4 corresponds to a real case study in the field of Retailing Management. Finally, in Section 5 we present our conclusions and possible future lines of research.

2 Theoretical Framework

Order-of-magnitude models are essential among the theoretical tools available for qualitative reasoning applied to physical systems [7, 21, 32]. The *classical orders-of-magnitude qualitative spaces* [33] are built from a set of ordered basic qualitative labels determined by a partition of the real line. A general algebraic structure, called Qualitative Algebra or Q-algebra, was defined based on this framework [34], providing a mathematical structure to unify sign algebra and interval algebra through a continuum of qualitative structures built from the roughest to the finest partition of the real line. Q-algebras and their algebraic properties have been extensively studied [27, 33].

In [29] a generalization of qualitative absolute orders-of-magnitude was proposed, something which served as the theoretical basis to develop a Measure Theory in this context. The classical orders-of-magnitude qualitative spaces verify the conditions of the generalized model introduced in [29].

2.1 Absolute Order-of-Magnitude Qualitative Spaces

Let's consider a finite set of *basic* labels, $\mathbb{S}_* = \{B_1, \ldots, B_n\}$, which is totally ordered as a chain: $B_1 < \ldots < B_n$. Usually, each basic label corresponds to a linguistic term, for instance "extremely bad'" <"very bad" < "bad" < "acceptable" < "good" < "very good"<"extremely good". However, it is not unusual for basic labels to be defined by a discretization of a real interval or the real line, given by a set $\{a_1, \ldots, a_{n+1}\}$ of real numbers as landmarks such as $B_i = [a_i, a_{i+1}]$, $i = 1, \ldots, n$.

Nevertheless, we consider a more general case in this chapter in which knowledge of landmark values is not required to introduce the basic labels.

The complete description universe for the Orders-of-Magnitude Space OM(n) with granularity n, is the set $\mathbb{S}_n$:

$$\mathbb{S}_n = \mathbb{S}_* \cup \{[B_i, B_j] \, | \, B_i, B_j \in \mathbb{S}_*, i < j\},$$

where the label $[B_i, B_j]$ with $i < j$ is defined as the set $\{B_i, B_{i+1}, \ldots, B_j\}$.

Consistent with the former example of linguistic labels, the label "moderately good" can be represented by ["acceptable", "good"], i.e., $[B_4, B_5]$. The label "don't know" is represented by ["extremely bad", "extremely good"], i.e., $[B_1, B_7]$. This least precise label is denoted by the symbol ?, i.e., $[B_1, B_n] \equiv ?$.

There is a partial order relation $\leq_P$ in $\mathbb{S}_n$, "to be more precise than", given by:

$$L_1 \leq_P L_2 \Longleftrightarrow L_1 \subset L_2. \tag{1}$$

This structure permits working with all different levels of precision from the basic labels to the ? label (see Figure 1).

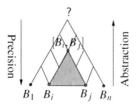

Fig. 1 The space $\mathbb{S}_n$.

To introduce the classical concept of entropy by means of qualitative order-of-magnitude spaces, the concept of measure is required. This concept seeks to generalize the concept of "length", "area"and "volume", understanding that these quantities do not necessarily correspond to their physical counterparts but that they may in fact represent others.

2.2 Qualitative Description Induced by an Evaluator

Let Λ be the set that represents a magnitude or a feature that is qualitatively described by means of the $\mathbb{S}_n$ labels. Since Λ can represent both a continuous magnitude such as position and temperature and a discrete feature such as salary, Λ will be:

$$\Lambda = \{a(t) = a_t \mid t \in I\},$$

where t is a continuous or discrete parameter, and I a set of indexes. An example would be $I = [t_0, t_1]$ in the case where $a(t)$ is a room temperature in a given instant t within this period of time or $I = \{1, \ldots, n\}$ in the case of the salary of n people to be qualitatively described.

This qualitative description is carried out by each evaluator and is represented by the function:

$$Q : \Lambda \to \mathbb{S}_n,$$

where $a_t \mapsto Q(a_t) = \mathscr{E}_t$ is the qualitative label with which the evaluator describes a_t. All the elements of the set $Q^{-1}(\mathscr{E}_t)$ are "representatives" of the label $\mathscr{E}_t$ or "are qualitatively described" by $\mathscr{E}_t$. From now on, this process of qualitative description will be referred to as the *qualitativization* process.

Function Q induces a partition of Λ by means of the equivalence relation:

$$a \sim_Q b \iff Q(a) = Q(b).$$

This partition will be denoted by $\Lambda / \sim_Q$, and its equivalence classes are the sets $Q^{-1}(Q(a_j)) = Q^{-1}(\mathscr{E}_j) \ \forall j \in J \subset I$. Each of these classes contains all the elements of Λ which are described by the same qualitative label (see Figure 2).

Example 1. Suppose there is a kettle heating water and we want to qualitivize the water temperature during a period of five minutes: $\Lambda = \{T(t) \mid t \in [0,5]\}$, assuming

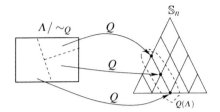

Fig. 2 The qualitativization of a set Λ by means of Q.

that there are three evaluators and the space of qualitative description is $\mathbb{S}_5$, with $B_1 = $ VERY COLD, $B_2 = $ COLD, $B_3 = $ WARM, $B_4 = $ HOT and $B_5 = $ VERY HOT. Let us consider the three following qualitativizations:

$$Q_1(T(t)) = \begin{cases} [B_1, B_2], & \text{if } t \in [0,2) \\ [B_3, B_4], & \text{if } t \in [2,4) \\ B_5, & \text{if } t \in [4,5] \end{cases} \qquad Q_2(T(t)) = \begin{cases} B_1, & \text{if } t \in [0,1) \\ B_2, & \text{if } t \in [1,2) \\ B_3, & \text{if } t \in [2,3) \\ B_4, & \text{if } t \in [3,4) \\ B_5, & \text{if } t \in [4,5] \end{cases}$$

and:

$$Q_3(T(t)) = [B_1, B_5] \text{ if } t \in [0,5].$$

Note that the qualitative description given by Q_2 is the most precise and that the description corresponding to Q_3 is the least. In addition, the intersection of the Q_1, Q_2 and Q_3 labels corresponding to each period of time is not empty. Thus, we can assume that there is some degree of consensus among the three evaluators.

The concepts of qualitative description precision and degree of consensus among a set of qualitativizations are formally introduced in the following sections.

2.3 The Algebraic Structure of a Set of Qualitative Descriptions

Let $\mathscr{Q} = \{Q \mid Q : \Lambda \to \mathbb{S}_n\}$ be the set of qualitativizations of Λ over $\mathbb{S}_n$ given by a group of evaluators.

Given $Q, Q' \in \mathscr{Q}$, two different operations are defined between them. Intuitively speaking, one is the result of *mixing* the two evaluations in a new evaluation that includes both opinions about each element of Λ, and the other one is the result of taking what is *common* between the two evaluations.

2.3.1 The Mix ⊔ Operation

Definition 1. Given two qualitativizations $Q, Q' \in \mathscr{Q}$, the operation $Q \sqcup Q'$ leads to a new qualitativization function $Q \sqcup Q' : \Lambda \to \mathbb{S}_n$ such that, for any $a_t \in \Lambda$,

$$(Q \sqcup Q')(a_t) = Q(a_t) \sqcup Q'(a_t),$$

where $\sqcup$ is the connex union of labels, i.e. the minimum label that contains both $Q(a_t)$ and $Q'(a_t)$:

$$[B_i, B_j] \sqcup [B_h, B_k] = [B_{\min\{i,h\}}, B_{\max\{j,k\}}],$$

using the convention $[B_i, B_i] = B_i$.

Note that if $Q(a_t) \cap Q'(a_t) \neq \emptyset$, then $Q(a_t) \sqcup Q'(a_t)$ is the simple union $Q(a_t) \cup Q'(a_t)$.

2.3.2 The Common $\cap$ Operation

The concept of consensus between two qualitativizations, Q and Q', is required in order to introduce the common operation:

Definition 2. Two qualitativizations Q, Q' are *in consensus*, $Q \rightleftarrows Q'$, iff

$$Q(a_t) \cap Q'(a_t) \neq \emptyset \quad \forall a_t \in \Lambda. \tag{2}$$

This last condition is equivalent to saying that, for all $a_t \in \Lambda$, $Q(a_t)$ and $Q'(a_t)$ are qualitatively equal, denoted $Q(a_t) \approx Q'(a_t)$.

It is clear that the relation $\rightleftarrows$ is symmetric and reflexive.

In general, a set $\{Q_i\}_{i \in I} \subset \mathscr{Q}$ of qualitativizations of Λ over $\mathbb{S}_n$ is *in consensus* iff

$$\bigcap_{i \in I} Q_i(a_t) \neq \emptyset \ \forall a_t \in \Lambda.$$

Note that, in this case, $Q \rightleftarrows Q'$ for all $Q, Q' \in \{Q_i\}_{i \in I}$.

As an example, the set $\{Q_1, Q_2, Q_3\}$ in Example 1 is in consensus.

Definition 3. Given two qualitativizations Q and Q' where $Q \rightleftarrows Q'$, the *common* $Q \cap Q'$ operation produces a new qualitativization function $Q \cap Q' : \Lambda \to \mathbb{S}_n$ such that

$$(Q \cap Q')(a_t) = Q(a_t) \cap Q'(a_t) \ \forall a_t \in \Lambda.$$

In general, if $\{Q_i\}_{i \in I} \subset \mathscr{Q}$ is in consensus, the operation *common* $\cap_{i \in I} Q_i$ produces a new qualitativization: $(\cap_{i \in I} Q_i)(a_t) = \cap_{i \in I} Q_i(a_t) \ \forall a_t \in \Lambda$.

2.3.3 The Algebraic Structure of the Set $\mathscr{Q}$

Definition 4. A weak partial lattice [15] is a set H with two binary partial operations $\wedge$ and $\vee$ satisfying, for all $a, b, c \in H$, the following statements and their dual, interchanging $\wedge$ and $\vee$:

(i) $a \wedge a$ exists and $a \wedge a = a$.
(ii) If $a \wedge b$ exists, then $b \wedge a$ exists, and $a \wedge b = b \wedge a$.

(iii) If $a \wedge b, (a \wedge b) \wedge c$ and $b \wedge c$ exist, then $a \wedge (b \wedge c)$ exists, and $(a \wedge b) \wedge c = a \wedge (b \wedge c)$.

If $b \wedge c, a \wedge (b \wedge c)$ and $a \wedge b$ exist, then $(a \wedge b) \wedge c$ exists, and $(a \wedge b) \wedge c = a \wedge (b \wedge c)$.

(iv) If $a \wedge b$ exists, then $a \vee (a \wedge b)$ exists, and $a = a \vee (a \wedge b)$.

The algebraic structure of the set $\mathcal{Q}$ and the $\sqcup$ and $\cap$ operations is given by the following proposition:

Proposition 1. $(\mathcal{Q}, \sqcup, \cap)$ *is a weak partial lattice.*

Proof. Demonstrating (see [15]) the following statements and their dual forms (obtained by changing $\sqcup$ by $\cap$) is sufficient. Note that given Q, Q', the $\sqcup$ operation is always defined, but the $\cap$ operation exists iff $Q \rightleftarrows Q'$.

1. $Q \sqcup Q = Q$.   2. $Q \sqcup Q' = Q' \sqcup Q$.
3. $(Q \sqcup Q') \sqcup Q'' = Q \sqcup (Q' \sqcup Q'')$.   4. $Q \cap (Q \sqcup Q')$ exists
                                                              and $Q \cap (Q \sqcup Q') = Q$.

And the dual ones:

1'. $Q \cap Q$ exists and $Q \cap Q = Q$.

2'. If $Q \cap Q'$ exist then $Q' \cap Q$ exists and $Q \cap Q' = Q' \cap Q$.

3'. If $Q \cap Q'$ and $(Q \cap Q') \cap Q''$ exist, then $Q' \cap Q''$ and $Q \cap (Q' \cap Q'')$ exist and $(Q \cap Q') \cap Q'' = Q \cap (Q' \cap Q'')$. If $Q' \cap Q''$ and $Q \cap (Q' \cap Q'')$ exist, then $Q \cap Q'$ and $(Q \cap Q') \cap Q''$ exist and $(Q \cap Q') \cap Q'' = Q \cap (Q' \cap Q'')$.

4'. If $Q \cap Q'$ exists, then $Q \sqcup (Q \cap Q') = Q$.

Statements 1, 2 and 3 are easily proved. Statement 4 is also true because $Q(a_t) \subset Q(a_t) \sqcup Q'(a_t)$ for any $a_t \in \Lambda$. Therefore $Q \rightleftarrows Q \sqcup Q'$ and $Q(a_t) \cap (Q(a_t) \sqcup Q'(a_t)) = Q(a_t)$. The dual statements are similarly proved. $\square$

From the general lattice theory [15, 2] applied to the weak partial lattice $(\mathcal{Q}, \sqcup, \cap)$ the following statements can be made:

- $Q \leq Q'$ iff $Q \sqcup Q' = Q$ defines a a *partial order relation*.
- $Q \sqcup Q' = \inf\{Q, Q'\}$.
- If $Q \rightleftarrows Q'$, then $Q \cap Q' = \sup\{Q, Q'\}$.
- $Q \leq Q'$ iff $Q \cap Q' = Q'$.

Note that in the case of $Q \leq Q'$, the qualitative description Q is less accurate than Q', because $Q'(a_t) \subset Q(a_t) \; \forall a_t \in \Lambda$, i.e., each element of set Λ is more precisely described by Q' than by Q :

$$Q \leq Q' \Leftrightarrow Q'(a_t) \leq_P Q(a_t) \; \forall a_t \in \Lambda.$$

Proposition 2. *Let $\mathcal{Q}_L$ be a subset of $\mathcal{Q}$ which is in consensus. Then $(\mathcal{Q}_L, \sqcup, \cap)$ is a distributive complete lattice.*

Proof. If the subset $\mathcal{Q}_L$ of $\mathcal{Q}$ is in consensus, then $(\mathcal{Q}_L, \sqcup, \cap)$ is a complete lattice, because the operation $\cap$ is defined in all cases.

When there is consensus the $\sqcup$ and $\cap$ operations on $\mathcal{Q}_L$ correspond exactly to the union and intersection of $\mathbb{S}_n$ labels respectively. Therefore, the distributive axioms, i.e., if $Q, Q', Q'' \in \mathcal{Q}_L$ then $Q \sqcup (Q' \cap Q'') = (Q \sqcup Q') \cap (Q \sqcup Q'')$ and $Q \cap (Q' \sqcup Q'') = (Q \cap Q') \sqcup (Q \cap Q'')$, are satisfied.   □

2.4 A Distance in a Complete Lattice $\mathcal{Q}_L$

Let us suppose that there exists a subset $\mathcal{Q}_L$ of $\mathcal{Q}$ which is in consensus (if this situation does not hold, in Section 3.2 a process to obtain consensus is presented). This subsection is devoted to define a distance between two qualitativizations $Q, Q' \in \mathcal{Q}_L$.

Definition 5. In the lattice $(\mathcal{Q}_L, \sqcup, \cap)$ the *null element* $0_{\mathcal{Q}_L}$ is defined as

$$0_{\mathcal{Q}_L} = \sqcup_{Q_i \in \mathcal{Q}_L} Q_i,$$

and the *universal element* $1_{\mathcal{Q}_L}$ is defined as

$$1_{\mathcal{Q}_L} = \cap_{Q_i \in \mathcal{Q}_L} Q_i.$$

The null element and the universal elements verify for all $Q \in \mathcal{Q}_L$ (see figure 3):

$$0_{\mathcal{Q}_L} \sqcup Q = 0_{\mathcal{Q}_L}, 0_{\mathcal{Q}_L} \cap Q = Q,$$

$$1_{\mathcal{Q}_L} \sqcup Q = Q, 1_{\mathcal{Q}_L} \cap Q = 1_{\mathcal{Q}_L},$$

and then

$$0_{\mathcal{Q}_L} \leq Q \leq 1_{\mathcal{Q}_L}.$$

Recall the definition of *chain*: a totally ordered ordered set of a poset.

By "*x covers y*" it is meant that $y < x$ and that $y < z < x$ is not satisfied by any z. A finite chain $x = a_1 < a_2 < \ldots < a_n = y$ is a *maximal chain* if each a_{i+1} covers a_i for $i = 1, \ldots, n-1$.

Let us assume that Λ is a finite set. Since $\mathbb{S}_n$ is also finite, then all the chains in $(\mathcal{Q}_L, \sqcup, \cap)$ are finite. Therefore, all finite maximal chains between fixed end points have the same length (Jordan-Dedekind theorem) [2].

Definition 6. If $Q, Q' \in \mathcal{Q}_L$, the length of a chain with end points Q and Q', $l([Q, Q'])$ is the cardinal of any maximal chain between Q and Q'. The length of $Q \in \mathcal{Q}_L$, $l(Q)$, is the length of $[0_{\mathcal{Q}_L}, Q]$ (see figure 3)

In the distributive lattice $(\mathcal{Q}_L, \sqcup, \cap)$ the following statement is satisfied for all Q and Q' in $\mathcal{Q}_L$:

$$l(Q) + l(Q') = l(Q \sqcup Q') + l(Q \cap Q'). \tag{3}$$

Lemma 1. *Since in $\mathcal{Q}_L$ the operation $\sqcup$ and $\cap$ are the infimum and supremum respectively then:*

$$(Q \cap Q') \sqcup (Q' \cap Q'') \geq Q' \tag{4}$$

$$Q' \geq (Q \sqcup Q') \cap (Q' \sqcup Q''). \tag{5}$$

Proof. It is a simply exercise using the definition of $\leq$ and properties of $\sqcup$ and $\cap$. □

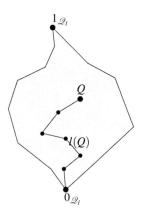

Fig. 3 The null and universal elements, and the length of a qualitativization.

The next theorem defines a distance in the lattice $(\mathscr{Q}_L, \sqcup, \cap)$:

Theorem 1. *In the lattice* $(\mathscr{Q}_L, \sqcup, \cap)$, *the function* $d : \mathscr{Q}_L \times \mathscr{Q}_L \to \mathbb{R}$ *defined as*

$$d(Q,Q') = l(Q \cap Q') - l(Q \sqcup Q'), \tag{6}$$

is a distance.

Proof. 1. Positive definiteness: Because $Q \sqcup Q' \leq Q \cap Q' \ \forall Q, Q'$ it is trivial to see that $l(Q \sqcup Q') \leq l(Q \cap Q')$, so $d(Q,Q') \geq 0$.
If $Q = Q'$ then $d(Q,Q') = 0$. Conversely,

$$d(Q,Q') = 0 \Rightarrow l(Q \sqcup Q') = l(Q \cap Q'),$$

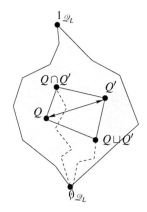

Fig. 4 The distance between two qualitativizations.

and this, together with the fact that $Q \sqcup Q' \leq Q \cap Q'$, and the Jordan-Dedekind theorem leads to $Q \sqcup Q' = Q \cap Q'$.

By the absorptive laws of lattices:

$$Q \cap (Q \sqcup Q') = Q \text{ and } Q \sqcup (Q \cap Q') = Q.$$

We have

$$Q = Q \cap (Q \sqcup Q') = Q \cap (Q \cap Q') = Q \cap Q',$$
$$Q' = Q' \cap (Q \sqcup Q') = Q' \cap (Q \cap Q') = Q \cap Q',$$

so $Q = Q'$.

2. Symmetry: Since $\sqcup$ and $\cap$ are commutative, $d(Q,Q') = d(Q',Q)$.
3. Triangle inequality: For all $Q,Q',Q'' \in \mathscr{Q}_L$

$$d(Q,Q') \leq d(Q,Q'') + d(Q'',Q').$$

We have

$$d(Q,Q'') + d(Q'',Q') = l(Q \cap Q'') + l(Q' \cap Q'') - (l(Q \sqcup Q'') + l(Q' \sqcup Q'')).$$

The two first summands can be expressed using the property (3)

$$l(Q \cap Q'') + l(Q' \cap Q'') = l((Q \cap Q'') \sqcup (Q' \cap Q'')) + l((Q \cap Q'') \cap (Q' \cap Q'')),$$

and then, by (4)

$$l(Q \cap Q'') + l(Q' \cap Q'') \geq l(Q'') + l((Q \cap Q' \cap Q'')).$$

Similarly from (3)

$$l(Q \sqcup Q'') + l(Q' \sqcup Q'') = l((Q \sqcup Q'') \sqcup (Q' \sqcup Q'')) + l((Q \sqcup Q'') \cap (Q'' \sqcup Q')),$$

and then, by (5):

$$l(Q \sqcup Q'') + l(Q' \sqcup Q'') \leq l(Q \sqcup Q' \sqcup Q'') + l(Q'').$$

So,

$$d(Q,Q'') + d(Q'',Q') \geq l(Q \cap Q' \cap Q'') - l(Q \sqcup Q' \sqcup Q'').$$

Now, using the fact that

$$Q \cap Q' \cap Q'' \geq Q \cap Q' \Rightarrow l(Q \cap Q' \cap Q'') \geq l(Q \cap Q')$$
$$Q \sqcup Q' \sqcup Q'' \leq Q \sqcup Q' \Rightarrow l(Q \sqcup Q' \sqcup Q'') \leq l(Q \sqcup Q'),$$

we conclude that

$$d(Q,Q'') + d(Q'',Q') \geq l(Q \cap Q') - l(Q \sqcup Q') = d(Q,Q'). \qquad \square$$

Thus far, we have introduced the basics of the qualitativization mathematical structure: the definition of qualitativization, the aggregation information for mix and common operations, and the algebraic structure of the set $\mathscr{Q}$. In the next sections these concepts are used to define the entropy of a given qualitativization and degree of consensus.

2.5   Entropy

The notion of information measures has been the subject of study for a long time and the interest in this research has been considerably renewed by the development of Zadeh's fuzzy sets [36], Possibility Theory and Shafer's Theory of Evidence [30]. An excellent overview of information measures in these fields can be found in [8, 9, 22]. Despite this interest, however, no work has been specifically dedicated to measuring the information within absolute order-of-magnitude qualitative spaces. Entropy, as defined in this chapter, is a measure of the information provided by an evaluator when he/she qualitativizes the set Λ. It can also be seen as the measure of the information the evaluator needs to assign a qualitative label to any element in Λ. The information used by the evaluator is the information that he/she has, possesses or knows about the elements in the set. The first step is to measure how much information is needed in order to map an element in Λ to a specific label. The following definitions are an extension of Shannon's Theory of Information [31].

2.5.1   Definition of a Measure in Absolute Order-of-Magnitude Spaces

The definition of a measure [16] in $\mathbb{S}_n$ is necessary to define the information of a label and the entropy.

Definition 7. Let X be a non-empty set and $\mathscr{C} \subset \mathscr{P}(X)$, with $\emptyset \in \mathscr{C}$. A *measure on* $\mathscr{C}$ is a map $\mu : \mathscr{C} \to [0, +\infty]$ satisfying the following conditions:

1. $\mu(\emptyset) = 0$.
2. For any sequence $(E_n)_{n=1}^{\infty}$ of pairwise disjoint sets of $\mathscr{C}$ such that $\cup_{n=1}^{+\infty} E_n \in \mathscr{C}$, then

$$\mu\left(\bigcup_{n=1}^{+\infty} E_n\right) = \sum_{n=1}^{+\infty} \mu(E_n).$$

Given that $\mathbb{S}_n$ is a finite set, the unions and sums in the definition are finite. Therefore, the general process to define a measure space via the σ-algebra generated by a semi-ring is not necessary. This process can be found in the case of the generalized order-of-magnitude spaces in [29].

For instance, a measure in the classical $\mathbb{S}_n$ that takes into account the lengths of the basic labels is given by: $\mu(B_i) = \mu([a_i, a_{i+1}]) = (a_{i+1} - a_i)/(a_n - a_1)$.

2.5.2 The Information of a Qualitative Label

The information of a label $\mathscr{E}$ is defined by a positive continuous real function of the measure of such label. It is denoted by $I(\mathscr{E})$. It is assumed that if a label $\mathscr{E}$ is more precise than a label $\mathscr{E}'$, then more information is needed to assign an element to $\mathscr{E}$ than to $\mathscr{E}'$:

$$\mathscr{E} \leq_P \mathscr{E}' \Rightarrow I(\mathscr{E}) \geq I(\mathscr{E}').$$

Another assumption about I function is that the information for the ? label is zero (no information is needed to assign the ? label to an element).

The following definition of I, inspired in the Shannon's Theory of information, verifies these assumptions:

Definition 8. Let μ be a normalized measure defined on $\mathbb{S}_n$, i.e., a measure such that $\mu(?) = 1$. The formula to define the information of a label $\mathscr{E} \in \mathbb{S}_n$ such that $\mu(\mathscr{E}) \neq 0$ is

$$I(\mathscr{E}) = \log \frac{1}{\mu(\mathscr{E})}.$$

Note that, for any $\mathscr{E}$, $\mu(\mathscr{E}) \leq 1$, and so $I(\mathscr{E}) \geq 0$.
Moreover, I decreases with respect $\leq_P$:

$$\mathscr{E} \leq_P \mathscr{F} \Rightarrow \mathscr{E} \subset \mathscr{F} \Rightarrow \mu(\mathscr{E}) \leq \mu(\mathscr{F}) \Rightarrow \log \frac{1}{\mu(\mathscr{E})} \geq \log \frac{1}{\mu(\mathscr{F})}$$

In addition, $I(?) = \log 1 = 0$.

Example 2. In the classical $\mathbb{S}_n$ model and by considering the measure $\mu(B_i) = \mu([a_i, a_{i+1}]) = (a_{i+1} - a_i)/(a_n - a_1)$, the information of a label is $I([a_i, a_{i+1}]) = \log\left(\frac{a_n - a_1}{a_{i+1} - a_i}\right)$.

Proposition 3. *For all labels $\mathscr{E}, \mathscr{F} \in \mathbb{S}_n$ it holds that:*

$$I(\mathscr{E} \sqcup \mathscr{F}) \leq \min\{I(\mathscr{E}), I(\mathscr{F})\} \leq I(\mathscr{E}) + I(\mathscr{F}).$$

Proof. It is sufficient to take into account that $\mathscr{E} \leq_P \mathscr{E} \sqcup \mathscr{F}$, $\mathscr{F} \leq_P \mathscr{E} \sqcup \mathscr{F}$ and that I is decreasing. $\square$

2.5.3 Definition of Entropy of a Qualitativization in $\mathbb{S}_n$

The entropy of a qualitativization, as mentioned above, is a measure of the information needed by the evaluator when qualitativizing set Λ. The most natural way to express this concept mathematically is to define the entropy H of a qualitativization Q as a weighted average of the information regarding the elements within the set Λ given by Q.

Let μ be a normalized measure defined on $\mathbb{S}_n$, and $\overline{\mu}$ be a normalized measure defined on the set Λ.

Definition 9. Entropy H of the set Λ given by Q is:

$$H(Q) = \sum_{\mathcal{E} \in \mathbb{S}_n, \mu(\mathcal{E}) \neq 0} \overline{\mu}(Q^{-1}(\mathcal{E}))I(\mathcal{E}), \tag{7}$$

where in the summation only the $\mathcal{E}$ such that $Q^{-1}(\mathcal{E}) \neq \emptyset$ appear.

Note that $H(Q) \geq 0$ for all qualitativizations Q, and $H(Q) = 0$ when the whole set Λ is described only by the label ?.

If $\Lambda / \sim_Q = \{X_i, i \in J\}$, that is, the set of equivalence classes of $\sim_Q$, then (7) can be expressed as

$$H(Q) = \sum_{i \in J} \overline{\mu}(X_i)I(Q(X_i)). \tag{8}$$

Our next proposition shows the monotonicity of the entropy with respect to the accuracy relation between qualitativizations.

Proposition 4. *Given two qualitativizations Q and Q', then*

$$Q \leq Q' \implies H(Q) \leq H(Q').$$

Proof. Let's assume that $\Lambda / \sim_Q = \{X_i \mid i \in M\}$, $\Lambda / \sim_{Q'} = \{Y_j \mid j \in N\}$, and $(\Lambda / \sim_Q) \cap (\Lambda / \sim_{Q'}) = \{X_i \cap Y_j \neq \emptyset \mid i \in M, j \in N\}$.

Then, for all X_i and Y_j, $X_i = \bigcup_{j \in N}(X_i \cap Y_j)$, $Y_j = \bigcup_{i \in M}(X_i \cap Y_j)$, where only the non-empty intersections $X_i \cap Y_j$ are written, and the unions are disjoint unions, because $\{Y_j\}_{j \in N}$ and $\{X_i\}_{i \in M}$ are classes of equivalence of Λ.

Therefore, the entropy of Q is:

$$H(Q) = \sum_{i \in M} \overline{\mu}(X_i)I(Q(X_i)) = \sum_{i \in M} \overline{\mu}\left(\bigcup_{j \in N}(X_i \cap Y_j)\right)I(Q(X_i)) =$$
$$= \sum_{i \in M}\left(\sum_{j \in N} \overline{\mu}(X_i \cap Y_j)\right)I(Q(X_i)) = \sum_{i \in M, j \in N} \overline{\mu}(X_i \cap Y_j)I(Q(X_i)).$$

Analogously, $H(Q') = \sum_{i \in M, j \in N} \overline{\mu}(X_i \cap Y_j)I(Q'(Y_j))$.

Given that $X_i \cap Y_j \neq \emptyset$, the hypothesis $Q \leq Q'$ implies that, for an element $a_t \in X_i \cap Y_j$, $Q'(a_t) = Q'(Y_j) \leq_P Q(a_t) = Q(X_i)$, that is, $Q'(Y_j) \subset Q(X_i)$. As such, $I(Q(X_i)) \leq I(Q'(Y_j))$ for every summand in $H(Q)$ and $H(Q')$, so the inequality $H(Q) \leq H(Q')$ is inferred. □

As a corollary to the last proposition we obtain entropy's subadditivity:

Corollary 1. *Given two qualitativizations Q and Q', then*

$$H(Q \sqcup Q') \leq \min\{H(Q), H(Q')\} \leq H(Q) + H(Q').$$

Proof. It suffices to take into account that $Q \sqcup Q' \leq Q, Q'$. □

The property of monotonicity together with the subadditivity are two of the main properties of information measures [8, 23].

2.6 Precision of a Qualitative Description

The entropy of set Λ when it is qualitativized by means of space $\mathbb{S}_n$ has a maximum value which allows us to define a measure of the precision of the qualitativizations given by the decision group.

Proposition 5. *Let* $\mathcal{E}_1^*, \ldots, \mathcal{E}_k^* \in \mathbb{S}_n$ *be the (basic) labels with minimum measure* μ, $m^* = \mu(\mathcal{E}_1^*) = \ldots = \mu(\mathcal{E}_k^*) \neq 0$. *Let us consider a qualitativization* $\widetilde{Q}$ *such that* $\widetilde{Q}(\Lambda) \subset \{\mathcal{E}_1^*, \ldots, \mathcal{E}_k^*\}$, *that is,* $\widetilde{Q}$ *maps the entire set* Λ *to the most precise labels.* *Then:* $H(Q) \leq H(\widetilde{Q}) = \log \frac{1}{m^*}$ $\forall Q$.

Proof. Since $m^* \leq \mu(\mathcal{E})$ $\forall \mathcal{E} \in \mathbb{S}_n$, then $I(\mathcal{E}) \leq \log(1/m^*)$. Thus, for any Q,

$$H(Q) = \sum_{\mathcal{E} \in \mathbb{S}_n} \overline{\mu}(Q^{-1}(\mathcal{E}))I(\mathcal{E}) \leq \log \frac{1}{m^*} \sum_{\mathcal{E} \in \mathbb{S}_n} \overline{\mu}(Q^{-1}(\mathcal{E})) = \log \frac{1}{m^*},$$

because $\{Q^{-1}(\mathcal{E})\}_{\mathcal{E} \in \mathbb{S}_n}$ is a partition of Λ and $\overline{\mu}$ is normalized. Moreover, since $\widetilde{Q}(\Lambda) \subset \{\mathcal{E}_1^*, \ldots, \mathcal{E}_k^*\}$, $H(\widetilde{Q}) = \sum_{\mathcal{E} \in \mathbb{S}_n} \overline{\mu}(\widetilde{Q}^{-1}(\mathcal{E})) \log \frac{1}{m^*} = \log \frac{1}{m^*}$. $\qquad\square$

According to this proposition the precision of a qualitativization is defined in the following way:

Definition 10. The *precision* of a qualitativization Q of set Λ, $h(Q)$, is the relative entropy respect to the maximum entropy $H(\widetilde{Q})$ for set Λ in $\mathbb{S}_n$:

$$h(Q) = \frac{H(Q)}{H(\widetilde{Q})}. \tag{9}$$

This quantity is a real number between 0 and 1; the more accurate the evaluator is, the closer $h(Q)$ is to 1. When Q maps the whole Λ to the most precise labels (basic labels with the smallest measure) then $h(Q) = 1$. In the opposite case, $h(Q) = 0$ when Q maps the whole Λ to the least precise label ?.

3 Consensus in the Group Decision

Measuring consensus has been tackled in the literature by several authors in different ways. The most studied approaches to measuring consensus use fuzzy linguistic information [5, 4, 26, 17]. In [6, 28] the degree of consensus is computed through an average, and in [10] it is related to a distance. However, there is a lack in the field of order-of magnitude qualitative reasoning.

The approach presented in this chapter, which is based on entropy as defined in the previous section, offers a new method with which to compute degrees of consensus in the frame of absolute order-of-magnitude models, where different levels of precision can be simultaneously considered. The way of dealing with consensus processes and measures presented is somewhat different to the usual one. In fact, the degree of consensus defined does not depend on the number of evaluators in the group, in the sense that if the number of evaluators that "think similarly" increases, the consensus degree does not increase.

3.1 Degree of Consensus

In order to introduce a definition for the degree of consensus, let us suppose that two evaluators qualitativize set Λ by means of Q and Q'. First of all, the degree of consensus can only be computed when consensus exists among them, i.e. if $Q \rightleftarrows Q'$.

If the two evaluators "think similarly", then the operation $\cap$ between Q, Q' which extracts their coincidences will produce a qualitativization similar to the qualitativization obtained by mixing them. In this case $H(Q \cap Q')$ will be quite similar to $H(Q \sqcup Q')$. Otherwise, $Q \cap Q'$ will be a qualitativization with a high degree of entropy, and $Q \sqcup Q'$ will have a low degree.

On the other hand, $H(Q \cap Q') \geq H(Q \sqcup Q')$ because $Q \cap Q' \geq Q \sqcup Q'$; thus the quotient $H(Q \sqcup Q')/H(Q \cap Q')$ is a real number between 0 and 1.

In order to generalize the quotient above to the case of group decisions with M evaluators, let us introduce the following notation:

Given a space $\mathbb{S}_n$, a finite nonempty set $\Lambda = \{a_1, \ldots, a_N\}$ and a group of evaluators $\mathbb{E} = \{\alpha_1, \ldots, \alpha_M\}$, the *group evaluation* of Λ is considered as the pair $(\Lambda, \mathcal{Q}_{\mathbb{E}})$, where $\mathcal{Q}_{\mathbb{E}} = \{Q_i : \Lambda \to \mathbb{S}_n \mid i \in \{1, \cdots M\}\}$, and Q_i is the evaluation of α_i.

Let's suppose that there is consensus among the group, i.e., $\cap_{i=1}^M Q_i(a_t) \neq \emptyset\ \forall a_t \in \Lambda$. The next definition regarding the degree of consensus thus measures the relation between the entropy of mix and common operations in the set of group qualitativizations:

Definition 11. Given a group evaluation $(\Lambda, \mathcal{Q}_{\mathbb{E}})$ in consensus, i.e., $\cap_{i=1}^M Q_i$ exists, the *degree of consensus* among the group, $\kappa(\mathcal{Q}_{\mathbb{E}})$, is

$$\kappa(\mathcal{Q}_{\mathbb{E}}) = \frac{H(\sqcup_{i=1}^M Q_i)}{H(\cap_{i=1}^M Q_i)} \tag{10}$$

(the only case in which κ is not well defined corresponds to the case $H(\cap_{i=1}^M Q_i) = 0$, that is, when all evaluators describe the elements of the full set Λ with the label ?).

This degree is a number between 0 and 1; the closer it is to 1, the closer the group is to being unanimous in its assessment. The next proposition shows that the degree of consensus within a group evaluation cannot be increased by adding a new evaluator to the group.

Proposition 6. *Consider a group evaluation* $(\Lambda, \mathcal{Q}_{\mathbb{E}})$ *in consensus. Let be* Q_{new} *a new evaluator of* Λ *such that* $\mathcal{Q}_{\mathbb{E}} \cup \{Q_{new}\}$ *is in consensus. Then*

$$\kappa(\mathcal{Q}_{\mathbb{E}} \cup \{Q_{new}\}) \leq \kappa(\mathcal{Q}_{\mathbb{E}}).$$

Proof. Based on the fact that $Q \sqcup Q' = \inf\{Q, Q'\}$ and $Q \cap Q' = \sup\{Q, Q'\}$:

$$(\sqcup_{i=1}^{M} Q_i) \sqcup Q_{\text{new}} \leq \sqcup_{i=1}^{M} Q_i \text{ and } \cap_{i=1}^{M} Q_i \leq (\cap_{i=1}^{M} Q_i) \cap Q_{\text{new}}.$$

Then, from Proposition 4:

$$H((\sqcup_{i=1}^{M} Q_i) \sqcup Q_{\text{new}}) \leq H(\sqcup_{i=1}^{M} Q_i) \text{ and } H(\cap_{i=1}^{M} Q_i) \leq H((\cap_{i=1}^{M} Q_i) \cap Q_{\text{new}}),$$

and hence $\kappa(\mathcal{Q}_{\mathbb{E}} \cup \{Q_{\text{new}}\}) \leq \kappa(\mathcal{Q}_{\mathbb{E}})$.   □

Therefore, the only way to increase the degree of consensus in a group is for some evaluator to reconsider the situation and his/her assessment.

3.2 Achieving Consensus

The necessary and sufficient condition for which there exists consensus is $\cap_{i=1}^{M} Q_i(a_t) \neq \emptyset \, \forall a_t \in \Lambda$. If this situation does not hold then a process has to be initiated to obtain consensus. In [6, 10, 25, 28], different approaches to this problem are found framed within fuzzy sets theory and aggregation operators. If two people disagree on some attribute and they want to reach an agreement, i.e., reach consensus, they can reconsider their positions and find points in common. When it is not possible, the process to obtain consensus can be done automatically.

The algorithm presented here is based on the following idea: to get positions closer by increasing the level of qualitative descriptions, i.e., by increasing the granularity of the space of orders-of-magnitude. For instance, let us consider an evaluation done over a space $\mathbb{S}_3$ where $B_1 = \text{LOW}, B_2 = \text{NORMAL}$, and $B_3 = \text{HIGH}$. If two evaluators' opinions about the same attribute are $B_2 = \text{NORMAL}$ and $B_3 = \text{HIGH}$, the consensus can be achieved in a space $\mathbb{S}_4$ where $B'_1 = \text{LOW}, B'_2 = \text{FAIRLY LOW}$, $B'_3 = \text{FAIRLY HIGH}$, and $B'_4 = \text{HIGH}$, by substituting the initial labels B_2 and B_3 by $[B'_2, B'_3] = [\text{FAIRLY LOW}, \text{FAIRLY HIGH}]$ and $[B'_3, B'_4] = [\text{FAIRLY HIGH}, \text{HIGH}]$ respectively. It can be understood as a process of automatic negotiation.

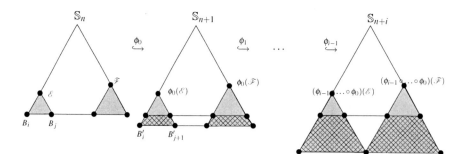

Fig. 5 The dive function.

Definition 12. Given a space $\mathbb{S}_n$ with basic labels $\mathscr{S} = \{B_1, \ldots, B_n\}$, and a space $\mathbb{S}_{n+1}$ with basic labels $\mathscr{S}' = \{B'_1, \ldots, B'_{n+1}\}$, the *dive function* is the map $\phi_0 : \mathbb{S}_n \rightarrow \mathbb{S}_{n+1}$, defined as follows:
For basic labels $B_i \in \mathbb{S}_n$, then

$$\phi_0(B_i) = [B'_i, B'_{i+1}],$$

and, for non-basic labels,

$$\phi_0([B_i, B_j]) = \bigcup_{k=i}^{j} \phi_0(B_k) = [B'_i, B'_{j+1}].$$

The dive function ϕ_0 is an injection of $\mathbb{S}_n$ into $\mathbb{S}_{n+1}$. With this function, each basic label in $\mathbb{S}_n$ is "split" into two new basic labels in $\mathbb{S}_{n+1}$. And in general, for each label $\mathscr{E}$ in $\mathbb{S}_n$, $\Phi_0(\mathscr{E})$ is obtained by adding a new basic label. In this same way, we can define $\phi_i : \mathbb{S}_{n+i} \rightarrow \mathbb{S}_{n+i+1}$, for $i \geq 1$ (see Figure 5), and the following chain can be considered:

$$\mathbb{S}_n \xrightarrow{\phi_0} \mathbb{S}_{n+1} \xrightarrow{\phi_1} \mathbb{S}_{n+2} \hookrightarrow \cdots \hookrightarrow \mathbb{S}_{n+m} \xrightarrow{\phi_m} \mathbb{S}_{n+m+1}$$

Then, given $\mathscr{E}, \mathscr{F} \in \mathbb{S}_n$ such that $\mathscr{E} \cap \mathscr{F} = \emptyset$, we can see that there exists a natural number $k \geq 1$ such that :

$$(\phi_{k-1} \circ \cdots \circ \phi_0)(\mathscr{E}) \cap (\phi_{k-1} \circ \cdots \circ \phi_0)(\mathscr{F}) \neq \emptyset.$$

Similarly and given $\mathscr{E}_1, \ldots, \mathscr{E}_M \in \mathbb{S}_n$ such that $\bigcap_{i=1}^{M} \mathscr{E}_i = \emptyset$, there exists $k \geq 1$ such that

$$\bigcap_{i=1}^{M} (\phi_{k-1} \circ \cdots \circ \phi_0)(\mathscr{E}_i) \neq \emptyset.$$

The following proposition allows us to extend the measure defined in $\mathbb{S}_n$ to the new space $\mathbb{S}_{n+1}$.

Proposition 7. *Let μ be a normalized measure defined on $\mathbb{S}_n$ and let us suppose that $\mathbb{S}_n$ is "dived" in $\mathbb{S}_{n+1}$. Then the measure μ can be extended to a normalized measure μ' in $\mathbb{S}_{n+1}$ defined, taking weights $0 < \lambda_1, \ldots, \lambda_n < 1$ and convex linear combinations, in the following way (see Figure 6):*

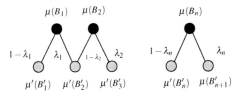

Fig. 6 The measure expansion.

$$\mu'(B_1') = (1 - \lambda_1)\mu(B_1)$$
$$\mu'(B_2') = \lambda_1\mu(B_1) + (1 - \lambda_2)\mu(B_2)$$
$$\vdots$$
$$\mu'(B_i') = \lambda_{i-1}\mu(B_{i-1}) + (1 - \lambda_i)\mu(B_i)$$
$$\vdots$$
$$\mu'(B_{n+1}') = \lambda_n\mu(B_n)$$

And for a non-basic label $\mathscr{E}' = [B_i', B_j'] \in \mathbb{S}_{n+1}$,

$$\mu'(\mathscr{E}') = \sum_{k=i}^{j}\mu'(B_k').$$

Proof. It is easy to check that μ' verifies the axioms of a measure and that it is normalized, i.e., $\mu'(B_1') + \ldots + \mu'(B_{n+1}') = \mu(B_1) + \ldots + \mu(B_n) = 1$. $\square$

The choice of the parameters λ_i might obviously affect to the achievement of consensus and it has be performed in each specific situation.

With the defined dive function and this extension of the measure, we can thus enact a process to reach consensus in a group evaluation $(\Lambda, \mathscr{Q}_\mathbb{E})$.

Let's suppose that the group is not in consensus, i.e., there exists a subset $\Gamma \subset \Lambda$ such that $\bigcap_{i=1}^{M} Q_i(a_t) = \emptyset \ \forall a_t \in \Gamma$. For each $a_t \in \Gamma$, let n_{a_t} be the first natural number such that

$$\bigcap_{i=1}^{M}(\phi_{n_{a_t}} \circ \cdots \circ \phi_0)(Q_i(a_t)) \neq \emptyset.$$

Considering $n^* = \max\{n_{a_t} | a_t \in \Gamma\}$, the group evaluation obtained, which is $\{\phi_{n^*} \circ \cdots \circ \phi_1 \circ \phi_0 \circ Q_i \mid i = 1, \ldots, M\}$, is in consensus in the space $\mathbb{S}_{n+n^*+1}$.

Now, we can calculate the degree of consensus κ within the group evaluation in which consensus has been obtained.

The following example shows the automatic process to achieve consensus among two evaluators. Once consensus has been achieved, the degree of the obtained consensus is computed.

Example 3. Let us suppose a committee consisting of two members $\mathbb{E} = \{\alpha_1, \alpha_2\}$ evaluates a candidate for a grant. Let us also assume that this candidate is evaluated in terms of three attributes $\Lambda = \{a_1, a_2, a_3\}$. Attribute a_1 is the quality of his/her CV, attribute a_2 is the quality of his/her publications, and a_3 is the quality of his/her research projects. The evaluation of each a_i is done over a space $\mathbb{S}_3$, where $B_1 =$ LOW, $B_2 =$ NORMAL, and $B_3 =$ HIGH.

Suppose that evaluator α_1 gives the candidate the following appraisal:

$$Q_1(a_1) = B_1, Q_1(a_2) = B_1, Q_1(a_3) = [B_2, B_3],$$

and committee member α_2 the following:

$$Q_2(a_1) = B_2, Q_2(a_2) = [B_1, B_2], Q_2(a_3) = B_3.$$

There is no consensus within the group because $Q_1(a_1) \cap Q_2(a_1) = \emptyset$.
The automatic negotiation only requires one step to achieve consensus:

$$(\phi_0 \circ Q_1)(a_1) = [B'_1, B'_2], \ (\phi_0 \circ Q_1)(a_2) = [B'_1, B'_2], \ (\phi_0 \circ Q_1)(a_3) = [B'_2, B'_4],$$
$$(\phi_0 \circ Q_2)(a_1) = [B'_2, B'_3], \ (\phi_0 \circ Q_2)(a_2) = [B'_1, B'_3], \ (\phi_0 \circ Q_2)(a_3) = [B'_3, B'_4],$$

and now $\phi_0 \circ Q_1 \rightleftarrows \phi_0 \circ Q_2$ in $\mathbb{S}_4$.

Let's take, in this example, $\mu(B_1) = \mu(B_2) = \mu(B_3) = 1/3, \overline{\mu}$ as the normalized counter measure, i.e., $\overline{\mu}(X_i) = \dfrac{\text{card}(X_i)}{\text{card}(\Lambda)}$, for each $X_i \subset \Lambda$, and $\lambda_1 = \lambda_2 = \lambda_3 = 1/2$.
So, $\mu'(B'_1) = \mu'(B'_4) = 1/6$ and $\mu'(B'_2) = \mu'(B'_3) = 2/6$.
The degree of consensus is:

$$\kappa(\{\phi_0 \circ Q_1, \phi_0 \circ Q_2\}) = \frac{H((\phi_0 \circ Q_1) \sqcup (\phi_0 \circ Q_2))}{H((\phi_0 \circ Q_1) \cap (\phi_0 \circ Q_2))}.$$

Since

$$((\phi_0 \circ Q_1) \sqcup (\phi_0 \circ Q_2))(\Lambda) = \{[B'_1, B'_3], [B'_1, B'_3], [B'_2, B'_4]\} \text{ and}$$
$$((\phi_0 \circ Q_1) \cap (\phi_0 \circ Q_2))(\Lambda) = \{B'_2, [B'_1, B'_2], [B'_3, B'_4]\},$$

using formula (8) we have:

$$\kappa(\{\phi_0 \circ Q_1, \phi_0 \circ Q_2\}) = \frac{\dfrac{2}{3}\log\dfrac{1}{5/6} + \dfrac{1}{3}\log\dfrac{1}{5/6}}{\dfrac{1}{3}\log\dfrac{1}{2/6} + \dfrac{1}{3}\log\dfrac{1}{3/6} + \dfrac{1}{3}\log\dfrac{1}{3/6}} = 0.22.$$

The main feature of this automatic negotiation process is that it is done by considering data obtained from the evaluators without having to interact with them again. This is useful when working on problems where it is difficult to contact the evaluators later. This situation can be found, for example, when processing data from surveys. In addition, the degree of consensus and the automatic negotiation process allow us to compare the internal coherence of different decision groups. A simple example of this is given below.

Example 4. Let $\mathbb{E}_1, \mathbb{E}_2$ and $\mathbb{E}_3$ be three committees from different areas of knowledge, selected to evaluate respective projects $\Lambda_1, \Lambda_2, \Lambda_3$ for an official announcement. Each committee consists of four evaluators, $\mathbb{E}_i = \{\alpha_1^i, \alpha_2^i, \alpha_3^i, \alpha_4^i\}$, and each project is characterized by two attributes $\Lambda_i = \{a_1^i, a_2^i\}$.

Qualitativization is done over the space $\mathbb{S}_5$ where $B_1 = \text{VERY BAD}, B_2 = \text{BAD}, B_3 = \text{REGULAR}, B_4 = \text{GOOD}$ and $B_5 = \text{VERY GOOD}$. The measure μ of all these basic labels is $1/5$, and the measure $\overline{\mu}$ in each Λ_i is the normalized cardinal measure. In Table 1 we summarize the qualitativizations of each project Λ_i given by the corresponding committee (the qualitativization of member α^i_j is done by means of the function Q^i_j), together with the results of the common and mix operations for each case and the degree of consensus for the third committee.

Table 1 Committees' evaluations.

$\mathbb{E}_1$	Λ_1		$\mathbb{E}_2$	Λ_2		$\mathbb{E}_3$	Λ_3	
	a^1_1	a^1_2		a^2_1	a^2_2		a^3_1	a^3_2
Q^1_1	B_3	B_4	Q^2_1	B_1	B_1	Q^3_1	$[B_2,B_3]$	$[B_1,B_2]$
Q^1_2	B_3	B_4	Q^2_2	B_2	B_2	Q^3_2	$[B_1,B_3]$	$[B_2,B_3]$
Q^1_3	B_3	B_4	Q^2_3	B_3	B_1	Q^3_3	$[B_1,B_2]$	$[B_1,B_2]$
Q^1_4	B_2	B_4	Q^2_4	B_1	$[B_1,B_2]$	Q^3_4	B_2	$[B_2,B_3]$
$\sqcup$	$[B_2,B_3]$	B_4	$\sqcup$	$[B_1,B_3]$	$[B_1,B_2]$	$\sqcup$	$[B_1,B_3]$	$[B_1,B_3]$
$\cap$	–	B_4	$\cap$	–	–	$\cap$	B_2	B_2
κ	–		κ	–		κ	0.32	

The consensus within the third committee is:

$$\kappa(\mathcal{Q}_{\mathbb{E}_3}) = \frac{-\log 3/5}{-\log 1/5} = 0.32$$

As there is no consensus within committees $\mathbb{E}_1$ and $\mathbb{E}_2$ when evaluating the respective projects Λ_1 and Λ_2, the dive function must be applied to these two committees' evaluations. In order to compare the degree of consensus among the three committees it is necessary to deal with the same granularity for all three. Since $\mathbb{E}_2$ needs two steps to reach consensus, the three degrees will be computed in $\mathbb{S}_7$; so, the dive function is applied twice to the three committees. The results are summarized in Table 2.

Table 2 Final consensus degrees, the dive function has been applied twice.

$\mathbb{E}_1$	Λ_1		$\mathbb{E}_2$	Λ_2		$\mathbb{E}_3$	Λ_3	
	a^1_1	a^1_2		a^2_1	a^2_2		a^3_1	a^3_2
Q^1_1	$[B''_3,B''_5]$	$[B''_4,B''_6]$	Q^2_1	$[B''_1,B''_3]$	$[B''_1,B''_3]$	Q^3_1	$[B''_2,B''_5]$	$[B''_1,B''_4]$
Q^1_2	$[B''_3,B''_5]$	$[B''_4,B''_6]$	Q^2_2	$[B''_2,B''_4]$	$[B''_2,B''_4]$	Q^3_2	$[B''_1,B''_5]$	$[B''_2,B''_5]$
Q^1_3	$[B''_3,B''_5]$	$[B''_4,B''_6]$	Q^2_3	$[B''_3,B''_5]$	$[B''_1,B''_3]$	Q^3_3	$[B''_1,B''_4]$	$[B''_1,B''_4]$
Q^1_4	$[B''_2,B''_4]$	$[B''_4,B''_6]$	Q^2_4	$[B''_1,B''_3]$	$[B''_1,B''_4]$	Q^3_4	$[B''_2,B''_4]$	$[B''_2,B''_5]$
$\sqcup$	$[B''_2,B''_5]$	$[B''_4,B''_6]$	$\sqcup$	$[B''_1,B''_5]$	$[B''_1,B''_4]$	$\sqcup$	$[B''_1,B''_5]$	$[B''_1,B''_5]$
$\cap$	$[B''_3,B''_4]$	$[B''_4,B''_6]$	$\cap$	B''_3	$[B''_2,B''_3]$	$\cap$	$[B''_2,B''_4]$	$[B''_2,B''_4]$
κ	0.58		κ	0.28		κ	0.37	

Taking all weights $\lambda_i = 1/2$, the measures of the basic labels of $\mathbb{S}_7$ are: $\mu(B_1'') = \mu(B_7'') = 1/20$, $\mu(B_2'') = \mu(B_6'') = 3/20$ and $\mu(B_3'') = \mu(B_4'') = \mu(B_5'') = 4/20$. As such:

$$\kappa(\mathcal{Q}_{\mathbb{E}_1}) = \frac{-\log 3/4 - \log 11/20}{-\log 2/5 - \log 11/20} = 0.58,$$

$$\kappa(\mathcal{Q}_{\mathbb{E}_2}) = \frac{-\log 4/5 - \log 3/5}{-\log 1/5 - \log 7/20} = 0.28,$$

$$\kappa(\mathcal{Q}_{\mathbb{E}_3}) = \frac{-\log 4/5}{-\log 11/20} = 0.37.$$

Notice that, although committee three, $\mathbb{E}_3$, was the only one achieving consensus at the beginning of the process, Λ_1's final degree of consensus is the greatest because, globally, the evaluations from its members where quite similar, and, therefore, the committee's final consensus is more accurate. The final degree of consensus in Λ_2 is the lowest, because the results of the mix operation in committee $\mathbb{E}_2$ are the less accurate labels.

4 Real Case Study

4.1 Description of Data

The previous methodology has been applied to the set of 44 features of a retailer firm. This extensive list was further reviewed in terms of the nature of the resource (legal, organizational, informational, human, etc). Details of the process of defining these variables are provided in [18] and measures evaluated are listed in table 3.

These features were assessed by eight managers. Managers evaluate each feature with respect to its importance for the success of the firm. In the evaluation of the measure, a $\mathbb{S}_7$ is used. The respondent is asked to indicate his subjective evaluation of the statement. The scale for importance is described as follows: Codes for importance go from $1 = B_1 = $ Absolutely Important to $7 = B_7 = $ Not Very Important. The measure of the basic labels is $\mu(B_i) = 1/7, i = 1,\ldots,7$. In table 3 there are the features and table 4 shows the values given by the managers.

4.2 The Computations

The set Λ consists of the 44 features of the table 3:

$$\Lambda = \{\text{Features}\}.$$

The set $\mathbb{E}$ is formed by the 8 managers:

$$\mathbb{E} = \{M_1,\ldots,M_8\},$$

Table 3 The 44 features considered

		Features
Physical Resource	Reach Ability	1. Number of customer visits
		2. Store location
Legal Resource	Reach Ability	3. The sales of private brand products
		4. Social responsibility
Human Resource	Human Management	5. Turnover
		6. Staff Training
Organizational Resource	Expansion Ability	7. Franchise system
		8. Store opening strategy
	Productivity	9. Sales per store
		10. Spend-per-visit rate
	General Management	11. Internal procedures
		12. Achievement of year-end goals
	Technology Management	13. Investments in technology development
	Management Organizational	14. Quality of data collection and process system
		15. Empowerment
		16. The listening ability of management
		17. Loss control
	Inventory Management	18. Inventory service level
	Marketing Management	19. Market positioning
		20. Store renovation/redecoration
	Management Financial	21. Cost control ability
		22. Percentage of partial-time staff
	Product Innovation	23. Shelf-life of new products
		24. The speed of new product development
	Loan Repay Ability	25. Past credit story
		26. Stockholder's background
	Diversification	27. Capital expenditures in internet channel
		28. Maintaining target customer in market diversification
Informational Resources	Market Segment	29. Following fashion trends
	Risk	30. Facing seasonal demands
	Strategic vision	31. Openness to criticism
		32. Willingness to innovate
Relational Resources	Stakeholder Relations	33. Customer complaints management
		34. Cost sharing with suppliers on promotions
		35. Joint venture opportunity
External Factors	The Actions from Outside	36. Changes in customer?s preferences
		37. Changes in supplier?s contract content
	Stakeholders	38. The innovation and imitation from competitor
	Political Enviromental	39. Change in government laws
		40. Stability of government
	Technological Environmental	41. Innovation of new technology equipment
		42. New management system software development
	Socio-culture Environmental	43. Change of population structure
		44. Change of lifestyle

and $\mathscr{Q}_{\mathbb{E}}$ is the table 4. Using this data the results about the precision are shown in table 5. It is clear that the initial data it is not in consensus, moreover in table 4 can be seen the wide range of the opinions of the evaluators, for example in the specific feature "Having a sound franchise system of stores" the evaluator M_8 gives a 6 to this feature and all the other evaluators evaluate that with a 1, and it will be necessary to apply the automatic negotiation process several times. That process, evidently will give a group in a low consensus because the number of $n^* = 6$:

$$\kappa(\phi^6 \circ \mathscr{Q}_{\mathbb{E}}) = 0.04.$$

The consensus degree can be computed by grouping the data by resources in order to determine if there is a big consensus in some subgroup of features, but again the wide range of the evaluators opinions give a low level of consensus after applying the automatic negotiation process several times, see table 6:

Table 4 Some of the features evaluated by the managers about the importance, and the degree of consensus for each feature.

	Media Saturn managers								
Feature	M_1	M_2	M_3	M_4	M_5	M_6	M_7	M_8	κ
Number of customer visits	7	7	6	7	6	6	7	7	0.53
Store location	3	6	7	5	6	7	7	5	0.03
The sales of private brand products	1	2	6	1	3	2	7	3	0.00
Social responsibility of the firm	3	4	2	3	3	3	4	2	0.20
Turnover of staff numbers	5	5	4	5	[3,5]	4	1	4	0.03
Staff training	6	7	6	7	5	6	7	6	0.08
Having a sound franchise system of stores	1	1	1	1	1	1	1	6	0.03
Store opening strategy	2	5	7	7	5	7	7	7	0.02
Sales per store	7	7	7	7	7	6	7	7	0.53
Customer spend-per-visit rate	6	1	5	4	7	5	6	7	0.00
Internal management procedures	3	5	5	2	6	5	7	6	0.03
Achievement of year-end goals	6	6	5	3	6	6	7	6	0.12
Investments in technology development	5	[4,5]	6	4	5	5	7	6	0.29
Quality of data collection and processing system	5	4	6	5	4	5	6	6	0.03
Empowerment of employees	7	7	4	7	5	5	7	6	0.29
The listening ability of management	5	6	6	5	7	5	7	6	0.03
Control of losses in stores	4	6	6	6	6	3	7	6	0.03
Inventory service level	6	6	5	6	6	3	7	6	0.29
Market positioning	7	5	6	7	7	6	6	7	0.05
Store renovation/redecoration	5	5	4	[3,4]	6	4	7	5	0.03
Cost control ability	3	6	7	6	6	5	7	6	0.03
Percentage of part-time staff	5	1	3	3	5 1	4	[5,6]	6	0.00
Shelf-life of new products	6	1	4	7	5	3	6	6	0.00
The speed of new product development	1	1	6	[1,7]	5	4	6	6	0.00
Past credit history of customers	1	1	1	1	1	2	6	[4,5]	0.00
Investors background	4	2	2	2	1	2	4	[4,5]	0.05
Capital expenditure on internet channel	6	5	7	5	3	3	6	7	0.03

Table 5 The precision of the eight managers.

	M_1	M_2	M_3	M_4	M_5	M_6	M_7	M_8
Precision	0.94	0.97	1	0.84	0.94	1	0.86	0.89

Table 6 The consensus by resources.

	Resources						Factor
	Physical	Legal	Human	Organizational	Informational	Relational	External
n^*	4	6	4	6	6	6	4
κ	0.07	0.08	0.03	0.01	0.01	0.04	0.04

4.3 Cluster

It can be studied the process of group in consensus formation. The next experiment has been performed choosing a subset of features of the original data. With the original set the group was not in consensus. Were needed two steps in the automatic negotiation process to get a two managers be in consensus. The third step gave four groups in consensus (table 7 shows that process):

$$G_1 = \{M_2, M_4, M_5\}, G_2 = \{M_1, M_3, M_4, M_7, M_8\},$$

$$G_3 = \{M_3, M_4, M_6, M_8\}, G_4 = \{M_4, M_5, M_7, M_8\}.$$

Table 7 The managers in consensus after an automatic negotiation.

	ϕ^2									ϕ^3							
M	1	2	3	4	5	6	7	8	M	1	2	3	4	5	6	7	8
1	⇄	-	-	-	-	-	-	-	1	⇄	-	⇄	⇄	-	-	⇄	⇄
2		⇄	-	-	-	-	-	-	2		⇄	-	⇄	⇄	-	-	-
3			⇄	-	-	-	-	-	3			⇄	⇄	-	⇄	⇄	⇄
4				⇄	-	-	⇄	-	4				⇄	⇄	⇄	⇄	⇄
5					⇄	-	-	-	5					⇄	-	⇄	⇄
6						⇄	-	-	6						⇄	-	⇄
7							⇄	-	7							⇄	⇄
8								⇄	8								⇄

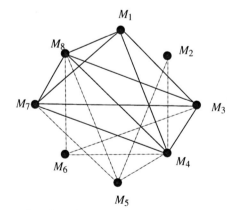

Fig. 7 The consensus graph and the cluster G_2.

The clusters can be seen in the graph in Figure 7, where the vertexes represent the managers, and edges (both continuous and discontinuous) join vertexes in consensus. The edges corresponding to the cluster G_2 are the continuous lines.

5 Conclusions and Future Research

In this chapter we present a methodology to measure precision and consensus in group decisions when the alternatives to be analyzed are represented in an absolute order-of-magnitude qualitative model. The operations considered to aggregate information provide this model a weak partial lattice structure. When there is consensus among the decision group, however, a distributive lattice structure is obtained.

The concept of entropy is introduced in this framework to measure the amount of information within a system when using order-of-magnitude descriptions to represent it. In addition, entropy allows us to measure consensus in group decision-making problems. The degree of consensus is introduced in order to obtain an objective measure of the decision group's reliability. If there is no consensus among the group, an automatic process is then initiated to achieve a global consensus. This process allows comparing the internal coherence of different decision groups.

As future research, and from a theoretical point of view, we plan to investigate in two directions. On one hand, both different degrees of consensus and different distances could be considered and compared, for instance, by defining conditional entropy and the information distance [1] derived from it in this framework. On the other hand, the clustering process obtained through automatic negotiation together with the distance presented will be the basis for a recommender system allowing recommendation to be derived from the nearest neighbor within a group of customers that is in consensus. A previous search of groups being in consensus with the user will make calculating a minimum distance easier and with lower cost.

In addition, this work and the related methodology could be applied to develop techniques to detect malfunctioning within an evaluation committee, i.e., finding incoherencies due to corruption or a lack of knowledge, and avoiding potential subjectivity caused by conflicts of interest regarding evaluators.

References

1. Bennet, C., Gács, P., Ming, L., Vitányi, M., Zurek, W.: Information distance. In: Proc. 25th ACM Symp. Theory of Comput., pp. 21–30 (1993)
2. Birkhoff, G.: Lattice Theory. American Mathematical Society (Colloquium Publications) (1967)
3. Boujelben, M.A., De Smet, Y., Frikha, A., Chabchoub, H.: Building a binary outranking relation in uncertain, imprecise and multi-experts contexts: The application of evidence theory. International Journal of Approximate Reasoning 50(8), 1259–1278 (2009)
4. Cabrerizo, F., Alonso, E., Herrera-Viedma, E.: A consensus model for group decision making problems with unbalanced fuzzy linguistic information. International Journal of Information Technology & Decision Making 8(1), 109–131 (2009)

5. Cabrerizo, F., Moreno, J., Pérez, J., Herrera-Viedma, E.: Analyzing consensus approaches in fuzzy group decision making: advantages and drawbacks. Soft Computing 14(5), 451–463 (2010)
6. Chiclana, F., Mata, F., Martínez, L., Herrera-Viedma, E., Alonso, S.: Integration of a consistency control module within a consensus model. International Journal of Uncertainty Fuzziness and Knowledge-Based Systems 16(1), 35–53 (2008)
7. Dague, P.: Numeric reasoning with relative orders of magnitude. In: AAAI Conference, Washington, pp. 541–547 (1993)
8. Dubois, D., Prade, H.: Properties of measures of information in evidence and possibility theories. Fuzzy Sets and Systems 100, 35–49 (1999)
9. Dubois, D., Prade, H.: Possibility theory, probability theory and multiple valued logics: A clarification. Annals of Mathematics and Artificial Intelligence 32, 35–66 (2001)
10. Eklund, P., Rusinowski, A., De Swart, H.: Consensus reaching in committees. European Journal of Operational Research 178, 185–193 (2007)
11. Herrera, F., Herrera-Viedma, E., Verdegay, J.: A model of consensus in group decision making under linguistic assessments. Fuzzy Sets and Systems (78) 73–87 (1996)
12. Herrera, F., Herrera-Viedma, E., Verdegay, J.: A rational consensus model in group decision making using linguistic assessments. Fuzzy Sets and Systems (81) 31–49 (1997)
13. Forbus, K.: Qualitative process theory. Artificial Intelligence 24, 85–158 (1984)
14. Forbus, K.: Qualitative Reasoning. CRC Hand-book of Computer Science and Engineering. CRC Press, Boca Raton (1996)
15. Grätzer, G.: General Lattice Theory. Birkhäuser, Basel (1998)
16. Halmos, P.R.: Measure Theory. Springer, Heidelberg (1974)
17. Herrera-Viedma, E., Alonso, S., Chiclana, F., Herrera, F.: A consensus model for group decision making with incomplete fuzzy preference relations. IEEE Transaccions on Fuzzy Systems 15(5), 863–877 (2007)
18. Hu, Y., Dawson, J., Ansell, J.: A framework for retail performance measurement. In: An application of resource advantage theory of competition. Tech. rep., School of Management and Echonomics, the University of Edimburgh (2007)
19. Jiang, Y.: An approach to group decision-making based on interval fuzzy preference relations. Journal of Systems Science and Systems Engineering 16(1), 113–120 (2007)
20. Kacprzyk, J., Fedrizzi, M., Nurmi, H.: Group decission making and consensus under fuzzy preferences and fuzzy majority. Fuzzy Sets and Systems 49, 21–31 (1992)
21. Kalagnanam, J., Simon, H., Iwasaki, Y.: The mathematical bases for qualitative reasoning. In: IEEE Expert, pp. 11–19. IEEE Computer Society Press, Los Alamitos (1991)
22. Klir, G.: Where do we stand on measures of uncertainty, ambiguity, fuzziness, and the like? Fuzzy Sets and Systems 24(2), 141–160 (1987)
23. Klir, G., Smith, R.: On measuring uncertainty and uncertainty-based information: Recent developments. Annals of Mathematics and Artificial Intelligence 32, 5–33 (2001)
24. Kuipers, B.: Making sense of common sense knowledge. Ubiquity 4(45) (2004)
25. Martínez, L., Montero, J.: Challenges for improving consensus reaching process in collective decisions. New Mathematics and Natural Computation 3(2), 203–217 (2007)
26. Mata, F., Martínez, L., Herrera-Viedma, E.: An adaptive consensus support model for group decision making problems in a multi-granular fuzzy linguistic context. IEEE Transaccions on Fuzzy Systems 17(2), 279–290 (2009)
27. Missier, A., Piera, N., Travé, L.: Order of magnitude algebras: a survey. Revue d'Intelligence Artificielle 3(4), 95–109 (1989)
28. Ngwenyama, O., Bryson, N., Mobolurin, A.: Supporting facilitation in group support systems: Techniques for analyzing consensus relevant data. Decision Support Systems 16, 155–168 (1996)

29. Roselló, L., Prats, F., Sánchez, M., Agell, N.: A definition of entropy based on qualitative descriptions. In: Bradley, E., Travé-Massuyés, L. (eds.) 22nd International Workshop on Qualitative Reasoning (QR 2008), University of Boulder, Colorado (2008)
30. Shafer, G.: A Mathematical Theory of Evidence. Princeton University Press, Princeton (1976)
31. Shannon, C.E.: A mathematical theory of communication. The Bell System Technical Journal 27, 379–423 (1948)
32. Struss, P.: Mathematical aspects of qualitative reasoning. AI in Engineering 3(3), 156–169 (1988)
33. Travé-Massuyès, L., Dague, P. (eds.): Modèles et raisonnements qualitatifs. Hermes Science Publications, Lavoisier (2003)
34. Travé-Massuyès, L., Piera, N.: The orders of magnitude models as qualitative algebras. In: 11th IJCAI, pp. 1261–1266 (1989)
35. Xie, G., Zhang, J., Lai, K., Yu, L.: Variable precision rough set for group decision making: An application. International Journal of Approximate Reasoning 49(2), 331–343 (2008)
36. Zadeh, L.: Fuzzy sets. Information and Control 8, 338–353 (1965)

On Consensus in Group Decision Making Based on Fuzzy Preference Relations

Meimei Xia and Zeshui Xu*

Abstract. In the process of decision making, the decision makers usually provide inconsistent fuzzy preference relations, and it is unreasonable to get the priority from an inconsistent preference relation. In this paper, we propose a method to derive the multiplicative consistent fuzzy preference relation from an inconsistent fuzzy preference relation. The fundamental characteristic of the method is that it can get a consistent fuzzy preference relation considering all the original preference values without translation. Then, we develop an algorithm to repair a fuzzy preference relation into the one with weak transitivity by using the original fuzzy preference relation and the constructed consistent one. After that, we propose an algorithm to help the decision makers reach an acceptable consensus in group decision making. It is worth pointing out that group fuzzy preference relation derived by using our method is also multiplicative consistent if all individual fuzzy preference relations are multiplicative consistent. Some examples are also given to illustrate our results.

Keywords: Group decision making; fuzzy preference relation; multiplicative consistency; weak transitivity; consensus.

1 Introduction

Preference relation is a very useful tool for providing information about the comparison of alternatives in decision making. Multiplicative preference relations (Saaty, 1980) and fuzzy preference relations (Tanino, 1984, 1988, 1990) are two of the most common preference relations. Over the last decades, many researchers have investigated multiplicative preference relations and achieved substantial

_____

Meimei Xia · Zeshui Xu
School of Economics and Management,
Southeast University, Nanjing, Jiangsu 211189, China
e-mail: xu_zeshui@263.net, meimxia@163.com

* Corresponding author.

E. Herrera-Viedma et al. (Eds.): Consensual Processes, STUDFUZZ 267, pp. 263–287.
springerlink.com   © Springer-Verlag Berlin Heidelberg 2011

results (Xu and Wei, 1999; Chiclana et al., 2001; Herrera et al., 2001; Xu, 2002, etc.). In recent years, more and more authors have been paying great attention to fuzzy preference relations (Xu, 2003; Herrera-Viedma et al., 2004; Xu and Da, 2005; Ma et al., 2006; Chiclana et al., 2008a, 2008b; Chicalna et al., 2009; Xu et al., 2009, etc.), most of them have mainly investigated how to get the priority of a fuzzy preference relation or to repair the inconsistency of a fuzzy preference relation or to estimate the unknown elements of an incomplete fuzzy preference relation. Xu (2003) presented an approach to improving consistency of a fuzzy preference relation and gave a practical iterative algorithm to derive a modified fuzzy preference relation with acceptable consistency by using the transformation formulas of fuzzy preference relation and multiplicative preference relation. Herrera-Viedma et al. (2004) presented a new characterization of consistency associated with the additive transitivity property of a fuzzy preference relation. They also gave a method to construct a consistent fuzzy preference relation from $n-1$ preference data. Xu and Da (2005) proposed a least deviation method to obtain a priority vector of a fuzzy preference relation based on the transformation relationship between fuzzy preference relation and multiplicative preference relation. Ma et al. (2006) presented a method to repair the inconsistency of a fuzzy preference relation to reach weak transitivity via a synthesis matrix which reflects the relationship between the fuzzy preference relation with additive consistency and the initial one given by a decision maker. Chiclana et al. (2008a) gave some methods to construct a consistent preference relation and estimated the missing values in an incomplete fuzzy preference relation which is based on the U-consistent criteria, i.e., the modeling of consistency of preferences via a self-dual almost continuous uninorm. Chiclana et al. (2009) put forward a functional equation to model the "cardinal consistency in the strength of preferences" of reciprocal preference relations. They pointed that the cardinal consistency with the conjunctive representable cross ratio uninorm is equivalent to Tanino (1984)'s multiplicative transitivity property. Although a lot of studies have been done about the consistency of fuzzy preference relations, some of them may produce the loss of the original decision information in the process of transformation or construction, while some only consider part of the original preference values which is unfair for other values. To overcome these issues, in this paper, a new method is given to construct the multiplicative consistent fuzzy preference relation from an inconsistent one preserving the original information as much as possible. An algorithm is also developed to repair the inconsistent fuzzy preference relation into the one with weak transitivity based on the constructed multiplicative consistent fuzzy preference relation and the original one.

In group decision making, the decision makers may come from different fields and have different background cultures which implies that they may have some divergent opinions. Thus how to reach group consensus is an interesting topic which has attracted great attention from many researchers (Inohara, 2000; Mohammed and Ringseis, 2001; Ben-Arieh and Chen, 2006; Ben-Arieh and Easton, 2007, etc.). Herrera-Viedma et al. (2002) proposed a consensus model for multi-person decision making problems with different preference structures based on two consensus criteria: 1) a consensus measure which indicates the agreement between decision makers' opinions and 2) a measure of proximity to find out how far the individual opinions are from the group opinion. Herrera-Viedma et al. (2005) presented a model of consensus support system to assist the decision makers in all phases of the consensus reaching process of group decision making problem with multi-granular linguistic preference relations. Herrera-Viedma et al. (2007) presented a consensus model for group decision making problems with incomplete fuzzy preference relations not only based on consensus measures but also on consistency measures, both of which are used to design a feedback mechanism that generates advice to the decision makers on how they should change and complete their fuzzy preference relations to obtain a solution with high consensus degree and maintaining a certain consistency level on their fuzzy preference relations. On the basis of additive weighted aggregation, Xu (2009) developed an automatic approach to reaching consensus among group opinions, which can avoid forcing the decision makers to modify their opinions. In this paper, an automatic approach is proposed to deal with the consensus of group fuzzy preferences based on the proposed multiplicative consistent fuzzy preference relation.

The remainder is constructed as follows: Section 2 gives an equivalent formula for the multiplicative consistency of fuzzy preference relation to identify whether a fuzzy preference relation is consistent or not, it goes further to give a method to construct a multiplicative consistent fuzzy preference relation from an inconsistent one. Section 3 proposes an algorithm to repair a fuzzy preference relation to the one with weak transitivity. In Section 4, we study the consensus of group decision making and develop an algorithm to reach an acceptable group consensus from individual fuzzy preference relations. Section 5 gives the concluding remarks.

2   The Construction of Multiplicative Consistent Fuzzy Preference Relations

Let $X = (x_1, x_2, \cdots, x_n)$ be a fixed set, then $R = (r_{ij})_{n \times n}$ is called a fuzzy preference relation (Orlovski, 1978) on $X \times X$ with the condition that:

$$r_{ij} \geq 0, \; r_{ij} + r_{ji} = 1, \; i, j = 1, 2, \cdots, n \qquad (1)$$

where r_{ij} denotes the degree that the alternative x_i is prior to the alternative x_j. i.e., $0.5 < r_{ij} < 1$ denotes that the alternative x_i is preferred to the alternative x_j, especially, $r_{ij} = 1$ denotes that the alternative x_i is absolutely preferred to the alternative x_j and $r_{ij} = 0.5$ denotes that there is no difference between the alternative x_i and the alternative x_j.

R is called an additive consistent fuzzy preference relation (Saaty, 1980; Mohammed and Ringseis, 2001; Ma et al., 2006), if it satisfies the additive transitivity property (Tanino, 1984):

$$r_{ij} = r_{ij} + r_{jk} - 0.5, \ i, j, k = 1, 2, \cdots, n \tag{2}$$

It is clear that the additive consistency property has some disadvantages, for example, if $r_{12} = 0.8$ and $r_{23} = 0.9$, then $r_{13} = 0.8 + 0.9 - 0.5 = 1.2 > 1$, which is not reasonable. Although it can be transformed into the value in $[0,1]$ by using Herrera-Viedma et al.'s method (2004), some preference information will be lost. If we use multiplicative consistency property, such a situation will never happen.

R is called a multiplicative consistent preference relation (Tanino, 1984, 1988, 1990), if it satisfies the multiplicative transitivity property:

$$r_{ij} r_{jk} r_{ki} = r_{ji} r_{kj} r_{ik}, \ i, j, k = 1, 2, \cdots, n \tag{2}$$

where $r_{ij} > 0$, for $i, j = 1, 2, \cdots, n$.

By the simple algebraic manipulation, Eq.(2) can be expressed as (Chiclana et al., 2009):

$$r_{ik} = \frac{r_{ij} r_{jk}}{r_{ij} r_{jk} + (1 - r_{ij})(1 - r_{jk})}, \ i, j, k = 1, 2, \cdots, n \tag{3}$$

Another important property of the fuzzy preference relation is the weak transitivity (Tanino, 1984, 1988) described as: If $r_{ij} \geq 0.5$ and $r_{jk} \geq 0.5$, then $r_{ik} \geq 0.5$, for $i, j, k = 1, 2, \cdots, n$.

Property 2.1 (Chiclana et al. (2009). *If a fuzzy preference relation $R = (r_{ij})_{n \times n}$ is multiplicative consistent, then it has weak transitivity.*

In this paper, we mainly discuss the multiplicative consistency of the fuzzy preference relation, thus it is necessary to assume that all the elements in the fuzzy preference relation R satisfy the condition that $r_{ij} > 0 (i, j = 1, 2, \cdots, n)$ in the remainder of this paper.

Based on the multiplicative transitivity, we first introduces a new characterization of the multiplicative consistent fuzzy preference relation, based on which, a method is given to construct the multiplicative consistent fuzzy preference relation from an inconsistent one.

Theorem 2.1. *For a fuzzy preference relation R, the following statements are equivalent:*

1) $r_{ik} = \dfrac{r_{ij} r_{jk}}{r_{ij} r_{jk} + (1 - r_{ij})(1 - r_{jk})}, \ i, j, k = 1, 2, \ldots, n$

$$2)\ r_{ik} = \frac{\sqrt[n]{\prod_{t=1}^{n}(r_{it}r_{tk})}}{\sqrt[n]{\prod_{t=1}^{n}(r_{it}r_{tk})} + \sqrt[n]{\prod_{t=1}^{n}((1-r_{it})(1-r_{tk}))}},\ i,k = 1,2,...,n$$

Proof. "1)$\Rightarrow$2)". Suppose

$$r_{ik} = \frac{r_{ij}r_{jk}}{r_{ij}r_{jk} + (1-r_{ij})(1-r_{jk})},\ i,j,k = 1,2,...,n \qquad (4)$$

$$\frac{r_{ik}}{1-r_{ik}} = \frac{(r_{it}r_{tk})/(r_{it}r_{tk} + (1-r_{it})(1-r_{tk}))}{1-(r_{it}r_{tk})/(r_{it}r_{tk} + (1-r_{it})(1-r_{tk}))}$$

$$= \frac{r_{it}r_{tk}}{(1-r_{it})(1-r_{tk})},\ i,k,t = 1,2,...,n \qquad (5)$$

then

$$r_{ik} = \frac{\sqrt[n]{(r_{ik}/1-r_{ik})^n}}{1 + \sqrt[n]{(r_{ik}/1-r_{ik})^n}}$$

$$= \frac{\sqrt[n]{(r_{it}r_{tk}/((1-r_{it})(1-r_{tk})))^n}}{1 + \sqrt[n]{(r_{it}r_{tk}/((1-r_{it})(1-r_{tk})))^n}}$$

$$= \frac{\sqrt[n]{(\prod_{t=1}^{n}(r_{it}r_{tk}))/(\prod_{t=1}^{n}((1-r_{it})(1-r_{tk})))}}{1 + \sqrt[n]{(\prod_{t=1}^{n}(r_{it}r_{tk}))/(\prod_{t=1}^{n}((1-r_{it})(1-r_{tk})))}}$$

$$= \frac{\sqrt[n]{\prod_{t=1}^{n}(r_{it}r_{tk})}}{\sqrt[n]{\prod_{t=1}^{n}(r_{it}r_{tk})} + \sqrt[n]{\prod_{t=1}^{n}((1-r_{it})(1-r_{tk}))}},\ i,k = 1,2,...,n \qquad (6)$$

"2)$\Rightarrow$1)". Suppose

$$r_{ij} = \frac{\sqrt[n]{\prod_{t=1}^{n}(r_{it}r_{tj})}}{\sqrt[n]{\prod_{t=1}^{n}(r_{it}r_{tj})} + \sqrt[n]{\prod_{t=1}^{n}((1-r_{it})(1-r_{tj}))}},\ i,j = 1,2,...,n \qquad (7)$$

$$r_{jk} = \frac{\sqrt[n]{\prod_{t=1}^{n}(r_{jt}r_{tk})}}{\sqrt[n]{\prod_{t=1}^{n}(r_{jt}r_{tk})} + \sqrt[n]{\prod_{t=1}^{n}((1-r_{jt})(1-r_{tk}))}},\ j,k = 1,2,...,n \qquad (8)$$

$$r_{ik} = \frac{\sqrt[n]{\prod_{t=1}^{n}(r_{it}r_{tk})}}{\sqrt[n]{\prod_{t=1}^{n}(r_{it}r_{tk})} + \sqrt[n]{\prod_{t=1}^{n}((1-r_{it})(1-r_{tk}))}}, \; i,k = 1,2,...,n \quad (9)$$

Let

$$S_{ij} = \sqrt[n]{\prod_{t=1}^{n}(r_{it}r_{tj})} + \sqrt[n]{\prod_{t=1}^{n}((1-r_{it})(1-r_{tj}))}, \; i,j = 1,2,...,n \quad (10)$$

$$T_{jk} = \sqrt[n]{\prod_{t=1}^{n}(r_{jt}r_{tk})} + \sqrt[n]{\prod_{t=1}^{n}((1-r_{jt})(1-r_{tk}))}, \; j,k = 1,2,...,n \quad (11)$$

then

$$\frac{r_{ij}r_{jk}}{r_{ij}r_{jk} + (1-r_{ij})(1-r_{jk})}$$

$$= \frac{\sqrt[n]{\prod_{t=1}^{n}(r_{it}r_{tj}r_{jt}r_{tk})}\big/ S_{ij}T_{jk}}{\sqrt[n]{\prod_{t=1}^{n}(r_{it}r_{tj}r_{jt}r_{tk})}\big/ S_{ij}T_{jk} + \left(1 - \sqrt[n]{\prod_{t=1}^{n}(r_{it}r_{tj})}\big/ S_{ij}\right)\left(1 - \sqrt[n]{\prod_{t=1}^{n}(r_{jt}r_{tk})}\big/ T_{jk}\right)}$$

$$= \frac{\sqrt[n]{\prod_{t=1}^{n}(r_{it}r_{tj}r_{jt}r_{tk})}}{\sqrt[n]{\prod_{t=1}^{n}(r_{it}r_{tj}r_{jt}r_{tk})} + \sqrt[n]{\prod_{t=1}^{n}((1-r_{it})(1-r_{tj})(1-r_{jt})(1-r_{tk}))}}$$

$$= \frac{\sqrt[n]{\prod_{t=1}^{n}(r_{it}r_{tk})}}{\sqrt[n]{\prod_{t=1}^{n}(r_{it}r_{tk})} + \sqrt[n]{\prod_{t=1}^{n}((1-r_{it})(1-r_{tk}))}} = r_{ik}, \; i,j,k = 1,2,...,n \quad (12)$$

and

$$r_{ik} + r_{ki} = \frac{\sqrt[n]{\prod_{t=1}^{n}(r_{it}r_{tk})}}{\sqrt[n]{\prod_{t=1}^{n}(r_{it}r_{tk})} + \sqrt[n]{\prod_{t=1}^{n}((1-r_{it})(1-r_{tk}))}}$$

$$+ \frac{\sqrt[n]{\prod_{t=1}^{n}(r_{kt}r_{ti})}}{\sqrt[n]{\prod_{t=1}^{n}(r_{kt}r_{ti})} + \sqrt[n]{\prod_{t=1}^{n}((1-r_{kt})(1-r_{ti}))}}$$

$$= \frac{\sqrt[n]{\prod_{t=1}^{n}(r_{it}r_{tk})}}{\sqrt[n]{\prod_{t=1}^{n}(r_{it}r_{tk})}+\sqrt[n]{\prod_{t=1}^{n}((1-r_{it})(1-r_{tk}))}}$$

$$+\frac{\sqrt[n]{\prod_{t=1}^{n}(1-r_{tk})(1-r_{it})}}{\sqrt[n]{\prod_{t=1}^{n}(1-r_{tk})(1-r_{it})}+\sqrt[n]{\prod_{t=1}^{n}(r_{it}r_{tk})}}=1,\ i,k=1,2,...,n \qquad (13)$$

which completes the proof of Theorem 2.1. □

By Theorem 2.1 and the definition of multiplicative consistent fuzzy preference relation, the following result can easily be given:

Corollary 2.1. *The fuzzy preference relation R is multiplicative consistent, if it satisfies the following:*

$$r_{ik}=\frac{\sqrt[n]{\prod_{t=1}^{n}(r_{it}r_{tk})}}{\sqrt[n]{\prod_{t=1}^{n}(r_{it}r_{tk})}+\sqrt[n]{\prod_{t=1}^{n}((1-r_{it})(1-r_{tk}))}}\ ,\ i,k=1,2,...,n \qquad (14)$$

However, the fuzzy preference relation given by the decision maker is usually inconsistent which may due to his/her not possessing a precise or sufficient level of knowledge of part of the problem, or because that the decision maker is unable to discriminate the degree to which some alternatives are better than the others. Thus it is very important how to provide the decision maker some useful tools to help them get a consistent fuzzy preference relation. Corollary 2.1 gives us an approach to constructing the multiplicative consistent fuzzy preference relation from an inconsistent one by considering all the elements in the original preference relation.

For a fuzzy preference relation R, the multiplicative consistent fuzzy preference relation of R is given as $\bar{R}=(\bar{r}_{ik})_{n\times n}$ where

$$\bar{r}_{ik}=\frac{\sqrt[n]{\prod_{t=1}^{n}(r_{it}r_{tk})}}{\sqrt[n]{\prod_{t=1}^{n}(r_{it}r_{tk})}+\sqrt[n]{\prod_{t=1}^{n}((1-r_{it})(1-r_{tk}))}}\ ,\ i,k=1,2,...,n \qquad (15)$$

If the original fuzzy preference relation is multiplicative consistent, then by Eq.(15), we can get a fuzzy preference relation which is the same as the original one; Otherwise the fuzzy preference relation derived by Eq.(15) is different from the original one.

Based on the above analysis, below we give an equivalent form of the definition of the traditional multiplicative consistent fuzzy preference relation which is based on Eq.(2) or Eq.(3).

Definition 2.1. *Let R be a fuzzy preference relation, and $\bar{R}$ be the fuzzy preference relation constructed by using Eq.(15). If $\bar{r}_{ik} = r_{ik}$, for all $i, k = 1,2,...,n$, then R is called a multiplicative consistent fuzzy preference relation; Otherwise, R is called an inconsistent fuzzy preference relation.*

In order to measure the consistent degree of a fuzzy preference relation, let

$$d(R,\bar{R}) = \frac{1}{n^2}\sum_{i=1}^{n}\sum_{j=1}^{n}\left|r_{ij} - \bar{r}_{ij}\right| \tag{16}$$

denote the deviation between the fuzzy preference relation $R = (r_{ij})_{n\times n}$ and the multiplicative consistent fuzzy preference relation $\bar{R} = (\bar{r}_{ij})_{n\times n}$ constructed by using Eq.(15). The smaller the value $d(R,\bar{R})$, the more consistent the fuzzy preference relation R . Especially, if $d(R,\bar{R}) = 0$, then R is multiplicative consistent.

From literature review, we can see many methods have been developed to construct the additive consistency preference relation based on the additive transitivity.

Example 2.1. *(Ma et al., 2006). Suppose that a decision maker gives a fuzzy preference relation on an alternative set $X = \{x_1, x_2, x_3, x_4\}$ as follows:*

$$R = \begin{pmatrix} 0.5 & 0.1 & 0.6 & 0.7 \\ 0.9 & 0.5 & 0.8 & 0.4 \\ 0.4 & 0.2 & 0.5 & 0.9 \\ 0.3 & 0.6 & 0.1 & 0.5 \end{pmatrix}$$

Ma et al. (2006)' method estimated the additive consistent fuzzy preference relation of R based on additive consistency as follows:

$$P = \begin{pmatrix} 0.5 & 0.325 & 0.475 & 0.600 \\ 0.675 & 0.5 & 0.650 & 0.775 \\ 0.525 & 0.350 & 0.5 & 0.625 \\ 0.400 & 0.225 & 0.375 & 0.5 \end{pmatrix}$$

Ma et al. (2006) denoted that the values of the elements may be less than 0 or greater than 1 in the fuzzy preference relations constructed. In such cases, they used some formulas to transform those exceeding values into the ones belonging to the interval $[0,1]$. However, such situations will never happen if our method is used which can preserve more original information.

If we use Eq.(15), then the constructed multiplicative consistent fuzzy preference relation of R can be given as:

$$\bar{R} = \begin{pmatrix} 0.5 & 0.2630 & 0.4164 & 0.6044 \\ 0.7370 & 0.5 & 0.6667 & 0.8107 \\ 0.5836 & 0.3333 & 0.5 & 0.6816 \\ 0.3956 & 0.1893 & 0.3184 & 0.5 \end{pmatrix}$$

Moreover

$$d(R,P) = 0.1563 > d(R,\bar{R}) = 0.1506$$

From an inconsistent fuzzy preference relation, Chiclana et al. (2008a) got an estimated preference relation using the global consistency based estimated values derived from many partial consistent values based on multiplicative consistency. Then they used the estimated preference relation to measure the consistency level of the original fuzzy preference relation. But the estimated preference relation is not consistent.

Example 2.2. *(Chiclana et al., 2008a). Let*

$$R = \begin{pmatrix} 0.5 & 0.55 & 0.7 & 0.95 \\ 0.45 & 0.5 & 0.65 & 0.9 \\ 0.3 & 0.35 & 0.5 & 0.75 \\ 0.05 & 0.1 & 0.25 & 0.5 \end{pmatrix}$$

Chiclana et al. (2008a) constructed the estimated fuzzy preference relation as:

$$UR = \begin{pmatrix} 0.5 & 0.62 & 0.78 & 0.9 \\ 0.38 & 0.5 & 0.7 & 0.89 \\ 0.22 & 0.3 & 0.5 & 0.86 \\ 0.01 & 0.11 & 0.14 & 0.5 \end{pmatrix}$$

Obviously, by Eq.(3), UR is not multiplicative consistent, i.e.,

$$\frac{ur_{12}ur_{23}}{ur_{12}ur_{23} + (1-ur_{12})(1-ur_{23})} = \frac{0.62 \times 0.7}{0.62 \times 0.7 + (1-0.62) \times (1-0.7)}$$

$$= 0.792 \neq ur_{23} = 0.78$$

If we use Eq.(15), then

$$\bar{R} = \begin{pmatrix} 0.5 & 0.5852 & 0.7484 & 0.9281 \\ 0.4148 & 0.5 & 0.6783 & 0.9015 \\ 0.2516 & 0.3217 & 0.5 & 0.8128 \\ 0.0719 & 0.0985 & 0.1872 & 0.5 \end{pmatrix}$$

We can find that $\bar{R}$ is a multiplicative consistent fuzzy preference relation. which is more suitably used to measure the consistency level (in this paper, we don't focus on this issue). Furthermore, we have

$$d(R,UR) = 0.0456 \geq d(R,\bar{R}) = 0.0248$$

Based on the multiplicative consistency, Chiclana et al. (2009) proposed a method to construct the consistent fuzzy preference relation from $n-1$ preference values such that $\{r_{i(i+1)} \mid i = 1,...,n-1\}$. If we apply it in Example 2.2, then the multiplicative consistent fuzzy preference relation of R is:

$$\hat{R} = \begin{pmatrix} 0.5 & 0.55 & 0.69 & 0.87 \\ 0.45 & 0.5 & 0.65 & 0.85 \\ 0.3 & 0.35 & 0.5 & 0.75 \\ 0.05 & 0.1 & 0.25 & 0.5 \end{pmatrix}$$

Although $\hat{R}$ is consistent, it is constructed by the set of the values, $\{0.55, 0.65, 0.75\}$, which does not consider other values of the fuzzy preference relation. Moreover, if we use $\hat{R}$ to measure the consistency levels of the values in R, then the values used to construct the consistent fuzzy preference relation always have the highest consistency levels.

From the above analysis, it can be concluded that our method can not only get the multiplicative consistent fuzzy preference relation considering all the elements in the original one, but also preserve more preference information than the existing methods based on additive transitivity for having no translations.

3   A Method for Repairing the Consistency of Fuzzy Preference Relations

If the fuzzy preference relation given by a decision maker is inconsistent, then it can be transformed into the multiplicative consistent one by using Eq.(15). However, there usually exist large deviations between the initial fuzzy preference relation and the transformed one. It is desirable that the modified fuzzy preference relation not only has weak transitivity, but also maintains the original preference information as much as possible (Ma et al., 2006). This section focuses on this issue and develops a method to repair the consistency of the fuzzy preference relations based on multiplicative transitivity property.

We first give a method to fuse two fuzzy preference relations $R_a = (r_{ija})_{n \times n}$ and $R_b = (r_{ijb})_{n \times n}$ into the fuzzy preference relation $R(\beta) = (r_{ij}(\beta))_{n \times n}$ $(0 \le \beta \le 1)$, where each element $r_{ij}(\beta)$ is defined as below:

$$r_{ij}(\beta) = \frac{(r_{ija})^{1-\beta}(r_{ijb})^{\beta}}{(r_{ija})^{1-\beta}(r_{ijb})^{\beta} + (1-r_{ija})^{1-\beta}(1-r_{ijb})^{\beta}} , \ i, j = 1, 2, ..., n \qquad (17)$$

where β in Eq.(17) is a controlling parameter, the smaller the value of β, the nearer $r_{ij}(\beta)$ is to r_{ija}, the bigger the value of β, the nearer $r_{ij}(\beta)$ is to r_{ijb}. Especially, $r_{ij}(0) = r_{ija}$ and $r_{ij}(1) = r_{ijb}$, $r_{ij}(\beta)$ is the value between r_{ija} and r_{ijb}, then the following theorem can be obtained:

Theorem 3.1. Let $R_a = (r_{ija})_{n \times n}$ and $R_b = (r_{ijb})_{n \times n}$ be two fuzzy preference relations, and $R(\beta) = (r_{ij}(\beta))_{n \times n}$ $(0 \le \beta \le 1)$ be their fusion by Eq.(17), then

1) $\min\{r_{ija}, r_{ijb}\} \le r_{ij}(\beta) \le \max\{r_{ija}, r_{ijb}\}$, $i, j = 1, 2, ..., n$.
2) $R(0) = R_a$, $R(1) = R_b$.
3) $R(\beta)$ is a fuzzy preference relation.

Proof. 2) is obvious. Now we prove 1) and 3), suppose

$$r_{ija} + r_{jia} = 1, \ r_{ijb} + r_{jib} = 1, \ r_{ija} \le r_{ijb}, \ i \le j, \ i, j = 1, 2, ..., n \qquad (18)$$

and

$$r_{ij}(\beta) = \frac{(r_{ija})^{1-\beta}(r_{ijb})^{\beta}}{(r_{ija})^{1-\beta}(r_{ijb})^{\beta} + (1-r_{ija})^{1-\beta}(1-r_{ijb})^{\beta}}$$

$$= \frac{1}{1 + (1/r_{ija} - 1)^{1-\beta}(1/r_{ijb} - 1)^{\beta}} , \ i \le j, \ i, j = 1, 2, ..., n \qquad (19)$$

Since

$$(1/\max\{r_{ija}, r_{ijb}\} - 1)^{1-\beta}(1/\max\{r_{ija}, r_{ijb}\} - 1)^{\beta}$$
$$\le (1/r_{ij}^{(a)} - 1)^{1-\beta}(1/r_{ij}^{(b)} - 1)^{\beta}$$

$$\leq (1/\min\{r_{ija}, r_{ijb}\} - 1)^{1-\beta} (1/\min\{r_{ija}, r_{ijb}\} - 1)^{\beta}, i, j = 1, 2, ..., n \quad (20)$$

then $\min\{r_{ija}, r_{ijb}\} \leq r_{ij}(\beta) \leq \max\{r_{ija}, r_{ijb}\}$. Similarly, for $j \leq i$, we can get the same result. Moreover,

$$r_{ij}(\beta) + r_{ji}(\beta)$$

$$= \frac{(r_{ija})^{1-\beta}(r_{ijb})^{\beta}}{(r_{ija})^{1-\beta}(r_{ijb})^{\beta} + (1-r_{ija})^{1-\beta}(1-r_{ijb})^{\beta}}$$

$$+ \frac{(r_{jia})^{1-\beta}(r_{jib})^{\beta}}{(r_{jia})^{1-\beta}(r_{jib})^{\beta} + (1-r_{jia})^{1-\beta}(1-r_{jib})^{\beta}}$$

$$= \frac{(r_{ija})^{1-\beta}(r_{ijb})^{\beta}}{(r_{ija})^{1-\beta}(r_{ijb})^{\beta} + (1-r_{ija})^{1-\beta}(1-r_{ijb})^{\beta}}$$

$$+ \frac{(1-r_{ija})^{1-\beta}(1-r_{ijb})^{\beta}}{(1-r_{ija})^{1-\beta}(1-r_{ijb})^{\beta} + (r_{ija})^{1-\beta}(r_{ijb})^{\beta}} = 1, i, j = 1, 2, ..., n \quad (21)$$

Thus, $R(\beta)$ is a fuzzy preference relation. □

From Theorem 3.1, we can establish a series of fuzzy preference relations between R_a and R_b with the change of the controlling parameter β in the interval $[0,1]$ according to the Eq.(17). Moreover, there are some other relations among the fuzzy preference relations R_a, R_b and $R(\beta)$ which can be described as:

Theorem 3.2. $R_a = (r_{ija})_{n \times n}$ and $R_b = (r_{ijb})_{n \times n}$ are multiplicative consistent if and only if $R(\beta) = (r_{ij}(\beta))_{n \times n}$ $(0 \leq \beta \leq 1)$ is multiplicative consistent.

Proof. Suppose that $R(\beta)$ is multiplicative consistent, $R(0) = R_a$, $R(1) = R_b$, then R_a and R_b are multiplicative consistent.

Conversely, assume that R_a and R_b are multiplicative consistent, and let

$$M_{ijka} = r_{ija} r_{jka} + (1-r_{ija})(1-r_{jka}), i, j, k = 1, 2, ..., n \quad (22)$$

$$N_{ijkb} = r_{ijb} r_{jkb} + (1-r_{ijb})(1-r_{jkb}), i, j, k = 1, 2, ..., n \quad (23)$$

$$G_{ij} = (r_{ija})^{1-\beta}(r_{ijb})^{\beta} + (1-r_{ija})^{1-\beta}(1-r_{ijb})^{\beta}, i, j = 1, 2, ..., n \quad (24)$$

$$H_{jk} = (r_{jka})^{1-\beta}(r_{jkb})^{\beta} + (1-r_{jka})^{1-\beta}(1-r_{jkb})^{\beta}, j, k = 1, 2, ..., n \quad (25)$$

then

$$r_{ik}(\beta) = \frac{(r_{ika})^{1-\beta}(r_{ikb})^{\beta}}{(r_{ika})^{1-\beta}(r_{ikb})^{\beta} + (1-r_{ika})^{1-\beta}(1-r_{ikb})^{\beta}}$$

$$= \frac{((r_{ija}r_{jka})/M_{ijka})^{1-\beta}((r_{ijb}r_{jkb})/N_{ijkb})^{\beta}}{((r_{ija}r_{jka})/M_{ijka})^{1-\beta}((r_{ijb}r_{jkb})/N_{ijkb})^{\beta} + (1-(r_{ija}r_{jka})/M_{ijka})^{1-\beta}(1-(r_{ijb}r_{jkb})/N_{ijkb})^{\beta}}$$

$$= \frac{(r_{ija}r_{jka})^{1-\beta}(r_{ijb}r_{jkb})^{\beta}}{(r_{ija}r_{jka})^{1-\beta}(r_{ijb}r_{jkb})^{\beta} + ((1-r_{ija})(1-r_{jka}))^{1-\beta}((1-r_{ijb})(1-r_{jkb}))^{\beta}}$$

$$i,j,k = 1,2,...,n \quad (26)$$

On the other hand,

$$\frac{r_{ij}(\beta)r_{jk}(\beta)}{r_{ij}(\beta)r_{jk}(\beta) + (1-r_{ij}(\beta))(1-r_{jk}(\beta))}$$

$$= \frac{((r_{ija}r_{ijb})^{1-\beta}(r_{jka}r_{jkb})^{\beta})/G_{ij}H_{jk}}{((r_{ija}r_{ijb})^{1-\beta}(r_{jka}r_{jkb})^{\beta})/G_{ij}H_{jk} + (1-((r_{ija})^{1-\beta}(r_{ijb})^{\beta})/G_{ij})(1-((r_{jka})^{1-\beta}(r_{jkb})^{\beta})/H_{jk})}$$

$$= \frac{(r_{ija}r_{ijb})^{1-\beta}(r_{jka}r_{jkb})^{\beta}}{(r_{ija}r_{ijb})^{1-\beta}(r_{jka}r_{jkb})^{\beta} + ((1-r_{ija})(1-r_{jka}))^{1-\beta}\left((1-r_{ijb})(1-r_{jkb})\right)^{\beta}}$$

$$i,j,k = 1,2,...,n \quad (27)$$

Therefore

$$r_{ij}(\beta) = \frac{(r_{ija})^{1-\beta}(r_{ijb})^{\beta}}{(r_{ija})^{1-\beta}(r_{ijb})^{\beta} + (1-r_{ija})^{1-\beta}(1-r_{ijb})^{\beta}}, \; i,j = 1,2,...,n \quad (28)$$

which implies that $R(\beta)$ is multiplicative consistent. $\square$

Based on Theorem 3.2, we can conclude that if both R_a and R_b are multiplicative consistent, then the fuzzy preference relations constructed by using Eq.(15) are also consistent, and generally, they contain not only the preference information of R_a but also the preference information of R_b.

Considering that the decision makers often provide the inconsistent fuzzy preference relations, Ma et al. (2006) gave an algorithm to repair the inconsistent fuzzy preference relation into the one with weak transitivity, but they are on the basis of additive consistency which has some defects mentioned in Section 2, thus the repaired fuzzy preference relation they finally got is sometimes unreasonable. In what follows, we utilize Eq.(15) to develop a new algorithm for repairing a fuzzy preference relation into the one with weak transitivity based on the constructed multiplicative consistent fuzzy preference relation and the original one.

Algorithm 3.1.

Let $R = (r_{ij})_{n \times n}$ be an initial fuzzy preference relation, p be the number of iterations, δ be the step size, and $0 \leq p\delta \leq 1$.

Step 1. Construct the multiplicative consistent fuzzy preference relation $\bar{R}$ from R by using Eq.(15).

Step 2. Construct the fused matrix $\hat{R} = (\hat{r}_{ik})_{n \times n}$ by using

$$\hat{r}_{ij} = \frac{r_{ij}^{1-p\delta} \bar{r}_{ij}^{p\delta}}{r_{ij}^{1-p\delta} \bar{r}_{ij}^{p\delta} + (1-r_{ij})^{1-p\delta}(1-\bar{r}_{ij})^{p\delta}}, \ i, j = 1, 2, ..., n \qquad (29)$$

Step 3. If $\hat{R}$ has weak transitivity according to Ma et al.'s method, then go to Step 5; Otherwise, go to the next step.

Step 4. Let $p = p + 1$, go to Step 2.

Step 5. Output $\hat{R}$.

In Algorithm 3.1, the fuzzy preference relation $\hat{R}$ is generated from the original fuzzy preference relation R and the multiplicative consistent fuzzy preference relation $\bar{R}$ under the parameter p. With the increase of the value of the parameter p, the generated fuzzy preference relation $\hat{R}$ can contain more and more information of the multiplicative consistent fuzzy preference relation $\bar{R}$, and ultimately, we will get the fuzzy preference relation $\hat{R}$ with weak transitivity.

The convergence property of Algorithm 3.1 can be shown as follows:

Theorem 3.3. *Algorithm 3.1 can derive a fuzzy preference relation* $\hat{R} = (\hat{r}_{ik})_{n \times n}$ *with weak transitivity from an inconsistent fuzzy preference relation* $R = (r_{ij})_{n \times n}$ *after a finite number of iterations.*

Proof. For a given natural number N, there exists a $\delta \in [0,1]$ such that $p\delta = 1$. Let $\delta = 1/N$, then after N iterations of calculation, we can obtain $\hat{R} = \bar{R}$, where $\bar{R}$ is the multiplicative consistent fuzzy preference relation constructed from R by using Eq.(15). Therefore, by Property 2.1, we know that $\hat{R}$ has weak transitivity.   □

In the following, we utilize the fuzzy preference relation:

$$R = \begin{pmatrix} 0.5 & 0.1 & 0.6 & 0.7 \\ 0.9 & 0.5 & 0.8 & 0.4 \\ 0.4 & 0.2 & 0.5 & 0.9 \\ 0.3 & 0.6 & 0.1 & 0.5 \end{pmatrix}$$

given by Ma et al. (2006) to illustrate Algorithm 3.1:

By Eq.(15), we first construct a multiplicative consistent fuzzy preference relation as follows:

$$\bar{R} = \begin{pmatrix} 0.5 & 0.2630 & 0.4164 & 0.6044 \\ 0.7370 & 0.5 & 0.6667 & 0.8107 \\ 0.5836 & 0.3333 & 0.5 & 0.6816 \\ 0.3956 & 0.1893 & 0.3184 & 0.5 \end{pmatrix}$$

Let $p = 1$, $\delta = 0.1$, then by Eq.(29), we have

$$R_1 = \begin{pmatrix} 0.5 & 0.1110 & 0.5820 & 0.6910 \\ 0.8890 & 0.5 & 0.7887 & 0.4453 \\ 0.4180 & 0.2113 & 0.5 & 0.8863 \\ 0.3090 & 0.5547 & 0.1137 & 0.5 \end{pmatrix}$$

According to Ma et al.'s method, R_1 does not have weak transitivity, then we let $p = 2$, and by Eq.(29), we have

$$R_2 = \begin{pmatrix} 0.5 & 0.1230 & 0.5639 & 0.6819 \\ 0.8770 & 0.5 & 0.7769 & 0.4916 \\ 0.4361 & 0.2231 & 0.5 & 0.8710 \\ 0.3181 & 0.5084 & 0.1290 & 0.5 \end{pmatrix}$$

which does not have weak transitivity. Then, we let $p = 3$, and by Eq.(29), we obtain

$$R_3 = \begin{pmatrix} 0.5 & 0.1362 & 0.5455 & 0.6727 \\ 0.8638 & 0.5 & 0.7647 & 0.5380 \\ 0.4545 & 0.2353 & 0.5 & 0.8540 \\ 0.3273 & 0.4620 & 0.1460 & 0.5 \end{pmatrix}$$

which has weak transitivity.

Therefore, our method obtains the repaired fuzzy preference relation with weak transitivity after three times of iterations. If Ma et al. (2006)'s method is used, then the fuzzy preference relation R_M with weak transitivity can be obtained after the same times of iteration as follows:

$$R_M = \begin{pmatrix} 0.5 & 0.1675 & 0.5625 & 0.6700 \\ 0.8325 & 0.5 & 0.7550 & 0.5125 \\ 0.4375 & 0.2450 & 0.5 & 0.8175 \\ 0.3300 & 0.4875 & 0.1825 & 0.5 \end{pmatrix}$$

and by Eq.(16), we calculate the deviations between the original preference relation R and the repaired ones, and get

$$d(R, R_M) = 0.0469 > d(R, R_3) = 0.0422$$

4 A Consensus Algorithm for Group Decision Making Based on Fuzzy Preference Relations

In group decision making, group consensus is an important issue which refers to how to obtain the maximum degree of consensus or agreement between the set of decision makers on the solution set of alternatives. A lot of work has been done on this topic and can be classified two classes: 1) Calculate the agreement amongst all the decision makers' opinions; 2) How far the individual opinions are from the group opinion. This paper focuses on 2) and proposes a method to reach group consensus based on fuzzy preference relations.

A group decision making can be described as follows: Suppose that m decision makers e_l $(l = 1, 2, \cdots, m)$ provide their individual fuzzy preference relations $R_l = (r_{ijl})_{n \times n}$ $(l = 1, 2, \cdots, m)$ over the alternatives $x_1, x_2, \cdots, x_n$, and $\lambda = (\lambda_1, \lambda_2, \cdots, \lambda_m)^T$ is the weight vector of the decision makers e_l $(l = 1, 2, \cdots, m)$ with the condition that $\sum_{l=1}^{m} \lambda_l = 1$ and $0 \le \lambda_l \le 1$, $l = 1, 2, \cdots, m$. To get the maximum group consensus, we first give an approach to fuse the individual fuzzy preference relations $R_l = (r_{ijl})_{n \times n}$ $(l = 1, 2, \cdots, m)$ into the group opinion:

Theorem 4.1. *Let $R_l = (r_{ijl})_{n \times n}$ $(l = 1, 2, \cdots, m)$ be m individual fuzzy preference relations, then their fusion $R = (r_{ij})_{n \times n}$ is also a fuzzy preference relation, where*

$$r_{ij} = \frac{\prod_{l=1}^{m}(r_{ijl})^{\lambda_l}}{\prod_{l=1}^{m}(r_{ijl})^{\lambda_l} + \prod_{l=1}^{m}(1-r_{ijl})^{\lambda_l}}, \ i,j=1,2,...,n \qquad (30)$$

Proof. It is obvious that

$$0 \le r_{ij} = \frac{\prod_{l=1}^{m}(r_{ijl})^{\lambda_l}}{\prod_{l=1}^{m}(r_{ijl})^{\lambda_l} + \prod_{l=1}^{m}(1-r_{ijl})^{\lambda_l}} \le 1, \ i,j=1,2,...,n \qquad (31)$$

Moreover

$$r_{ij} + r_{ji} = \frac{\prod_{l=1}^{m}(r_{ijl})^{\lambda_l}}{\prod_{l=1}^{m}(r_{ijl})^{\lambda_l} + \prod_{l=1}^{m}(1-r_{ijl})^{\lambda_l}} + \frac{\prod_{l=1}^{m}(r_{jil})^{\lambda_l}}{\prod_{l=1}^{m}(r_{jil})^{\lambda_l} + \prod_{l=1}^{m}(1-r_{jil})^{\lambda_l}}$$

$$= \frac{\prod_{l=1}^{m}(r_{ijl})^{\lambda_l}}{\prod_{l=1}^{m}(r_{ijl})^{\lambda_l} + \prod_{l=1}^{m}(1-r_{ijl})^{\lambda_l}} + \frac{\prod_{l=1}^{m}(1-r_{ijl})^{\lambda_l}}{\prod_{l=1}^{m}(1-r_{ijl})^{\lambda_l} + \prod_{l=1}^{m}(r_{ijl})^{\lambda_l}} = 1,$$

$$i,j=1,2,...,n \quad (32)$$

which completes the proof of Theorem 4.1. □

In Theorem 4.1, Eq.(30) is an extension of Eq.(17) on m dimensions, i.e., if $m=2$, then Eq.(30) reduces to Eq.(17). It is similar to the weighted averaging operator or the weighted geometric operator. Furthermore, based on the idea of the ordered weighted averaging operator (Yager, 1988) or the ordered weighted geometric operator (Chiclana et al., 2001; Xu and Da, 2002), Eq.(17) can also be extended to the following form:

$$r_{ij} = \frac{\prod_{\bar{l}=1}^{m}(r_{ij\bar{l}})^{\lambda_l}}{\prod_{\bar{l}=1}^{m}(r_{ij\bar{l}})^{\lambda_l} + \prod_{\bar{l}=1}^{m}(1-r_{ij\bar{l}})^{\lambda_l}}, \ i,j=1,2,...,n \qquad (33)$$

where $r_{ij\bar{l}}$ is the $\bar{l}$ th largest of $r_{ij\bar{l}}$ $(l=1,2,\cdots,m)$.

Based on Theorem 4.1 and Eq.(3), we can get the following interesting result:

Theorem 4.2. *If all individual fuzzy preference relations* $R_l = (r_{ijl})_{n \times n}$ $(l=1,2,\cdots,m)$ *are multiplicative consistent, then their fusion* $R = (r_{ij})_{n \times n}$ *is also multiplicative consistent.*

Proof. Assume that $R_l = (r_{ijl})_{n \times n}$ $(l=1,2,\cdots,m)$ is multiplicative consistent, and let

$$Y_{ijkl} = r_{ijl}r_{jkl} + (1-r_{ijl})(1-r_{jkl}), \ i,j,k=1,2,...,n \qquad (34)$$

$$E_{ij} = \prod\nolimits_{l=1}^{m}(r_{ijl})^{\lambda_l} + \prod\nolimits_{l=1}^{m}(1-r_{ijl})^{\lambda_l}, \ i,j=1,2,...,n \qquad (35)$$

$$F_{jk} = \prod\nolimits_{l=1}^{m}(r_{jkl})^{\lambda_l} + \prod\nolimits_{l=1}^{m}(1-r_{jkl})^{\lambda_l}, \ j,k=1,2,...,n \quad (36)$$

then we have

$$
\begin{aligned}
r_{ik} &= \frac{\prod_{l=1}^{m}(r_{ikl})^{\lambda_l}}{\prod_{l=1}^{m}(r_{ikl})^{\lambda_l} + \prod_{l=1}^{m}(1-r_{ikl})^{\lambda_l}} \\
&= \frac{\prod_{l=1}^{m}((r_{ijl}r_{jkl})/Y_{ijkl})^{\lambda_l}}{\prod_{l=1}^{m}((r_{ijl}r_{jkl})/Y_{ijkl})^{\lambda_l} + \prod_{l=1}^{m}(1-(r_{ijl}r_{jkl})/Y_{ijkl})^{\lambda_l}} \\
&= \frac{\prod_{l=1}^{m}(r_{ijl}r_{jkl})^{\lambda_l}}{\prod_{l=1}^{m}(r_{ijl}r_{jkl})^{\lambda_l} + \prod_{l=1}^{m}((1-r_{ijl})(1-r_{jkl}))^{\lambda_l}}, \ i,j,k=1,2,...,n \quad (37)
\end{aligned}
$$

On the other hand

$$
\begin{aligned}
r_{ik} &= \frac{r_{ij}r_{jk}}{r_{ij}r_{jk} + (1-r_{ij})(1-r_{jk})} \\
&= \frac{\left(\prod_{l=1}^{m}(r_{ijl})^{\lambda_l}\prod_{l=1}^{m}(r_{jkl})^{\lambda_l}\right)/(E_{ij}F_{jk})}{\left(\prod_{l=1}^{m}(r_{ijl})^{\lambda_l}\prod_{l=1}^{m}(r_{jkl})^{\lambda_l}\right)/(E_{ij}F_{jk}) + (1-\prod_{l=1}^{m}(r_{ijl})^{\lambda_l}/E_{ij})(1-\prod_{l=1}^{m}(r_{jkl})^{\lambda_l}/F_{jk})} \\
&= \frac{\prod_{l=1}^{m}(r_{ijl})^{\lambda_l}\prod_{l=1}^{m}(r_{jkl})^{\lambda_l}}{\prod_{l=1}^{m}(r_{ijl})^{\lambda_l}\prod_{l=1}^{m}(r_{jkl})^{\lambda_l} + \prod_{l=1}^{m}(1-r_{ijl})^{\lambda_l}\prod_{l=1}^{m}(1-r_{jkl})^{\lambda_l}} \\
&= \frac{\prod_{l=1}^{m}(r_{ijl}r_{jkl})^{\lambda_l}}{\prod_{l=1}^{m}(r_{ijl}r_{jkl})^{\lambda_l} + \prod_{l=1}^{m}((1-r_{ijl})(1-r_{jkl}))^{\lambda_l}}, \ i,j,k=1,2,...,n \quad (38)
\end{aligned}
$$

Therefore

$$r_{ik} = \frac{r_{ij}r_{jk}}{r_{ij}r_{jk} + (1-r_{ij})(1-r_{jk})}, \ i,j,k=1,2,...,n \qquad (39)$$

i.e., R is multiplicative consistent. $\qquad\qquad\qquad\qquad\qquad\qquad\qquad\square$

For example, if we fuse the multiplicative consistent fuzzy preference relations obtained in Examples 2.1-2.3 by using Eq.(30), then we have

1) If $\lambda = (0.2, 0.3, 0.5)^{\mathrm{T}}$, then

$$R = \begin{pmatrix} 0.5 & 0.5224 & 0.6569 & 0.7856 \\ 0.4776 & 0.5 & 0.6364 & 0.7701 \\ 0.3431 & 0.3636 & 0.5 & 0.6568 \\ 0.2144 & 0.2299 & 0.3432 & 0.5 \end{pmatrix}$$

2) If $\lambda = (1/3, 1/3, 1/3)^{\mathrm{T}}$, then

$$R = \begin{pmatrix} 0.5 & 0.4978 & 0.6470 & 0.8075 \\ 0.5022 & 0.5 & 0.6490 & 0.8089 \\ 0.3530 & 0.3510 & 0.5 & 0.6959 \\ 0.1925 & 0.1911 & 0.3041 & 0.5 \end{pmatrix}$$

3) If $\lambda = (0.7, 0.2, 0.1)^{\mathrm{T}}$, then

$$R = \begin{pmatrix} 0.5 & 0.5252 & 0.6904 & 0.8816 \\ 0.4748 & 0.5 & 0.6685 & 0.8706 \\ 0.3096 & 0.3315 & 0.5 & 0.7695 \\ 0.1184 & 0.1294 & 0.2305 & 0.5 \end{pmatrix}$$

Clearly, all the above fuzzy preference relations derived by taking different weight vectors are multiplicative consistent.

Based on the above analysis and the idea of Xu (2009), we develop an automatic algorithm to reach group consensus from individual fuzzy preference relations. The details are as follows:

Algorithm 4.1.
Step 1. *Construct the multiplicative consistent fuzzy preference relation* $\overline{R} = (\overline{r}_{ij})_{n \times n}$ *from* $R_l = (r_{ijl})_{n \times n}$ $(l = 1, 2, \cdots, m)$ *by using Eq.(15).*

Step 2. *Fuse all individual preference relations* $\overline{R}_l = (\overline{r}_{ijl})_{n \times n}$ *into a group fuzzy preference relation* $\overline{R} = (\overline{r}_{ij})_{n \times n}$ *by the Eq.(30). For convenience, let* $\overline{R}_l^{(0)} = (\overline{r}_{ijl}^{(0)})_{n \times n} = \overline{R}_l = (\overline{r}_{ijl})_{n \times n}$, $\overline{R}^{(0)} = (\overline{r}_{ij}^{(0)})_{n \times n} = \overline{R} = (\overline{r}_{ij})_{n \times n}$, *and* $s = 0$.

Step 3. *Calculate the deviation degree between each individual fuzzy preference relation* $\overline{R}_l^{(s)}$ *and the group fuzzy preference relation* $\overline{R}^{(s)}$ *by using Eq.(16), i.e.,*

$$d(\overline{R}_l^{(s)}, \overline{R}^{(s)}) = \frac{1}{n^2} \sum_{i=1}^{n} \sum_{j=1}^{n} \left| r_{ijl}^{(s)} - r_{ij}^{(s)} \right| \qquad (40)$$

Suppose that $\rho = \rho^*$ *is the dead line of acceptable deviation between each individual fuzzy preference relation and the group fuzzy preference relation. If* $d(\overline{R}_l^{(s)}, \overline{R}^{(s)}) \le \rho^*$, *for all* $l = 1, 2, \cdots, m$, *then go to Step 5; Otherwise, go to the next step.*

Step 4. *Let* $\overline{R}_l^{(s+1)} = (\overline{r}_{ijl}^{(s+1)})_{n \times n}$ *and if* $l \in \{t | d(\overline{R}_t^{(s)}, \overline{R}^{(s)}) \le \rho^*\}$, *then*

$$\overline{r}_{ijl}^{(s+1)} = \frac{(\overline{r}_{ijl}^{(s)})^{1-\eta} (\overline{r}_{ij}^{(s)})^{\eta}}{(\overline{r}_{ijl}^{(s)})^{1-\eta} (\overline{r}_{ij}^{(s)})^{\eta} + (1-\overline{r}_{ijl}^{(s)})^{1-\eta} (1-\overline{r}_{ij}^{(s)})^{\eta}}, \ i, j = 1, 2, ..., n \quad (41)$$

and else, $\overline{R}_l^{(s+1)} = \overline{R}_l^{(s)}$. *Then let* $\overline{R}^{(s+1)} = (\overline{r}_{ij}^{(s+1)})_{n \times n}$, *where*

$$r_{ij}^{(s+1)} = \frac{\prod_{l=1}^{m} (r_{ijl}^{(s+1)})^{\lambda_l}}{\prod_{l=1}^{m} (r_{ijl}^{(s+1)})^{\lambda_l} + \prod_{l=1}^{m} \prod_{l=1}^{m} (1-r_{ijl}^{(s+1)})^{\lambda_l}}, \ i, j = 1, 2, ..., n \quad (42)$$

let $s = s+1$, *return to step 3.*

Step 5. *Output* $\overline{R}_l^{(s)}$ *and* $\overline{R}^{(s)}$.

Obviously, the prominent characteristic of Algorithm 4.1 is that it can automatically modify the diverging individual fuzzy preference relations so as to reach an acceptable consensus, and thus can avoid forcing the decision makers to modify their preferences, which makes the decision more scientifically and efficiently. The algorithm is very suitable for the cases where it is urgent to obtain a solution of consensus, or the decision makers can not or are unwilling to revaluate the alternatives. In the following, we give an example to illustrate Algorithm 4.1.

Example 4.1. *Suppose that there are four decision makers* $e_l \ (l = 1, 2, , 3, 4)$, *who provide their fuzzy preference relations about the four alternatives* x_1, x_2, x_3 *and* x_4 *as follows:*

$$R_1 = \begin{pmatrix} 0.5 & 0.3 & 0.7 & 0.1 \\ 0.7 & 0.5 & 0.6 & 0.6 \\ 0.3 & 0.4 & 0.5 & 0.2 \\ 0.9 & 0.4 & 0.8 & 0.5 \end{pmatrix}, \ R_2 = \begin{pmatrix} 0.5 & 0.4 & 0.6 & 0.2 \\ 0.6 & 0.5 & 0.7 & 0.4 \\ 0.4 & 0.3 & 0.5 & 0.1 \\ 0.8 & 0.6 & 0.9 & 0.5 \end{pmatrix}$$

$$R_3 = \begin{pmatrix} 0.5 & 0.5 & 0.7 & 0.1 \\ 0.5 & 0.5 & 0.8 & 0.4 \\ 0.3 & 0.2 & 0.5 & 0.2 \\ 0.9 & 0.6 & 0.8 & 0.5 \end{pmatrix}, \quad R_4 = \begin{pmatrix} 0.5 & 0.4 & 0.7 & 0.8 \\ 0.6 & 0.5 & 0.4 & 0.3 \\ 0.3 & 0.6 & 0.5 & 0.1 \\ 0.2 & 0.7 & 0.9 & 0.5 \end{pmatrix}$$

Without loss of generality, here we let $\rho^* = 0.05$. Now we use Algorithm 4.1 to reach acceptable group consensus, which involves the following steps:

Step 1. Construct the multiplicative consistent fuzzy preference relations $\overline{R}_l$ $(l = 1,2,3,4)$ from R_l $(l = 1,2,3,4)$ by using Eq.(15):

$$\overline{R}_1 = \begin{pmatrix} 0.5 & 0.2761 & 0.5276 & 0.2069 \\ 0.7239 & 0.5 & 0.7454 & 0.4061 \\ 0.4724 & 0.2546 & 0.5 & 0.1893 \\ 0.7931 & 0.5939 & 0.8107 & 0.5 \end{pmatrix},$$

$$\overline{R}_2 = \begin{pmatrix} 0.5 & 0.3639 & 0.6262 & 0.2069 \\ 0.6361 & 0.5 & 0.7454 & 0.3132 \\ 0.3738 & 0.2546 & 0.5 & 0.1347 \\ 0.7931 & 0.6868 & 0.8653 & 0.5 \end{pmatrix}$$

$$\overline{R}_3 = \begin{pmatrix} 0.5 & 0.3583 & 0.6382 & 0.2084 \\ 0.6417 & 0.5 & 0.7595 & 0.3204 \\ 0.3618 & 0.2405 & 0.5 & 0.1299 \\ 0.7916 & 0.6796 & 0.8701 & 0.5 \end{pmatrix}$$

$$\overline{R}_4 = \begin{pmatrix} 0.5 & 0.6612 & 0.7534 & 0.5106 \\ 0.3388 & 0.5 & 0.6101 & 0.3483 \\ 0.2466 & 0.3899 & 0.5 & 0.2546 \\ 0.4894 & 0.6517 & 0.7454 & 0.5 \end{pmatrix}$$

Step 2. Fuse all individual fuzzy preference relations $\overline{R}_l$ $(l = 1,2,3,4)$ into group fuzzy preference relation $\overline{R}$ by Eq.(30). For convenience, let $s = 0$, $\overline{R}_l^{(0)} = (\overline{r}_{ijl}^{(0)})_{n \times n} = \overline{R}_l = (\overline{r}_{ijl})_{n \times n}$, and $\overline{R}^{(0)} = (\overline{r}_{ij}^{(0)})_{n \times n} = \overline{R} = (\overline{r}_{ij})_{n \times n}$, $l = 1,2,3,4$, then

$$\bar{R}^{(0)} = \begin{pmatrix} 0.5 & 0.4112 & 0.6405 & 0.2699 \\ 0.5888 & 0.5 & 0.7184 & 0.3462 \\ 0.3595 & 0.2816 & 0.5 & 0.1718 \\ 0.7301 & 0.6538 & 0.8282 & 0.5 \end{pmatrix}$$

Step 3. Calculate the deviation degree between each individual preference relation $\bar{R}_l^{(0)}$ and the group preference relation $\bar{R}^{(0)}$ by Eq.(40):

$$d(\bar{R}^{(0)}, \bar{R}_1^{(0)}) = 0.0519, \ d(\bar{R}^{(0)}, \bar{R}_2^{(0)}) = 0.0277,$$
$$d(\bar{R}^{(0)}, \bar{R}_3^{(0)}) = 0.0282, \ d(\bar{R}^{(0)}, \bar{R}_4^{(0)}) = 0.0996$$

thus

$$d(\bar{R}, \bar{R}_1) = 0.0519 > 0.05, \ d(\bar{R}, \bar{R}_4) = 0.0996 > 0.05$$

and then, go to Step 4.

Step 4. Suppose $\eta = 0.5$, then we recalculate $\bar{R}_l^{(1)} = (\bar{r}_{ijl}^{(1)})_{n \times n}$, $l = 1, 4$ and $\bar{R}^{(1)} = (\bar{r}_{ij}^{(1)})_{n \times n}$, where $\bar{R}_l^{(1)} = \bar{R}_l^{(0)}$, $l = 2, 3$, and

$$\bar{r}_{ijl}^{(1)} = \frac{(\bar{r}_{ijl}^{(0)})^{1-\eta} (\bar{r}_{ij}^{(0)})^{\eta}}{(\bar{r}_{ijl}^{(0)})^{1-\eta} (\bar{r}_{ij}^{(0)})^{\eta} + (1-\bar{r}_{ijl}^{(0)})^{1-\eta} (1-\bar{r}_{ij}^{(0)})^{\eta}}, \ i, j, l = 1, 4$$

$$r_{ij}^{(1)} = \frac{\prod_{l=1}^{m} (r_{ijl}^{(1)})^{\lambda_l}}{\prod_{l=1}^{m} (r_{ijl}^{(1)})^{\lambda_l} + \prod_{l=1}^{m} (1-r_{ijl}^{(1)})^{\lambda_l}}, \ i, j = 1, 2, 3, 4$$

thus

$$\bar{R}_1^{(1)} = \begin{pmatrix} 0.5 & 0.3404 & 0.5852 & 0.2370 \\ 0.6596 & 0.5 & 0.7321 & 0.3757 \\ 0.4148 & 0.2679 & 0.5 & 0.1804 \\ 0.7630 & 0.6243 & 0.8196 & 0.5 \end{pmatrix}$$

$$\bar{R}_4^{(1)} = \begin{pmatrix} 0.5 & 0.5387 & 0.7000 & 0.3831 \\ 0.4613 & 0.5 & 0.6665 & 0.3472 \\ 0.3000 & 0.3335 & 0.5 & 0.2102 \\ 0.6169 & 0.6528 & 0.7898 & 0.5 \end{pmatrix}$$

$$\bar{R}^{(1)} = \begin{pmatrix} 0.5 & 0.3985 & 0.6385 & 0.2533 \\ 0.6015 & 0.5 & 0.7272 & 0.3387 \\ 0.3615 & 0.2728 & 0.5 & 0.1611 \\ 0.7467 & 0.6613 & 0.8389 & 0.5 \end{pmatrix}$$

Then, let $s = 1$, and return to Step 3, i.e., we need to recalculate the deviation degree between each individual preference relation $\bar{R}_l^{(1)}$ and the group preference relation $\bar{R}^{(1)}$ by Eq.(40):

$$d(\bar{R}^{(1)}, \bar{R}_1^{(1)}) = 0.0236, \ d(\bar{R}^{(1)}, \bar{R}_2^{(1)}) = 0.0204,$$
$$d(\bar{R}^{(1)}, \bar{R}_3^{(1)}) = 0.0209, \ d(\bar{R}^{(1)}, \bar{R}_4^{(1)}) = 0.0562$$

Since $d(\bar{R}^{(1)}, \bar{R}_4^{(1)}) = 0.0562 > 0.05$, then go to Step 4, we get $\bar{R}_1^{(2)} = R_1^{(1)}$, $\bar{R}_2^{(2)} = R_2^{(1)}$, $\bar{R}_3^{(2)} = R_3^{(1)}$ and

$$\bar{R}_4^{(2)} = \begin{pmatrix} 0.5 & 0.4679 & 0.6699 & 0.3146 \\ 0.5321 & 0.5 & 0.6977 & 0.3430 \\ 0.3301 & 0.3023 & 0.5 & 0.1844 \\ 0.6854 & 0.6570 & 0.8156 & 0.5 \end{pmatrix}$$

$$\bar{R}^{(2)} = \begin{pmatrix} 0.5 & 0.3816 & 0.6304 & 0.2393 \\ 0.6184 & 0.5 & 0.7343 & 0.3376 \\ 0.3969 & 0.2657 & 0.5 & 0.1557 \\ 0.7607 & 0.6624 & 0.8443 & 0.5 \end{pmatrix}$$

In this case, by Eq.(40), we have

$$d(\bar{R}^{(2)}, \bar{R}_1^{(2)}) = 0.0192, \ d(\bar{R}^{(2)}, \bar{R}_2^{(2)}) = 0.0139,$$
$$d(\bar{R}^{(2)}, \bar{R}_3^{(2)}) = 0.0163, \ d(\bar{R}^{(2)}, \bar{R}_4^{(2)}) = 0.0340$$

i.e., all of the above deviations $d(\bar{R}^{(2)}, \bar{R}_l^{(2)}) \ (l = 1, 2, 3, 4)$ are less than 0.05, therefore, the acceptable group consensus is reached.

5 Concluding Remarks

In this paper, we have proposed a new characterization for a multiplicative consistent fuzzy preference relation, based on which a method has been developed to construct the multiplicative consistent fuzzy preference from an inconsistent one. The method can not only consider all the elements in the fuzzy preference relation, but also preserve the original preference information as much as possible. In the cases where the deviation between the constructed multiplicative consistent fuzzy preference relation and the original one is large, we have developed an algorithm to repair the fuzzy preference relation into the one with weak transitivity. Some examples have been utilized to compare our method with the existing ones.

Moreover, we have suggested an automatic algorithm to get group consensus by repairing the individual preferences so as to reduce the deviations between the individual fuzzy preference relations and the group's one. We have also shown that if all individual fuzzy preference relations are multiplicative consistent, then the group fuzzy preference relation that derived by using our method is multiplicative consistent.

In addition, our method can be reasonably used to estimate the unknown elements in an incomplete fuzzy preference relation, which is an interesting issue for future work.

Acknowledgment

The work was partly supported by the National Science Fund for Distinguished Young Scholars of China (No.70625005), the National Natural Science Foundation of China (No.71071161), and the Program Sponsored for Scientific Innovation Research of College Graduate in Jiangsu Province (No.CX10B_059Z).

References

1. Ben-Arieh, D., Chen, Z.F.: Linguistic-labels aggregation and consensus measure for autocratic decision making using group recommendations. IEEE Transactions on Systems, Man, and Cybernetics-Part A 36, 558–568 (2006)
2. Ben-Arieh, D., Easton, T.: Multi-criteria group consensus under linear cost opinion elasticity. Decision Support Systems 43, 713–721 (2007)
3. Chiclana, F., Herrera, F., Herrera-Viedma, E.: Integrating three representation models in fuzzy multipurpose decision making based on fuzzy preference relations. Fuzzy Sets and Systems 97, 33–48 (1998)
4. Chiclana, F., Herrera, F., Herrera-Viedma, E.: Integrating multiplicative preference relations in a multipurpose decision-making based on fuzzy preference relations. Fuzzy Sets and Systems 122, 277–291 (2001)
5. Chiclana, F., Herrera-Viedma, E., Alonso, S., Herrera, F.: A note on the estimation of missing pair wise preference values: a U-consistency based method. International Journal of Uncertainty, Fuzziness and Knowledge-Based Systems 16, 19–32 (2008a)
6. Chiclana, F., Herrera-Viedma, E., Alonso, S., Herrera, F.: Cardinal consistency of reciprocal preference relation: a characterization of multiplicative transitivity. IEEE transactions on fuzzy systems 17, 14–23 (2009)
7. Chiclana, F., Mata, F., Alonso, S., Martinez, L., Herrera-Viedma, E., Alonso, S.: Integration of a consistency control module within a consensus model. International Journal of Uncertainty, Fuzziness and Knowledge-Based Systems 16, 35–53 (2008b)

8. Herrera, F., Herrera-Viedma, E., Chiclana, F.: Multiperson decision-making based on multiplicative preference relations. European Journal of Operational Research 129, 372–385 (2001)
9. Herrera-Viedma, E., Herrera, F., Chiclana, F.: A consensus model for multiperson decision making with different preference structures. IEEE transactions on systems, Man and Cybernetics-Part A 32, 394–402 (2002)
10. Herrera-Viedma, E., Martínez, L., Mata, F., Chiclana, F.: A consensus support systems model for group decision making problems with multigranular linguistic preference relations. IEEE Transactions on Fuzzy Systems 13, 644–658 (2005)
11. Herrera-Viedma, E., Martínez, L., Mata, F., Chiclana, F.: A consensus model for group decision making with incomplete fuzzy preference relations. IEEE Transactions on Fuzzy Systems 15, 863–877 (2007)
12. Inohara, T.: On consistent coalitions in group decision making with flexible decision makers. Applied Mathematics and Computation 109, 101–119 (2000)
13. Ma, J., Fan, Z.P., Jiang, Y.P., Mao, J.Y., Ma, L.: A method for repairing the inconsistency of fuzzy preference relations. Fuzzy Sets and Systems 157, 20–33 (2006)
14. Mohammed, S., Ringseis, E.: Cognitive diversity and consensus in group decision making: The role of inputs, processes, and outcomes. Organizational Behavior and Human Decision Processes 85, 310–335 (2001)
15. Saaty, T.L.: The Analytic Hierarchy Process. McGraw-Hill, New York (1980)
16. Tanino, T.: Fuzzy preference orderings in group decision-making. Fuzzy Sets and Systems 12, 117–131 (1984)
17. Tanino, T.: Fuzzy preference relations in group decision making. In: Kacprzyk, J., Roubens, M. (eds.) Non-Conventional Preference Relations in Decision-Making, pp. 54–71. Springer, Heidelberg (1988)
18. Tanino, T.: On group decision-making under fuzzy preferences. In: Kacprzyk, J., Fedrizzi, M. (eds.) Multiperson Decision-Making Using Fuzzy Sets and Possibility Theory, pp. 172–185. Kluwer Academic Publishers, Dordrecht (1990)
19. Xu, Y.J., Li, Q.D., Liu, L.H.: Normalizing rank aggregation method for priority of a fuzzy preference relation and its effectiveness. International Journal of Approximate Reasoning 50, 1287–1297 (2009)
20. Xu, Z.S.: On consistency of the weighted geometric mean complex judgement matrix in AHP. European Journal of Operational Research 126, 683–687 (2002)
21. Xu, Z.S.: An Approach to Improving Consistency of Fuzzy Preference Matrix. Fuzzy Optimization and Decision Making 2, 3–21 (2003)
22. Xu, Z.S.: An automatic approach to reaching consensus in multiple attribute group decision making. Computers & Industrial Engineering 56, 1369–1374 (2009)
23. Xu, Z.S., Da, Q.L.: The ordered weighted geometric averaging operators. International Journal of Intelligent Systems 17, 709–716 (2002)
24. Xu, Z.S., Da, Q.L.: A least deviation method to obtain a priority vector of a fuzzy preference relation. European Journal of Operational Research 164, 206–216 (2005)
25. Xu, Z.S., Wei, C.P.: A consistency improving method in the Analytic Hierarchy Process. European Journal of Operational Research 116, 443–449 (1999)
26. Yager, R.R.: On ordered weighted averaging aggregation operators in multi- criteria decision making. IEEE Transactions on Systems, Man, and Cybernetics 18, 183–190 (1988)

Supporting Consensus Reaching Processes under Fuzzy Preferences and a Fuzzy Majority via Linguistic Summaries and Action Rules *

Sławomir Zadrożny, Janusz Kacprzyk, and Zbigniew W. Raś

Abstract. We deal with the classic approach to the evaluation of of a degree of consensus due to Kacprzyk and Fedrizzi [21, 22, 23] in which a soft degree of consensus has been introduced as a degree to which, for instance, "most of the important individuals agree as to almost all of the relevant options". The fuzzy majority is equated with a fuzzy linguistic quantifiers (most, almost all, ...) and handled via Zadeh's [64] classic calculus of linguistically quantified propositions and Yager's [63] OWA (ordered weighted average) operators. The consensus reaching process is run by a moderator who may need a support which is provided by a novel combination of: first, the use of the a soft degree of consensus alone along the lines of Fedrizzi, Kacprzyk and Zadrożny [12], Fedrizzi, Kacprzyk, Owsiński and Zadrożny [11], Kacprzyk and Zadrożny [31, 34]. Second, the linguistic data summaries in the sense of Yager [62], Kacprzyk and Yager [26], Kacprzyk, Yager and Zadrożny [27], in particular in its protoform based version proposed by Kacprzyk and Zadrożny [33], [35], are are employed to indicate in a natural language some interesting relations between individuals and options to help the moderator identify crucial (pairs of) individuals and/options which pose some threats to the reaching of consensus. Third, using results obtained in our recent paper (Kacprzyk, Zadrożny and Raś [41], we additionally use a novel data mining tool, a so-called action rule proposed by Raś

Sławomir Zadrożny · Janusz Kacprzyk
Systems Research Institute, Polish Academy of Sciences
ul. Newelska 6, 01-447 Warsaw, Poland
e-mail: {kacprzyk,zadrozny}@ibspan.waw.pl

Zbigniew W. Raś
Computer Science Department, College of Computing and Informatics
University of North Carolina at Charlotte, Charlotte, NC 28223, USA and
Institute of Computer Science, Polish Academy of Sciences
ul.Ordona 20, 01-237 Warsaw, Poland
e-mail: ras@uncc.edu

* This work was supported by the Ministry of Science and Higher Education in Poland under Grant N N519 404734.

E. Herrera-Viedma et al. (Eds.): Consensual Processes, STUDFUZZ 267, pp. 289–314.
springerlink.com © Springer-Verlag Berlin Heidelberg 2011

and Wieczorkowska [48], which are meant in our context to find best concessions to be offered to the individuals for changing their preferences to increase the degree of consensus.

Keywords: consensus, consensus reaching support, fuzzy preference, fuzzy majority, fuzzy logic, linguistic quantifier, linguistic summary, OWA (ordered weighted averaging) operator, action rule.

1 Introduction

We deal with the following consensus reaching setting (process). There is a (finite) group of individuals and a (finite) set of options. The individuals provide their preferences as to the particular pairs of options. Normally, these preferences (testimonies) are different in the beginning, and in a step-by-step consensus reaching process run by a moderator, they are being changed, by some argumentation, mutual concessions, etc., possibly in the direction of increasing consensus.

Our point of departure is a novel human-consistent approach to the definition of a soft degree of consensus introduced by Kacprzyk and Fedrizzi [21, 22, 23] who proposed a degree of consensus as a degree to which, e.g., "*most* of the *relevant* (knowledgeable, expert, . . .) individuals agree as to *almost all* of the *important* options".

Then, we use a general setting for a moderator-run group decision support system for supporting consensus reaching proposed by Fedrizzi, Kacprzyk and Zadrożny [12], and then further developed by Fedrizzi, Kacprzyk, Owsiński and Zadrożny [11], and Kacprzyk and Zadrożny [31, 34] that is based on the soft degrees of consensus mentioned above. Moreover, some additional information on the interplay between the preferences of the individuals is shown to be useful.

As it was shown by the authors in our former papers (Kacprzyk and Zadrożny [32, 37]), the use of linguistic data summaries in the sense of Yager [62], or maybe rather in its extended and implementable version of Kacprzyk and Yager [26] or Kacprzyk, Yager and Zadrożny [27], may be useful to provide insight and information as to what is "going wrong" in the consensus reaching process, what is to be paid attention to, which pairs of individuals/options may pose some problems, etc. Kacprzyk and Zadrożny's [37] approach to use ontologies in this context may also be useful but will not be dealt with here.

We assume as the main elements the *individual* and *social fuzzy preference relations* – cf., e.g., Nurmi [45], and a fuzzy majority to model group decision making and consensus reaching as proposed first by Kacprzyk [19], [20]; for a comprehensive review, cf. Kacprzyk, Zadrożny, Fedrizzi and Nurmi [39], [40]. The fuzzy majority may be represented in a natural way via the so-called *linguistic quantifiers*: *most, almost all, much more than a half,* . . . which cannot be handled by conventional logical calculi. Fuzzy logic can provide here simple and efficient tools, notably Zadeh's [64] calculus of linguistically quantified proposition. One can also use Yager's [63] OWA (ordered weighted average) operators which provide a much needed generality and flexibility.

Fuzzy majority is commonly used by the humans, and an often cited example in a biological context is due to Loewer and Laddaga [43]:

> ...It can correctly be said that there is a *consensus* among biologists that Darwinian natural selection is an important cause of evolution though there is currently *no consensus* concerning Gould's hypothesis of speciation. This means that there is a *widespread agreement* among biologists concerning the first matter but *disagreement* concerning the second ...

and it is clear that a rigid majority as, e.g., more than 75% would not be adequate here.

The use of fuzzy linguistic quantifiers has been proposed by the authors to introduce a fuzzy majority for measuring (a degree of) consensus and deriving new solution concepts in group decision making (cf. Kacprzyk [19, 20], Kacprzyk and Fedrizzi [21, 22, 23]); cf. also works on generalized choice functions under fuzzy and non-fuzzy majorities by Kacprzyk and Zadrożny [29, 31, 35]. For a comprehensive review, see Kacprzyk, Zadrożny, Fedrizzi and Nurmi [39, 40].

The soft degrees of consensus proposed in those works have proved to be conceptually powerful and useful for an implemented decision support system for consensus reaching proposed by Fedrizzi, Kacprzyk and Zadrożny [12], and then further developed by Fedrizzi, Kacprzyk, Owsiński and Zadrożny [11] or Kacprzyk and Zadrożny [31, 34], and Zadrożny and Kacprzyk [67].

This soft degree of consensus is meant to overcome some "rigidness" of the conventional concept of consensus in which (full) consensus occurs only when "*all* the individuals agree as to *all* the issues"; this is unrealistic is practice. The new degree of consensus can be therefore equal to 1, which stands for full consensus, when, for instance, "*most* of the (important) individuals agree as to *almost all* (of the relevant) options". This new degree of consensus has been proposed by Kacprzyk and Fedrizzi [21, 22, 22] using Zadeh's [64] calculus of linguistically quantified propositions. Then, Fedrizzi, Kacprzyk and Nurmi [10] have proposed to use the OWA (ordered weighted average) operators instead of Zadeh's calculus. It works well though some deeper works on the semantics of the OWA operators is relevant as shown by Zadrożny and Kacprzyk [68].

Basically, in all the above works the individuals provide their testimonies concerning options in question as *fuzzy preference relations* expressing preferences given in pairwise comparisons of options. A decision is to be taken after reaching *consensus*. Since in the beginning the individuals are usually far from consensus, a discussion – run by a moderator – proceeds in the group to clarify points of view, exchange information, advocate opinions, persuade stubborn individuals to be more flexible, etc., which should imply some changes in the preferences, possibly getting them closer to consensus.

To facilitate the work of the moderator, and make it more effective and efficient, some support is provided to him or her with some hints as to most promising directions of a further discussion, focusing discussion in the group, troublesome options and/or individuals, etc.

One of the first and more successful approach is due to Kacprzyk and Zadrożny [32] in which linguistic data summaries were used to derive some additional measures of what is going on in the group with respect to the uniformity of their preferences; its extension is given in Kacprzyk and Zadrożny [37]. Linguistic data summaries have been used therein. For instance, various linguistic summaries may be used concerning the individuals and alternatives, exemplified by "Most individuals definitely prefer option o_{i1} to option o_{i2}, moderately prefer o_{i3} to o_{i4}, ...", "Most individuals definitely preferring o_{i1} to o_{i2} also definitely prefer o_{i3} to o_{i4}", "Most individuals choose options o_{i1}, o_{i2}, ...", "Most individuals reject options o_{i1}, o_{i2}, ...", "Most options are dominated by option o_i in opinion of individual e_k", "Most options are dominated by option o_i in opinion of individual e_{k1}, e_{k2}, ...", "Most options dominating alternative o_i in opinion of individual e_{k1} also dominate option o_i in opinion of individual e_{k2}","Most options are preferred by individual e_k", etc.

Such summaries may help the moderator and the group gain a deeper understanding and insight of even intricate and sophisticated relations within the individuals and their testimonies (preferences). In the derivation of those valuable summaries Kacprzyk and Zadrożny's [28, 29, 30] works on general choice functions in group decision making have played a significant role. Among other approaches one should also cite Herrera-Viedma, Mata, Martínez and Pérez [16, 17].

A conceptually new approach to providing the moderator with some additional tool was proposed by Kacprzyk, Zadrożny and Raś [41] in which the use of so-called *action rules*, proposed by Raś and Wieczorkowska [48] in the context of Pawlak's [47] *information systems*, was advocated.

Basically, in that new approach the (traditional) fuzzy preference relations may be extended to the Atanassov's intuitionistic fuzzy preferences relations which specify for each pair of individuals a degree of preference of the former alternative over the latter one as a pair that consists of the degree of membership and non-membership, which do not need to sum up to 1, hence a hesitation margin (the difference between 1 and the sum of them) exists. The intuitionistic fuzzy preference relation may provide a better and more flexible framework to deal with lack of information and bipolar information, and also in a sense are better suited for the use of action rules to be briefly explained below.

The concept of an *action rule* was proposed by Raś and Wieczorkowska [48], in the context of Pawlak's [47] *information systems*, and then has been extensively further studied and developed by Raś and his collaborators [49, 50, 51, 58, 59, 60]. The purpose of an action rule is to show how a subset of flexible attributes should be changed to obtain an expected change of the decision attribute for a subset of objects characterized by some values of the subset of stable attributes. For example, in a bank context, an action rule may, e.g., indicate that offering a 20% reduction in a monthly bank account fee instead of a 10% reduction to a middle-aged customer is expected to increase his or her spendings from medium to high. The action rules are sought so that the cost of change of an attribute value, which may be different for different attributes, be the "cheapest" provided they support the expected change of the decision attribute.

It is clear a great potential of action rules in our context of the supporting of consensus reaching processes. Namely, we can imagine that we wish to find some "concessions" to eventually be offered to some individuals so that they change their preferences, and to choose those "concessions" that would bear the lowest cost.

The structure of the paper is the following. First, we will briefly present the calculus of linguistically quantified propositions, then the essence of action rules. Next, we will briefly present the concept of a soft degree of consensus, and the essence of a consensus reaching process. We will provide first an exposition on how linguistic data summaries can be used to help gain insight into the very essence of consensus reaching, and then we will show how the use of action rules can help gain that even more.

2   Linguistic Quantifiers and a Fuzzy Logic Based Calculus of Linguistically Quantified Propositions

A *linguistically quantified proposition*, exemplified by "most individuals are convinced", may be generally written as

$$Qy\text{'s are } S \tag{1}$$

or, with importance R added, exemplified by "most of the important individuals are convinces", as

$$QRy\text{'s are } S \tag{2}$$

or more conveniently written as

$$Qy\text{'s are } (R, S) \tag{3}$$

Then, T, i.e. its truth (validity), directly corresponds to the truth value of (1) or (2) calculated by using Zadeh's [64] calculus of linguistically quantified statements, or the OWA operators (cf. Yager [63]) which will not be discussed here.

Assuming a (proportional, nondecreasing) linguistic quantifier Q to be a fuzzy set in $[0, 1]$, for instance

$$\mu_{"most"}(x) = \begin{cases} 1 & \text{for } x \geq 0.8 \\ 2x - 0.6 & \text{for } 0.3 < x < 0.8 \\ 0 & \text{for } x \leq 0.3 \end{cases} \tag{4}$$

the truth values (from $[0, 1]$) of (1) and (2) are calculated, respectively, as:

$$\text{truth}(Qy\text{'s are } S) = \mu_Q[\frac{1}{n}\sum_{i=1}^{n}\mu_S(y_i)] \tag{5}$$

$$\text{truth}(QRy\text{'s are } S) = \mu_Q[\frac{\sum_{i=1}^{n}(\mu_R(y_i) \wedge \mu_S(y_i))}{\sum_{i=1}^{n}\mu_R(y_i)}] \tag{6}$$

where "$\wedge$" (minimum) can be replaced by, e.g., a t-norm.

The fuzzy predicates S and R are assumed atomic, i.e. referring to just one attribute, but can readily be extended to cover more sophisticated summaries involving some confluence of various attribute values as, e.g, *young and well paid*, or even the most interesting summarizers (concepts) as, e.g.: productive workers, difficult orders, etc. to be defined by a a complicated combination of attributes, a hierarchy (not all attributes are of the same importance for a concept in question), the attribute values are ANDed and/or ORed, k out of n, most, ... of them should be accounted for, etc.

3   A Soft Degree of Consensus under Fuzzy Preferences and a Fuzzy Majority

We have a set of n options, $O = \{o_1, \ldots, o_n\}$, and a set of m individuals, $E = \{e_1, \ldots, e_m\}$. Each individual e_k provides his or her *individual fuzzy preference relation*, P_k, given by its membership function $\mu_{P_k} : O \times O \to [0, 1]$ which, if card O is small enough, may be represented by a matrix $[r_{ij}^k]$ such that $r_{ij}^k = \mu_{P_k}(o_i, o_j)$; $i, j = 1, \ldots, n; k = 1, \ldots, m; r_{ij}^k + r_{ji}^k = 1$.

The (soft) degree of consensus is now derived in three steps: (1) for each pair of individuals we derive a degree of agreement as to their preferences between all the pair of options, (2) we aggregate these degrees to obtain a degree of agreement of each pair of individuals as to their preferences between $Q1$ (a fuzzy linguistic quantifier as, e.g., *most, almost all, much more than 50%, ldots*) pairs of options, and (3) we aggregate these degrees to obtain a degree of agreement of $Q2$ (another fuzzy linguistic quantifier) pairs of individuals as to their preferences between $Q1$ pairs of options which is meant to be the degree of consensus sought.

Let us denote for brevity by $\mathrm{AGG}_Q(.)$ the linguistic quantifier driven aggregation operations due to (5). We start with the degree of (strict) agreement between individuals e_{k1} and e_{k2} as to their preferences between options o_i and o_j

$$v_{ij}(k1, k2) = \begin{cases} 1 \text{ if } r_{ij}^{k1} = r_{ij}^{k2} \\ 0 \text{ otherwise} \end{cases} \tag{7}$$

where: $k1 = 1, \ldots, m - 1; k2 = k1 + 1, \ldots, m; i = 1, \ldots, n - 1;$ and $j = i + 1, \ldots, n$.

The degree of agreement between individuals $k1$ and $k2$ as to their preferences between all the pairs of options is

$$v(k1, k2) = \frac{2}{n(n-1)} \sum_{i=1}^{n-1} \sum_{j=i+1}^{n} v_{ij}(k1, k2) \tag{8}$$

The degree of agreement between individuals $k1$ and $k2$ as to their preferences between $Q1$ pairs of options is

$$v_{Q1}(k1, k2) = \mathrm{AGG}_{Q1}(\{v_{ij}(k1, k2)\}_{1 \le i < j \le n}) \tag{9}$$

where $\mathrm{AGG}_{Q1}(.)$ is the aggregation of $v_{ij}(k1,k2)$'s with respect to $Q1$ due to (5), where F is meant as "is high", i.e. close to 1.

In turn, the degree of agreement of all the pairs of individuals as to their preferences between $Q1$ pairs of options is

$$v_{Q1} = \frac{2}{m(m-1)} \sum_{k1=1}^{m-1} \sum_{k2=k1+1}^{m} (v_{Q1}(k1,k2)) \qquad (10)$$

and, finally, the degree of agreement of $Q2$ pairs of individuals as to their preferences between $Q1$ pairs of options, called the *degree of Q1/Q2 - consensus* is

$$\mathrm{con}(Q1,Q2) = \mathrm{AGG}_{Q2}(\{v_{Q1}(k1,k2)\}_{1 \leq k1 < k2 \leq m}) \qquad (11)$$

where $\mathrm{AGG}_{Q2}(.)$ is defined similarly as $\mathrm{AGG}_{Q1}(.)$.

Since the strict agreement (7) may be viewed too rigid, we can use the degree of *sufficient agreement* (at least to degree $\alpha \in [0,1]$) of individuals e_{k1} and e_{k2} as to their preferences between options o_i and o_j, as well as the the degree of strong agreement of individuals $k1$ and $k2$ as to their preferences between options o_i and o_j, obtaining the degree of $\alpha/Q1/Q2$ - consensus and $s/Q1/Q2$ - consensus, respectively (cf. Kacprzyk and Fedrizzi [21, 22, 23]).

An important issue is the addition of the importance of individuals and the relevance of options which may be tricky from the conceptual point of view as it may be viewed as imposing some authoritarian or subjective differentiation of the individuals. Formally, it is handles as for the importance of the options.

As we have already mentioned, for the use of the action rules, we extend this basic fuzzy preference modeling approach assumed above by, first, adopting a bipolar view of preferences (cf. Dubois and Prade [8], and in particular their modeling via *Atanassov's intuitionistic fuzzy sets* (IFSs) [3]. Second, instead of using numeric membership (and non-membership) degrees we use *linguistic terms*.

An intuitionistic fuzzy set X, denoted for brevity A-IFS, is represented by a pair of membership, μ_X, and non-membership, ν_X, functions, such that:

$$\mu_X(x) + \nu_X(x) \leq 1. \qquad (12)$$

An *intuitionistic fuzzy preference relation* $R_k^{\text{A-IFS}}$, denoted for brevity as an A-IFS preference relation, of individual e_k, is an A-IFS in $S \times S$, which is thus defined by its membership function $\mu_{R_k^{\text{A-IFS}}}(o_i,o_j)$ and non-membership function $\nu_{R_k^{\text{A-IFS}}}(o_i,o_j)$. The former is meant as the degree of preference (intensity) of o_i over o_j, and latter as the degree to which o_i is *not* preferred over o_j, here. To simplify, the latter may be interpreted in the spirit of the reciprocity as the intensity of preference of o_j over o_i which implies:

$$\nu_{R_k^{\text{A-IFS}}}(o_i,o_j) = \mu_{R_k^{\text{A-IFS}}}(o_j,o_i). \qquad (13)$$

The use of A-IFSs provides for a more flexible representation, notably for taking into account pro and con arguments while determining the preferences which is a widely advocated approach (cf., e.g., [37, 5]). We will omit the superscript A-IFS in $R_m^{\text{A-IFS}}$ in what follows, to simplify notation. For more details on the A-IFS preference relations and their use in group decision making and consensus reaching, cf. Szmidt and Kacprzyk [55, 56].

For our purposes a discretization of the universe in which degrees of preference are expressed is needed so that instead of using the interval $[0, 1]$ we adopt the ordinal linguistic approach [7], and represent preference degrees using an ordered set of linguistic terms (linguistic labels). We assume that an individual is specifying for each pair (o_i, o_j) only the membership degrees $\mu_R(o_i, o_j)$ and $\mu_R(o_j, o_i)$ using the following linguistic terms set $\mathcal{L}$:

$$definitely \succ strongly \succ moderately \succ weakly \succ not\_at\_all \quad (14)$$

which in general will be denoted as:

$$\mathcal{L} = \{l_T, l_{T-1}, \ldots, l_1, l_0\} \quad (15)$$

$$l_T \succ l_{T-1} \succ \ldots \succ l_1 \succ l_0 \quad (16)$$

where T is an even number.

Due to the assumed order (16), the infimum and supremum operators may be applied to sets as well as to multisets of linguistic terms:

$$\inf(\{l_{k_1}, l_{k_2}, \ldots, l_{k_n}\}) = l_{k_i} \text{ if } k_i \in \{k_1, \ldots, k_n\} \wedge \neg\exists_{j \in \{k_1,\ldots,k_n\}} l_j \succ l_{k_i} \quad (17)$$

$$\sup(\{l_{k_1}, l_{k_2}, \ldots, l_{k_n}\}) = l_{k_i} \text{ if } k_i \in \{k_1, \ldots, k_n\} \wedge \neg\exists_{j \in \{k_1,\ldots,k_n\}} l_j \succ l_{k_i} \quad (18)$$

Moreover, an *antonym* operator *ant*, which is an involutive order-reversing operator, is assumed on $\mathcal{L}$:

$$ant : \mathcal{L} \longrightarrow \mathcal{L} \qquad ant(l_k) = l_{T-k} \quad (19)$$

The use of such an operator effectively assumes that the linguistic terms form an interval scale.

An A-IFS preference relation, generalized using linguistic terms instead of numerical membership and non-membership degrees, is an *Atanassov's intuitionistic L-fuzzy set* [4], to be denoted as an A-ILFS set, and has to satisfy a counterpart of the property (12), which is expressed as:

$$\neg(\mu_R(o_i, o_j) \succ ant(\nu_R(o_i, o_j))) \quad (20)$$

which can be rewritten as:

$$\neg(\mu_R(o_i, o_j) \succ ant(\mu_R(o_j, o_i))) \quad (21)$$

as we have $\nu_R(o_i, o_j) = \mu_R(o_j, o_i)$. We will refer to such an R as an A-ILFS preference relation.

Thus, for example (referring to the linguistic terms set (14)), for $\mu_R(o_i, o_j) = definitely$ the value of $\mu_R(o_j, o_i)$ is determined to be not_at_all, what means that alternative o_i is definitely (fully) preferred to the alternative o_j. Other two interesting cases are where $\mu_R(o_i, o_j) = \mu_R(o_j, o_i) = moderately$ and $\mu_R(o_i, o_j) = \mu_R(o_j, o_i) = not\_at\_all$. In both cases an individual may be seen as indifferent to the choice between the alternatives o_i and o_j, but due to the semantics of the A-IFS/A-ILFS sets the first case may be seen as a genuine indifference, while the second corresponds to the situation where an individual is undecided, unable to make a choice due to, e.g., lack of information.

This also shows a need for a "linguistic counterpart" of the *hesitation margin* $\pi(x)$, which is defined in the A-IFS set theory for regular "numerical" membership degrees as

$$\pi(x) = 1 - \mu(x) - \nu(x) \tag{22}$$

In case of a simple lattice defined by (15)–(18), its counterpart for the A-ILFS sets can be defined in the following way. Let $\mu(x) = l_u$, $\nu(x) = l_w$ and $ant(l_u) = l_t$. Then:

$$\pi(x) = l_z, \text{ where } z = t - w \tag{23}$$

More details on the use of A-IFS fuzzy preference relations, amnd the related issues, in the setting considered here, can be found in Kacprzyk, Zadrożny and Raś [41].

4   The Concept of an Action Rule

The concept of an *action rule* was proposed by Raś and Wieczorkowska [48], in the context of Pawlak's [47] *information systems*, i.e. triples $IS = \{O, A, V\}$, where O is a finite set of objects, A is a set of its attributes and $V = \bigcup_{a \in A} V_a$, with V_a being a domain of attribute a. If one of the attributes $d \in A$ is distinguished and called a decision, then an information system is called a *decision system*. The set of attributes A may be further partitioned into subsets of *stable* and *flexible* attributes, denoted as A_{St} and A_{Fl} [48]. Thus $A = A_{St} \cup A_{Fl} \cup \{d\}$. This partitioning is crucial as the essence of an action rule is to show how a subset of flexible attributes should be changed to obtain expected change of the decision attribute for a subset of objects characterized by some values of the subset of stable attributes. For example, let objects $o \in O$ be bank customers characterized by such stable (from a bank perspective) attributes as age, profession, etc. and flexible attributes such as the type of an account, reduction of the monthly fee, etc., and the decision attribute is the customer's total monthly spendings. Then, an action rule may, e.g., indicate that offering a 20% reduction in monthly fee instead of a 10% reduction to a middle-aged customer is expected to increase his or her spendings from medium to high.

More formally, let us define first some auxiliary concepts [50]. An *atomic action term* is $(a, x \rightarrow y)$, where $a \in A$ is an attribute and $x, y \in V_a$ are values belonging to its domain. An *action term* t is a set of atomic action terms: $t = \{(a_1, x_1 \rightarrow y_1), \ldots, (a_n, x_n \rightarrow y_n)\}$, $a_i \in A$, $a_i \neq a_j$ for $i \neq j$ and $x_i, y_i \in V_{a_i}$. The *domain* of an action term t, denoted by $Dom(t)$, is a set of all attributes in t, i.e., $Dom(t) = \{a_1, \ldots, a_n\}$.

Finally, an *action rule* is $r = [t_1 \Rightarrow t_2]$, where t_1 is an action term and t_2 is an atomic action term referring to the decision attribute, i.e., $Dom(t_2) = \{d\}$. So, if the bank customers are characterized by `age`, `reduction` (monthly fee reduction) and `spendings`, then an action rule may take the following form:

$$[\{(\texttt{age}, middleaged \rightarrow middleaged),$$
$$(\texttt{reduction}, 10\% \rightarrow 20\%)\} \Rightarrow$$
$$(\texttt{spendings}, medium \rightarrow high)]$$

The measures of *support* (*supp*) and *confidence* (*conf*) are used to evaluate the action rules for a given information system $IS = \{O, A, V\}$. For an action term $t = \{(a_1, x_1 \rightarrow y_1), \ldots, (a_n, x_n \rightarrow y_n)\}$ let us denote by $N_S(t)$ the following pair of sets:

$$N_S(t) = [X, Y] = [\bigcap_{1 \leq i \leq n} \{o \in O : a_i(o) = x_i\}, \bigcap_{1 \leq i \leq n} \{o \in O : a_i(o) = y_i\}].$$

Further, for an action rule $[t_1 \Rightarrow t_2]$, let $N_S(t_1) = [X_1, Y_1]$ and $N_S(t_2) = [X_2, Y_2]$. Then the measures of support and confidence are defined as follows:

$$supp(r) = card(X_1 \cap X_2) \tag{24}$$

$$conf(r) = \frac{card(X_1 \cap X_2)}{card(X_1)} \frac{card(Y_1 \cap Y_2)}{card(Y_1)} \tag{25}$$

and 0 for the for denominators equal 0.

There is also an important issue of the cost of change of an attribute value which may be different for different attributes. The goal is thus to find the "cheapest" rules supporting the expected change of the decision attribute.

By an *association action rule* we mean any expression $[t_1 \Rightarrow t_2]$, where t_1 and t_2 are action terms.

Let $(a, x_1 \rightarrow y_1)$ is an atomic action term. We assume that the cost of changing the value of attribute a from x_1 to y_1 is denoted by $cost_{IS}((a, x_1 \rightarrow y_1))$ as introduced in [59]. For simplicity, the subscript IS is usually omitted. Let $t_1 = (a, x_1 \rightarrow y_1)$, $t_2 = (b, x_2 \rightarrow y_2)$ be two atomic action terms. We say that t_1, t_2 are positively correlated if change represented by t_1 implies change represented by t_2 and vice versa. Similarly, t_1 and t_2 are negatively correlated if change represented by t_1, i.e., $x_1 \rightarrow y_1$ implies change opposite to the one represented by t_2, i.e., $y_2 \rightarrow x_2$ and vice versa.

Now, assume that action term t is constructed from atomic action terms $T = \{t_1, t_2, \ldots, t_m\}$. We introduce a binary relation $\simeq$ on T defined as: $t_i \simeq t_j$ iff t_i and t_j are positively correlated. Relation $\simeq$ is an equivalence relation and it partitions

T into m equivalence classes $(T = T_1 \cup T_2 \cup \ldots \cup T_m)$, for some m. In each equivalence class T_i, an atomic action term $a(T_i)$ of the lowest cost is identified. The cost of t is defined as: $cost(t) = \sum \{cost(a(T_i)) : 1 \leq i \leq m\}$.

Now, if $r = [t_1 \Rightarrow t]$ is an association action rule, then r is simple if $cost(t_1 \cup t) = cost(t_1)$. The cost of r is $cost(t_1)$.

The user provides three thresholds, λ_1 - minimum support, λ_2 - minimum confidence, λ_3 - maximum cost. Let t be a frequent action set in S and t_1 is its subset. Any association action rule $r \in AAR_S(\lambda_1, \lambda_2)$ is called an association action rule of an acceptable cost if $cost(r) \leq \lambda_3$. Similarly, a frequent action set t is called a frequent action set of an acceptable cost if $cost(t) \leq \lambda_3$.

In order to construct simple association action rules of the lowest cost, we build frequent action sets of an acceptable cost following the classical Agrawal strategy (see [1]) for generating frequent sets enhanced by additional constraint which requires to verify the cost of frequent action sets. Any frequent action set which cost is higher than λ_3, is removed. Now, if t is a frequent action set of an acceptable cost and $\{a(T_i) : i \leq m\}$ is a collection of atomic action sets constructed by this strategy, then $\bigcup \{a(T_i) : i \leq m\} \Rightarrow [t - \{a(T_i) : i \leq m\}]$ is a simple association action rule of acceptable cost assuming that its confidence is not greater than λ_2.

For details related to action rules discovery, see [49, 50, 51].

5   A Consensus Reaching Process

We consider the setting of the consensus reaching process along the lines of Fedrizzi, Kacprzyk and Zadrożny [12], Fedrizzi, Kacprzyk, Owsiński and Zadrożny [11], Kacprzyk and Zadrożny [31, 34], and Zadrożny and Furlani [66]. For a different perspective, but within this copnceptual framework the works of the Spanish researchers, notably, alphabetically, Chiclana, Herrera, Herrera-Viedma, Martinez, Mata, etc. [44, 14]

Basically, there is a group of participating individuals and a moderator who is meant to effectively and efficiently run the consensus reaching session. The individuals and the moderator exchange information and opinions, provide argumentation, operating in a network as shown in Figure 1.

The consensus reaching is a dynamic process meant as follows. In the beginning, at $t = 0$, the individuals present their testimonies, i.e. their initial fuzzy (A-ILFS, to be more specific) preference relations, which may differ from each other to a large extent. The moderator tries to persuade them to change their preference relations using some argumentation, concessions, etc. By assumption, the individuals are rationally committed to reaching consensus, which is a *sine qua non* condition for any serious consensus reaching process, and the group should possibly get closer to consensus.

We have a set of m individuals $E = \{e_1, \ldots, e_m\}$ whose testimonies concerning a set of n options (alternatives) $O = \{o_1, \ldots, o_n\}$ are *individual A-ILFS preference relations*. A moderator stimulates an exchange of information, rational argument, discussion, creative thinking, clarification of positions, etc. These should eventually

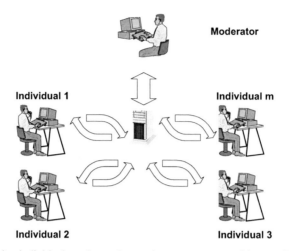

Fig. 1 Particiating individuals and a moderator in a consensus reaching session

lead to a change of the individual fuzzy preference relations. Some individuals, even
if not willing to change their original preferences, can accept consensual preferences
of the group provided their arguments has been heard and discussed. Thus, their
acceptance of consensus may be viewed as a change of their preferences. This is
repeated until the group gets sufficiently close to consensus, i.e. until the individual
A-ILFS preference relations become similar enough, or until some time limit is
reached.

The moderator's job may be however difficult, for instance due to many individuals and options. He or she should be somehow supported, for instance via an
effective and efficient human-computer interface, enhanced communication capabilities, advanced presentation tools for the visualization or verbalization of results
obtained, etc.

In this work we propose a new approach. Its first part is a novel, natural language based support – suggested in Kacprzyk and Zadrożny [38] – that is based on
the verbalization of results obtained by using linguistic summaries of data in the
sense of Yager [62], but in their implementable and extended version proposed by
Kacprzyk and Yager [26] and Kacprzyk Yager and Zadrożny [27]. Even more so, we
use here linguistic data summaries in the sense of the recent papers by Kacprzyk and
Zadrożny [33], in which a protoform based analysis was presented, and in Kacprzyk
and Zadrożny [38], in which an extremely powerful and far reaching relation to natural language generation (cf. Reiter and Dale [52]) was shown. The second part is
the use of action rules as an additional tool as proposed by Kacprzyk, Zadrożny and
Raś [41].

The rationale for such a hybrid approach to consensus reaching support is that
the process may be long and its support may be greatly enhanced by all kinds of additional indicators summarizing various aspects of the process, notably preferences
of the individuals during a discussion, cf. Zadrożny [65].

6 A Concept of a Linguistic Data Summary

A linguistic data summary is meant as a natural language like sentence that subsumes the very essence (from a certain point of view) of a (numeric) set of data, too large to be comprehensible by humans. The original Yagers approach to the linguistic summaries (cf. Yager [62], Kacprzyk and Yager [26], Kacprzyk, Yager and Zadrożny [27] and Kacprzyk and Zadrożny [33]) may be expressed as follows: $Y = \{y_1, \ldots, y_n\}$ is a set of objects, $A = \{A_1, \ldots, A_m\}$ is a set of attributes characterizing objects from Y, and $A_j(y_i)$ denotes a value of attribute A_j for object y_i.

A linguistic summary of Y consists of:

- a summarizer S, i.e. an attribute together with a linguistic value (label) defined on the domain of attribute A_j;
- a quantity in agreement Q, i.e. a linguistic quantifier (e.g. most);
- truth (validity) T of the summary, i.e. a number from the interval $[0, 1]$ assessing the truth (validity) of the summary (e.g., 0.7),

and, optionally, a qualifier R may occur, i.e. another attribute together with a linguistic value (label) defined on the domain of attribute A_k determining a (fuzzy subset) of Y.

In our context we may identify objects with individuals and their attributes with their preferences over various pairs of options. Then, the linguistic summary may be exemplified by

$$T(Most \text{ of individuals prefer option } o_1 \text{ to } o_2) = 0.7 \qquad (26)$$

A richer form of the summary may include a qualifier as in, e.g.,

$$T(Most \text{ of } important \text{ individuals prefer option } o_1 \text{ to } o_2) = 0.7 \qquad (27)$$

These truth values can be calculated by using Zadeh's calculus of linguistically quantified propositions as prsesented in Section 2.

7 Helping the Moderator Run a Consensus Reaching Session Using Linguistic Data Summaries

An important component of a consensus reaching support system is a set of indicators assessing how far the group is from consensus, what are the obstacles in reaching consensus, which preference matrix may be a candidate for a consensual one, etc. These indicators may be treated as some data summaries.

The original definition of a degree of consensus, i.e. the degree to which "*Most* of the *important* individuals agree in their preferences as to *almost all* of the *important* options", may be more formally expressed as follows:

$$Qh\text{'s are } (B', Qq\text{'s are } (I', \text{sim}(p_q^{h_1}, p_q^{h_2})) \qquad (28)$$

where: $h \in E \times E$ is a pair of individuals, B' represents importance of a pair of individuals (related to B, an importance of particular individuals), $q \in O \times O$ is a pair of options, I' represents importance of a pair of options (related to I, an importance of particular options), $p_q^{h_i}$ is a preference degree of individual i of pair h for pair of options q, and sim($\cdot$, $\cdot$) is a measure of similarity between two preference degrees.

This definition is an example of a nested linguistic summary defined for the space of pairs of individuals and options. The summarizer S and qualifier R are composed of features of either individuals or options (depending on the perspective adopted, to be discussed later) and fuzzy values (labels) expressing degree of preferences or importance weights of individuals/options.

7.1 Individuals as Objects

The objects of a linguistic summary may be identified with individuals and their attributes are preference degrees for the particular pairs of options and importance degrees of the individuals. Formally, referring to Section 6, we have:

$$Y = E \tag{29}$$

and

$$A = \{\mathcal{P}_{ij}\} \cup \{\mathcal{B}\} \tag{30}$$

where attributes $\mathcal{P}_{ij}$ correspond to preference degrees over pairs of options (o_i, o_j) and $\mathcal{B}$ represents the importance.

Then, the following types of summaries may be useful for consensus reaching session guidance.

Consensus indicating/building summaries

They corresponds to a flexible definition of consensus (cf. (28)) that states that *most of the individuals express similar preferences*, for instance "Most individuals definitely prefer o_{i1} to o_{i2}, moderately prefer o_{i3} to o_{i4}, ...", etc. formally written as

$$Qe_k(p_{i1,i2}^k = definite) \wedge (p_{i3,i4}^k = moderate) \wedge \ldots \tag{31}$$

If the list of conjuncts is long enough, then the truth of (31) means that there is a consensus among the individuals as to their preferences.

Clearly, this type of a linguistic summary may be used as another definition of consensus. Similarly to (28), importance weights of individuals and/or options may be added.

If the list of conjuncts is short, such a summary may be treated as a suggestion for building consensus as it indicates opinions that are shared by the group of individuals. Thus, they may be either further discussed to extend the common understanding in the group or assumed as agreed upon making it possible to proceed to other issues.

Discussion targeting summaries

They may be used to direct a further discussion in the group, and may disclose some patterns of understanding. For instance:

Most individuals definitely preferring o_{i1} to o_{i2} also definitely prefer o_{i3} to o_{i4}

to be formally expressed as

$$Qe_k(p_{i1,i2}^k = definite, p_{i3,i4}^k = moderate) \tag{32}$$

The discovery of association expressed with such a summary may trigger a further discussion enabling a better understanding of the decision problem.

Option choice oriented summaries

Thus far we assumed that the goal is to agree upon the content of the preference matrices. Usually, however, the aim of the discussion is to select either an option or a set of options preferred by the group, and then an agreement as to the preferences in respect to *all* or even *most* pairs of options may be unnecessary. To generate summaries taking that into account we have to assume a constructive definition of an option preferred by an individual as implied by his or her fuzzy preference relation, and we can apply here some *choice functions* considered by Kacprzyk and Zadrożny [29], [31], [35]. They are based on the concept of the classical choice function, C, that may be defined in a slightly simplified general form as:

$$C(S, P) = o_0, \quad o_0 \subseteq S \tag{33}$$

that may be exemplified by

$$C(S, P) = \{o_i \in O : \forall_{i \neq j} P(o_i, o_j)\} \tag{34}$$

where P denotes a classical crisp preference relation.

In the case of a fuzzy preference relation, P, we assume C to be a fuzzy set of chosen options defined as:

$$\mu_C(o_i) = \min_j \mu_P(o_i, o_j) \tag{35}$$

which may lead to a more flexible formula by replacing the strict min operator with a linguistic quantifier Q (e.g., "most") yielding:

$$\mu_C(o_i) = T(Qo_j \ P(o_i, o_j)) \tag{36}$$

For our further discussion the specific form of the choice function is not important, and we assume (36). In fact, it is possible that each individual adopts a different choice function, and hence a choice function assigned to each individual is denoted as C_k. For a related deep analysis of choice function in our setting, cf. Kacprzyk and Zadrożny [29], [31], [35].

Now, we can define a linguistic summary selecting a set of collectively preferred options, for instance as:

Most individuals choose options $o_{i1}, o_{i2}, \ldots$

to be formally expressed as, e.g.,

$$Qe_k \left(\mu_{C_k}(o_{i1}) = high \right) \wedge \left(\mu_{C_k}(o_{i2}) = very\ high \right) \wedge \ldots \tag{37}$$

where membership degrees to a choice set are discretized and expressed using linguistic labels.

The options referred to in such a summary qualify as a consensus solution if the goal of the group is to arrive at a subset of collectively preferred options. Therefore, such a summary is an alternative indicator of consensus.

On the other hand, a summary exemplified by

Most individuals reject options $o_{i1}, o_{i2}, \ldots$

to be formally expressed as, e.g.,

$$Qe_k \left(\mu_{C_k}(o_{i1}) = low \right) \wedge \left(\mu_{C_k}(o_{i2}) = very\ low \right) \wedge \ldots \tag{38}$$

make it possible to exclude the options concerned from a further consideration.

Therefore, by using the concept of a choice function we can get a constructive and practical definition of a consensus degree. Namely, both (28) and (31) refer to the preferences of the individuals over all pairs of options, possibly with importance weights. Since these importance weights are set independently of the current "standing" of the options implied by preference relations, a more rational definition should put more emphasis on preferences related to the options preferred by individuals and less on those rejected by them. Thus, the importance weights of pairs of options in (28) may be assumed as:

$$\mu_{B'_{kl}}(o_i, o_j) = f(\mu_{C_k}(o_i), \mu_{C_l}(o_i), \mu_{C_k}(o_j), \mu_{C_l}(o_j)) \tag{39}$$

that is, importance weights of pairs of options are specific for each pair of individuals. Function f may be exemplified by a simple arithmetic average.

7.2 Options as Objects

Objects of linguistic summaries may also be equated with options and, then, their attributes are preference degrees over other options as expressed by particular individuals adding, possibly, importance degrees of the options. Formally, we have:

$$Y = O \tag{40}$$

and

$$A = \{\mathcal{P}_{ij}^k\} \cup \{\mathcal{I}\} \tag{41}$$

where attributes $\mathcal{P}_{ij}^k$ correspond to preference degrees over other options and $\mathcal{I}$ represents importance.

This perspective may give an additional insight into the structure of preferences of both the entire group and particular individuals. For example, a summary:

Most options are dominated by option o_i in opinion of individual k

formally expressed as, e.g.,

$$Qo_j\ p_{ij}^k = definite \tag{42}$$

directly corresponds to the choice function mentioned earlier. Namely, if such a summary is valid, then it means that option o_i belongs to the choice set of individual k. On the other hand, a summary like:

Most options are dominated by option o_i in opinion of individual $k1, k2, \ldots$

formally expressed as, e.g.,

$$Qo_j\ (p_{ij}^{k1} = definite) \wedge (p_{ij}^{k2} = definite) \wedge \ldots \tag{43}$$

indicates option o_i as a candidate for a consensual solution.

Interesting patterns in the group may be grasped via linguistic summaries exemplified by:

Most options dominating option o_i in opinion of individual $k1$ also dominate option o_i in opinion of individual $k2$

to be formally expressed as, e.g.,

$$Qo_j\ (p_{ji}^{k1} = definite, p_{ji}^{k2} = definite) \tag{44}$$

Such a summary indicates a similarity of preferences of individuals $k1$ and $k2$. This similarity is here limited to just a pair of options but may be much more convincing in case of :

$$Qo_j\ (p_{ji_1}^{k1} = definite \wedge p_{ji_2}^{k1} = definite \wedge \ldots ,$$
$$p_{ji_1}^{k2} = definite \wedge p_{ji_2}^{k2} = definite \wedge \ldots)$$

Another perspective may be obtained assuming a different set of attributes for options. Namely, we can again employ the concept of a choice set and characterize each option o_i by a vector:

$$[\mu_{C_1}(o_i), \mu_{C_2}(o_i), \ldots, \mu_{C_m}(o_i)] \tag{45}$$

Then, a summary like

Most options are preferred by individual e_k

formally expressed as, e.g.,

$$Qo_i \, \mu_{C_k}(o_i) = high \tag{46}$$

indicates individual e_k as being rather indifferent in his/her preferences, while a summary like

Most options are rejected by individual e_l

formally expressed as, e.g.,

$$Qo_i \, \mu_{C_l}(o_i) = low \tag{47}$$

suggests that individual e_l exposes a clear preference towards a limited subset of options.

The second representation of options as objects may be seen as a kind of a compression of the first. Namely, for a given option o_i all p_{ij}^k's related to individual e_k which represent o_i in (41) are compressed into one number $\mu_{C_k}(o_i)$ in (45), i.e.,

$$[p_{i1}^k, p_{i2}^k, \ldots, p_{in}^k] \longrightarrow \mu_{C_k}(o_i) \tag{48}$$

Another compression is possible by aggregating, for a given option o_i, all p_{ij}^k's related to option o_j which represent o_i in (41) into one number, i.e.,

$$[p_{ij}^1, p_{ij}^2, \ldots, p_{ij}^m] \longrightarrow \text{aggregation}(p_{ij}^k)_{k=1,m} \tag{49}$$

The aggregation operator may take various forms, including a linguistic quantifier guided aggregation. The representation of options as objects obtained thus far may be used to generate summaries with interpretations similar to (45), but with slightly different semantics. The difference is related to the *direct* and *indirect* approaches to group decision making as discussed in Kacprzyk [19], [20], and Zadrożny [65].

This subsumes some basic possible verbalized types of an additional information, which is based on linguistic summaries, that can be of a great help in supporting the moderator to effectively and efficiently run a consensus reaching session. Among other approaches in a related spirit one should also cite Herrera-Viedma, Mata, Martinez and Pérez [16, 17].

8 Helping the Moderator Run a Consensus Reaching Session Using Action Rules

As the main driving force for the consensus reaching process is an exchange of arguments during the discussion, the system has to provide the moderator and the whole group with some advice (feedback information) on how far the group is from consensus, what are the most controversial issues (options), whose preferences are in the highest disagreement with the rest of the group, how their change would influence the consensus degree, etc. The latter can be done via the action rules.

8.1 Individuals Treated as Objects

If we identify the set of objects O with the set of individuals E, then the set of attributes A is, similarly as in Subsction 7.1, composed of (cf. Kacprzyk, Zadrożny and Raś [41]):

1. preference degrees (membership and non-membership expressed using linguistic terms) of a given individual for the particular pairs of alternatives,
2. an importance degree of an individual,
3. a personal consensus degree PCD, for a given individual and $DPCD$s, for a given individual and all pairs of the alternatives,
4. a choice set implied by the preference relation of a given individual.

where the *personal consensus degree*, $PCD(e_k)$ is defined as the truth value of the proposition:

> "Preferences of the expert e_k as to *most relevant* pairs of alternatives are in agreement with the preferences of *most* important experts" (50)

while the *detailed personal consensus degree*, $DPCD(e_k, s_i, s_j)$ is defined as the truth value of the proposition:

> "Preference of the expert e_k as to a pair of options (s_i, s_j) is in agreement with the preferences of *most important* experts" (51)

For more information on the definitions and properties of PCD and $DPCD$, cf. Kacprzyk, Zadrożny and Raś [41].

From the perspective of the action rules, we consider these groups of attributes as flexible group, stable group and the decision attributes groups. Thus, while mining action rules we pick up one from the last group of attributes and then start one of the algorithms mentioned in [60, 48, 50, 51, 49, 58]. A typical scenario may be the following. If the consensus degree in the group is too low, then the PCDs and $DPCD$s are computed. Next, we look for the rules which suggest how some individuals should change their preferences so as to change their PCD value from low to high, e.g.:

$$[\{(\text{importance}, important \rightarrow important),$$
$$(\mu_R(o_i, o_j), not\_at\_all \rightarrow moderately)$$
$$(\mu_R(o_j, o_i), not\_at\_all \rightarrow not\_at\_all)\} \Rightarrow$$
$$(\text{PCD}, medium \rightarrow high)]$$

suggesting that for important individuals it is enough to change their preferences concerning a given pair of options to get an increase of the personal consensus degree (PCD).

First, in order to produce such rules we need to discretize the values of PCD, using, e.g., another set of linguistic terms $\{very\ high, high, medium, low, very\ low\}$. Second, one has to be careful while generating the action rules so as not to suggest changes in preferences violating the consistency of the A-ILFS preference relations (21). The simplest solution is to treat both the membership values $\mu_R(o_i, o_j)$ and $\mu_R(o_j, o_i)$ as one atomic value, from the point of view of the action rules. Third, the special role of the hesitation margin $\pi(x)$ should be noted. Namely its value is a function of two other degrees, thus its direct use as an attribute in the description of an individual and its further use in the action rules does not make any difference. However, the hesitation margin may be used to assess the cost of given action rule since it may be assumed that the cost of changing the preference degree for which the hesitation margin is high should be lower. Also the importance may be seen as contributing to the cost evaluation: the higher importance of an individual the higher the cost of change.

Finally, it should be emphasized that we mean the action rules as a recommendation only. Thus, the changes suggested by the generated action rules are presented for consideration to the relevant individuals and they decide if and how to take them into account. The suggestions provided by an action rule are meant to trigger a discussion by showing some patterns in the group's preferences. It does not have to be necessarily the case that immediate implementation of these changes secures the increase of the agreement in the group.

Similar action rules may be generated with respect to a specific individual and specific pair of options, using the $DPCD$ indicator as the decision attribute.

The fourth group of attributes which may be associated with an individual consists of the membership and non-membership degrees of particular options to a choice set induced by his or her preference relation. These attributes may appear in actions rules of the following type:

$$[\{(\texttt{importance}, very\ important \rightarrow very\ important),$$
$$(\mu_R(o_i, o_j), not\_at\_all \rightarrow moderately)$$
$$(\mu_R(o_i, o_k), not\_at\_all \rightarrow strongly)\} \Rightarrow$$
$$(\mu_{C(S,R)}(o_i), weakly \rightarrow strongly)]$$

that is, for very important individuals if they change preferences as to an option o_i with respect to o_j and o_k as shown, then the membership of o_i to the choice sets of these individuals should be promoted from "weakly" to "strongly".

Such an action rule, again, should be seen first of all as providing possibly interesting information for the individuals. Namely, in the example given above, the individuals learn that there is some tight relation between the overall status of option o_i (represented by its membership degree to the individual's choice sets) and its position with respect to o_j and o_k. This may trigger a further discussion and provide much clarification.

Finally, it should be observed that the association action rules, mentioned in Section 4 may provide even more valuable information as they point out a set of changes that are expected to occur simultaneously provided that a number of preferences degrees changes are implemented.

8.2 Options Treated as Objects

Action rules may also be generated with respect to the set of options S instead as to the set of objects O. The set of attributes A is then composed of:

1. preference degrees with respect to the rest of the alternatives as expressed by all individuals,
2. a relevance degree of an option,
3. the option consensus degree OCD,
4. the choice sets membership and non-membership degrees of an option.

where the *option consensus degree*, $OCD(s_i)$ is defined as the truth value of the proposition (cf. Kacprzyk, Zadrożny and Raś [41]):

> "*Most important* pairs of experts
> agree in their preferences with re-   (52)
> spect to the alternative s_i"

The generation and use of the action rules in this case is similar to the one for individuals playing the role of objects. In particular, the attributes of the first group are treated as flexible, and may be expressed for option o_i as a sequence of pairs of the membership and non-membership degrees of the A-ILFS preference relation:

$$\mu_{R_1}(o_1, o_i), \nu_{R_1}(o_1, o_i), \ldots, \mu_{R_1}(o_i, o_N), \nu_{R_1}(o_i, o_N),$$
$$\mu_{R_2}(o_1, o_i), \nu_{R_2}(o_1, o_i), \ldots, \mu_{R_2}(o_i, o_N), \nu_{R_2}(o_i, o_N),$$
$$\ldots$$
$$\mu_{R_M}(o_1, o_i), \nu_{R_M}(o_1, o_i), \ldots, \mu_{R_M}(o_i, o_N), \nu_{R_M}(o_i, o_N),$$

assuming $1 < i < N$. The attributes in this group are thus indexed by a number of an individual and by a number of an optoon, different from o_i The second attribute is stable, and the third and fourth serve as decision attributes.

Thus, for example, we can obtain action rules stating that for relevant options, if individuals e_k and e_m change their preferences (membership degrees) from "weakly" to "strongly" then the option consensus degree (OCD) change from "low" to "high" (we assume, as in case of the PCD's, that the range of values of the OCD is discretized using a set of linguistic terms).

The fourth group of attributes which may be associated with the options consists of the membership and non-membership degrees of a given option to the choice sets induced by the preference relations of particular individuals. For option o_i these form a vector:

$$[\mu_{C(S,R_1)}(o_i), \nu_{C(S,R_1)}(o_i), \ldots, \mu_{C(S,R_M)}(o_i), \nu_{C(S,R_M)}(o_i)] \qquad (53)$$

Then, the following action rule may be generated:

$$[\{(\texttt{relevance}, relevant \rightarrow relevant),$$
$$(\mu_{R_k}(\cdot, o_j), not\_at\_all \rightarrow strongly)\} \Rightarrow$$
$$(\mu_{C(S,R_k)}, weakly \rightarrow strongly)]$$

stating that for a relevant option, if the preferences of individual e_k concerning this option and option o_j change from lack of preference to strong preference, then the membership of this option to the choice set of individual R_k is expected to change from weak to strong.

An extended analysis, by taking into account the cost an of action rules can be performed but will not be shown in this paper. For details we refer the readers to our paper Kacprzyk, Zadrożny and Raś [41].

is not decided in his preferences then he or she should be open for

9 Concluding Remarks

The purpose of the paper was to present an extended, and a more unified and comprehensive approach to generate linguistic summaries of the "state of the matter" in the consensus reaching process run in a group of individuals by a moderator. We had shown some other linguistically quantified propositions that are linguistic summaries of various relations between the individuals and their preferences over the set of options. These linguistic summaries may give an extraordinary insight into what is the present "state of the mind" of the group, and which paths (related to changes of testimonies of various individuals with respect to various options) may be promising for getting closer to consensus.

It should be noted that we have used our protoform based approach to linguistic data summaries as shown in Kacprzyk and Zadrożny [33] which provides a powerful general framework and also, as recently shown in Kacprzyk and Zadrożny [38] can make the use of tools and software developed in natural language generation (NLG) possible which may greatly simplify implementations.

We have also outlined another set of tools that could help the moderator run a consensus reaching session, namely those employing the concept of a action rule that may essentially make possible to find with respect to which individual and option a concession may be made so that specific individuals be willing to change their preferences concerning specific pairs of options getting closer to consensus.

The combination of the use of linguistic summaries and action rules, which is a novel result presented in this paper, seems to add an enriched power of a set of linguistic summaries based set of tools to support the moderator run consensus reaching process proposed by the authors in previous papers.

References

1. Agrawal, R., Srikant, R.: Fast algorithms for mining association rules. In: Bocca, J.B., Jarke, M., Zaniolo, C. (eds.) Proc. 20th International Conference on Very Large Databases, pp. 487–499. Morgan Kaufmann Publishers Inc., San Francisco (1994)
2. Alonso, S., Cabrerizo, F.J., Chiclana, F., Herrera, F., Herrera-Viedma, E.: Group decision making with incomplete fuzzy linguistic preference relations. Int. J. Intell. Syst. 24(2), 201–222 (2009)
3. Atanassov, K.T.: Intuitionistic fuzzy sets. Fuzzy Sets and Systems 20, 87–96 (1986)
4. Atanassov, K.T.: Intuitionistic Fuzzy Sets. Physica Verlag, Heidelberg (1999)
5. Bonnefon, J.-F., Dubois, D., Fargier, H., Leblois, S.: Qualitative heuristics for balancing the pros and cons. Theory and Decision 65, 71–95 (2008)
6. Cabrerizo, F.J., Alonso, S., Herrera-Viedma, E.: A consensus model for group decision making problems with unbalanced fuzzy linguistic information. International Journal of Information Technology and Decision Making 8(1), 109–131 (2009)
7. Delgado, M., Verdegay, J.L., Vila, M.A.: On aggregation operations of linguistic labels. IJIS 8, 351–370 (1993)
8. Dubois, D., Prade, H.: An introduction to bipolar representations of information and preference. International Journal of Intelligent Systems 23(8), 866–877 (2008)
9. Fedrizzi, M., Kacprzyk, J., Zadrożny, S.: An interactive multi-user decision support system for consensus reaching processes using fuzzy logic with linguistic quantifiers. Decision Support Systems 4(3), 313–327 (1988)
10. Fedrizzi, M., Kacprzyk, J., Nurmi, H.: Consensus degrees under fuzzy majorities and fuzzy preferences using OWA (ordered weighted average) operators. Control and Cybernetics 22, 71–80 (1993)
11. Fedrizzi, M., Kacprzyk, J., Owsiński, J.W., Zadrożny, S.: Consensus reaching via a GDSS with fuzzy majority and clustering of preference profiles. Annals of Operations Research 51, 127–139 (1994)
12. Fedrizzi, M., Kacprzyk, J., Zadrożny, S.: An interactive multi - user decision support system for consensus reaching processes using fuzzy logic with linguistic quantifiers. Decision Support Systems 4, 313–327 (1988)
13. Herrera, F., Herrera-Viedma, E., Verdegay, J.L.: A model of consensus in group decision making under linguistic assessments. Fuzzy Sets and Systems 78, 73–88 (1996)
14. Herrera, F., Herrera-Viedma, E.: Choice Functions for Linguistic Preference Relations. In: 6th International Conference on Information Processing and Management of Uncertainty in Knowledge-Bases Systems IPMU 1998, Paris, France, pp. 152–157 (1998)
15. Herrera, F., Herrera-Viedma, E.: Choice functions and mechanisms for linguistic preference relations. European Journal of Operational Research 120, 144–161 (2000)
16. Herrera-Viedma, E., Mata, F., Martinez, L., Pérez, L.G.: An adaptive module for the consensus reaching process in group decision making problems. In: Proc. MDAI, pp. 89–98 (2005)
17. Herrera-Viedma, E., Martinez, L., Mata, F., Chiclana, F.: A consensus support system model for group decision-making problems with multi-granular linguistic preference relations. IEEE Trans. on Fuzzy Systems 13(5), 644–658 (2005)
18. Kacprzyk, J.: On some fuzzy cores and 'soft' consensus measures in group decision making. In: Bezdek, J.C. (ed.) He Analysis of Fuzzy Information, vol. 2, pp. 119–130. CRC Press, Boca Raton (1987)
19. Kacprzyk, J.: Group decision - making with a fuzzy majority via linguistic quantifiers. Part I: A consensory - like pooling; Part II: A competitive - like pooling. Cybernetics and Systems: an Int. Journal 16, 119–129 (Part I), 131 - 144, Part II. (1985)

20. Kacprzyk, J.: Group decision making with a fuzzy linguistic majority. Fuzzy Sets and Systems 18, 105–118 (1986)
21. Kacprzyk, J., Fedrizzi, M.: 'Soft' consensus measures for monitoring real consensus reaching processes under fuzzy preferences. Control and Cybernetics 15, 309–323 (1986)
22. Kacprzyk, J., Fedrizzi, M.: A 'soft' measure of consensus in the setting of partial (fuzzy) preferences. European Journal of Operational Research 34, 315–325 (1988)
23. Kacprzyk, J., Fedrizzi, M.: A 'human - consistent' degree of consensus based on fuzzy logic with linguistic quantifiers. Mathematical Social Sciences 18, 275–290 (1989)
24. Kacprzyk, J., Fedrizzi, M., Nurmi, H.: Group decision making and consensus under fuzzy preferences and fuzzy majority. Fuzzy Sets and Systems 49, 21–31 (1992)
25. Kacprzyk, J., Nurmi, H.: On fuzzy tournaments and their solution concepts in group decision making. European Journal of Operational Research 51, 223–232 (1991)
26. Kacprzyk, J., Yager, R.R.: Linguistic summaries of data using fuzzy logic. IJGS 30, 33–154 (2001)
27. Kacprzyk, J., Yager, R.R., Zadrożny, S.: A fuzzy logic based approach to linguistic summaries of databases. IJAMCS 10, 813–834 (2000)
28. Kacprzyk, J., Zadrożny, S.: Computing with words in decision making through individual and collective linguistic choice rules. Int. J. Uncertain. Fuzziness Knowl.-Based Syst. 9(Supplement), 89–102 (2001)
29. Kacprzyk, J., Zadrożny, S.: Collective choice rules in group decision making under fuzzy preferences and fuzzy majority: a unified OWA operator based approach. CC 31, 937–948 (2002)
30. Kacprzyk, J., Zadrożny, S.: Dealing with imprecise knowledge on preferences and majority in group decision making: towards a unified characterization of individual and collective choice functions. Bulletin of the Polish Academy of Sciences (Tech. Sci.) 51, 279–302 (2003)
31. Kacprzyk, J., Zadrożny, S.: An Internet-based group decision support system. Management VII(28), 2–10 (2003)
32. Kacprzyk, J., Zadrożny, S.: Linguistically quantified propositions for consensus reaching support. In: Proc. of the IEEE International Conference on Fuzzy Systems, Budapest, Hungary, July 25-29, pp. 1135–1140 (2004)
33. Kacprzyk, J., Zadrożny, S.: Computing with words in intelligent database querying: standalone and Internet-based applications. Information Sciences 134, 71–109 (2005)
34. Kacprzyk, J., Zadrożny, S.: On a concept of a consensus reaching process support system based on the use of soft computing and Web techniques. In: Ruan, D., Montero, J., Lu, J., Martinez, L., D'hondt, P., Kerre, E.E. (eds.) Computational Intelligence in Decision and Control, pp. 859–864. World Scientific, Singapore (2008)
35. Kacprzyk, J., Zadrożny, S.: Towards a general and unified characterization of individual and collective choice functions under fuzzy and nonfuzzy preferences and majority via the ordered weighted average operators. International Journal of Intelligent Systems 24(1), 4–26 (2009)
36. Kacprzyk, J., Zadrożny, S.: Protoforms of linguistic database summaries as a human consistent tool for using natural language in data mining. International Journal of Software Science and Computational Intelligence 1(1), 100–111 (2009)
37. Kacprzyk, J., Zadrożny, S.: Soft computing and Web intelligence for supporting consensus reaching. Soft Computing 14, 833–846 (2010)
38. Kacprzyk, J., Zadrożny, S.: Computing with words is an implementable paradigm: fuzzy queries, linguistic data summaries and natural language generation. IEEE Transactions on Fuzzy Systems 18, 461–472 (2010)

39. Kacprzyk, J., Zadrożny, S., Fedrizzi, M., Nurmi, H.: On group decision making, consensus reaching, voting and voting paradoxes under fuzzy preferences and a fuzzy majority: a survey and a granulation perspective. In: Pedrycz, W., Skowron, A., Kreinovich, V. (eds.) Handbook of Granular Computing, pp. 906–929. Wiley, Chichester (2008)
40. Kacprzyk, J., Zadrożny, S., Fedrizzi, M., Nurmi, H.: On group decision making, consensus reaching, voting and voting paradoxes under fuzzy preferences and a fuzzy majority: a survey and some perspectives. In: Bustince, H., Herrera, F., Montero, J. (eds.) Fuzzy Sets and Their Extensions: Representation, Aggregation and Models, pp. 263–295. Springer, Heidelberg (2008)
41. Kacprzyk, J., Zadrożny, S., Raś, Z.W.: How to support consensus reaching using action rules: a novel approach. International Journal of Uncertainty, Fuzziness and Knowledge-Based Systems 18(4), 451–470 (2010)
42. Kacprzyk, J., Zadrożny, S., Wilbik, A.: Linguistic summarization of some static and dynamic features of consensus reaching. In: Reusch, B. (ed.) Computational Intelligence, Theory and Applications, pp. 19–28. Springer, Heidelberg (2006)
43. Loewer, B., Laddaga, R.: Destroying the consensus. Special Issue on Consensus
44. Mata, F., Martínez, L., Herrera-Viedma, E.: An Adaptive Consensus Support Model for Group Decision-Making Problems in a Multigranular Fuzzy Linguistic Context. IEEE Transactions on Fuzzy Systems 17(2), 279–290 (2009)
45. Nurmi, H.: Approaches to collective decision making with fuzzy preference relations. Fuzzy Sets and Systems 6, 187–198 (1981)
46. Nurmi, H., Kacprzyk, J., Fedrizzi, M.: Probabilistic, fuzzy and rough concepts in social choice. European Journal of Operational Research 95, 264–277 (1996)
47. Pawlak, Z.: Information systems theoretical foundations. Inf. Syst. Journal 6(3), 205–218 (1981)
48. Raś, Z.W., Wieczorkowska, A.A.: Action-rules: How to increase profit of a company. In: Zighed, D.A., Komorowski, J., Żytkow, J.M. (eds.) PKDD 2000. LNCS (LNAI), vol. 1910, pp. 587–592. Springer, Heidelberg (2000)
49. Dardzińska, A., Raś, Z.W.: Extracting Rules from Incomplete Decision Systems: System ERID. In: Studies in Computational Intelligence on Foundations and Novel Approaches in Data Mining, pp. 143–153. Springer, Heidelberg (2006)
50. Raś, Z.W., Dardzińska, A.: Action Rules Discovery without Pre-existing Classification Rules. In: Chan, C.-C., Grzymala-Busse, J.W., Ziarko, W.P. (eds.) RSCTC 2008. LNCS (LNAI), vol. 5306, pp. 181–190. Springer, Heidelberg (2008)
51. Raś, Z.W., Tsay, L.-S., Dardzińska, A.: Tree-Based Algorithms for Action Rules Discovery. In: Mining Complex Data, Studies in Computational Intelligence, pp. 153–163. Springer, Heidelberg (2009)
52. Reiter, E., Dale, R.: Building Natural Language Generation Systems. Cambridge University Press, Cambridge (2000)
53. Szmidt, E., Kacprzyk, J.: Intuitionistic fuzzy sets in group decision making. Notes on Intuitionistic Fuzzy Sets 2, 15–32 (1996)
54. Szmidt, E., Kacprzyk, J.: Intuitionistic fuzzy relations and consensus formations. Notes on Intuitionistic Fuzzy Sets 6, 1–10 (2000)
55. Szmidt, E., Kacprzyk, J.: Using intuitionistic fuzzy sets in group decision making. Control and Cybernetics 31, 1055–1057 (2002)
56. Szmidt, E., Kacprzyk, J.: A consensus-reaching process under intuitionistic fuzzy preference relations. International Journal of Intelligent Systems 18(7), 837–852 (2003)
57. Szmidt, E., Kacprzyk, J.: A new concept of a similarity measure for intuitionistic fuzzy sets and its use in group decision making. In: Torra, V., Narukawa, Y., Miyamoto, S. (eds.) MDAI 2005. LNCS (LNAI), vol. 3558, pp. 272–282. Springer, Heidelberg (2005)

58. Tzacheva, A., Raś, Z.W.: Action rules mining. International Journal of Intelligent Systems 20(7), 719–736 (2005)
59. Tzacheva, A., Raś, Z.W.: Constraint based action rule discovery with single classification rules. In: An, A., Stefanowski, J., Ramanna, S., Butz, C.J., Pedrycz, W., Wang, G. (eds.) RSFDGrC 2007. LNCS (LNAI), vol. 4482, pp. 322–329. Springer, Heidelberg (2007)
60. Tsay, L.-S., Raś, Z.W., Wieczorkowska, A.: Tree-based Algorithm for Discovering Extended Action-Rules (System DEAR2). Intelligent Information Systems, 459–464 (2004)
61. Xu, Z., Yager, R.R.: Intuitionistic and interval-valued intuitionistic fuzzy preference relations and their measures of similarity for the evaluation of agreement within a group. Fuzzy Optimization and Decision Making 8, 123–139 (2009)
62. Yager, R.R.: A new approach to the summarization of data. IS 28, 69–86 (1982)
63. Yager, R.R.: On ordered weighted averaging operators in multicriteria decision making. IEEE Transactions onf Systems, Man and Cybernetics SMC 18, 183–190 (1988)
64. Zadeh, L.A.: A Computational Approach to Fuzzy Quantifiers in Natural Languages. Computers and Mathematics with Applications 9, 149–184 (1983)
65. Zadrożny, S.: An approach to the consensus reaching support in fuzzy environment. In: Kacprzyk, J., Nurmi, H., Fedrizzi, M. (eds.) Consensus under Fuzziness, pp. 83–109. Kluwer, Boston (1997)
66. Zadrożny, S., Furlani, P.: Modelling and supporting of the consensus reaching process using fuzzy preference relations. Control and Cybernetics 20, 135–154 (1991)
67. Zadrożny, S., Kacprzyk, J.: An Internet-based group decision and consensus reaching support system. In: Yu, X., Kacprzyk, J. (eds.) Applied Decision Support with Soft Computing, pp. 263–275. Springer, Heidelberg (2003)
68. Zadrożny, S., Kacprzyk, J.: Issues in the practical use of the OWA operators in fuzzy querying. Journal of Intelligent Information Systems 33, 307–325 (2009)

Part IV: Modern Trends in Consensus Reaching Support, and Applications

Consensual Processes Based on Mobile Technologies and Dynamic Information

I.J. Pérez, F.J. Cabrerizo, M.J. Cobo, S. Alonso, and E. Herrera-Viedma

Abstract. The aim of this contribution is to present a prototype of decision support system based on mobile technologies and dynamic information. Users can run the system on their own mobile devices in order to provide their preferences at anytime and anywhere. The system provides consensual and selection support to deal with dynamic decision making situations. Furthermore, the system incorporates a mechanism that allows to manage dynamic decision situations in which some information about the problem is not constant through the time, it gives more realism to decision processes with high or dynamic set of alternatives, focussing the discussion in a subset of them that changes in each stage of the process. The experts' preferences are represented using a linguistic approach. In such a way, we provide a new linguistic framework, that is mobile and dynamic, to deal with group decision making problems.

1 Introduction

Group Decision Making (GDM) arises from many real world situations [28, 42]. Thus, the study of decision making is necessary and important not only in Decision Theory but also in areas such as Management Science, Operations Research,

I.J. Pérez · M.J. Cobo · E. Herrera-Viedma
Dept. of Computer Science and Artificial Intelligence, University of Granada, Spain
e-mail: ijperez@decsai.ugr.es, mjcobo@decsai.ugr.es,
    viedma@decsai.ugr.es

F.J. Cabrerizo
Dept. of Software Engineering and Computer Systems,
Distance Learning University of Spain (UNED), Madrid, Spain
e-mail: cabrerizo@issi.uned.es

S. Alonso
Dept. of Software Engineering, University of Granada, Spain
e-mail: zerjioi@ugr.es

E. Herrera-Viedma et al. (Eds.): Consensual Processes, STUDFUZZ 267, pp. 317–337.
springerlink.com   © Springer-Verlag Berlin Heidelberg 2011

Politics, Social Psychology, Artificial Intelligence, Soft Computing, and so on. In these situations, there is a problem that can be solved in different ways and a group of experts trying to achieve a consensual solution. To do this, experts have to express their preferences by means of a set of assessments over a set of alternatives.

In the last years, the interaction human-technology has had several significant advances. The spread of mobile devices has increased accessibility to data and, in turn, influenced the time and the way in which users make decisions. Users can make real-time decisions based on the most up-to-date data accessed via wireless devices, such as portable computers, mobile phones, and personal digital assistants (PDAs). So, the application of the latest technologies extends opportunities and allows to carry out consensual processes in new frameworks. We assume that if the communications are improved the decisions will be upgraded, because people can focuss on the problem with less wasted time on unimportant issues [31, 43].

Several authors have provided interesting results on GDM with the help of fuzzy theory [10, 17, 27, 28, 29, 30, 37]. There are decision situations in which the experts' preferences cannot be assessed precisely in a quantitative form but may be in a qualitative one, and thus, the use of a *linguistic approach* is necessary [3, 4, 5, 16, 19, 26, 45, 50]. The *linguistic approach* is an approximate technique which represents qualitative aspects as linguistic values by means of *linguistic variables*, that is, variables whose values are not numbers but words or sentences in a natural or artificial language [15].

In this chapter we present a prototype of mobile decision support system (DSS) to deal automatically with linguistic GDM problems based on mobile technologies. This mobile DSS allows to develop dynamic consensual processes. In fact, at every stage of the decision process, users, in order to reach a common solution, receive recommendations to help them to change their preferences and they are able to send their updated preferences at any moment. Additionally, to better simulate real decision making processes, the mobile DSS includes a tool to manage dynamic sets of alternatives [38], that is, not only dynamic addition of new alternatives that, due to some dynamic external factors, can appear during the decision process, but also deleting some of them considered good alternatives at the beginning of the process but not so later on or are unavailable at the time.

In order to do this, the paper is set out as follows. Some preliminary aspects about GDM models, linguistic approach and mobile technologies usage in GDM problems are presented in Section 2. Section 3 defines the prototype of a mobile DSS and Section 4 includes a practical experiment. Finally, in Section 5 we point out our conclusions.

2   Preliminaries

In this section we present some considerations about GDM problems, the fuzzy linguistic approach and the use of mobile technologies in consensual processes.

2.1 GDM Problems

In a GDM problem we have a finite set of feasible alternatives. $X = \{x_1, x_2, \ldots, x_n\}$, $(n \geq 2)$ and the best alternatives from X have to be identified according to the information given by a set of experts, $E = \{e_1, e_2, \ldots, e_m\}$, $(m \geq 2)$.

Resolution methods for GDM problems are usually composed by two different processes [19] (see Figure 1):

1. *Consensus process:* Clearly, in any decision process, it is preferable that the experts reach a high degree of consensus on the solution set of alternatives. Thus, this process refers to how to obtain the maximum degree of consensus or agreement among the experts on the solution alternatives.
2. *Selection process:* This process consists in how to obtain the solution set of alternatives from the opinions on the alternatives given by the experts. Furthermore, the selection process is composed of two different phases:

   a. Aggregation phase: This phase uses an aggregation operator in order to transform the individual preferences on the alternatives into a collective preference.
   b. Exploitation phase: This phase transforms the collective preference into a partial ranking of alternatives that helps to make the final decision.

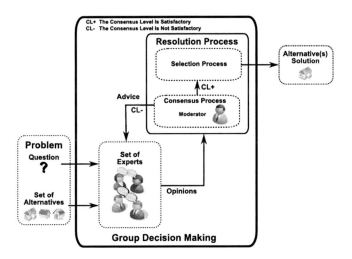

Fig. 1 *Resolution process of a GDM*

2.2 Fuzzy Linguistic Approach

There are situations in which the information cannot be assessed precisely in a quantitative form but may be in a qualitative one. For example, when attempting to qualify phenomena related to human perception, we are often led to use words in natural language instead of numerical values, e.g. when evaluating quality of a restaurant, terms like *good, medium* or *bad* can be used. In other cases, precise quantitative information cannot be stated because either it is unavailable or the cost for its computation is too high and an "approximate value" can be applicable, eg. when evaluating the speed of a car, linguistic terms like *fast, very fast* or *slow* can be used instead of numeric values [3, 13]. The use of Fuzzy Sets Theory has given very good results for modelling qualitative information [49].

Fuzzy linguistic modelling is a tool based on the concept of linguistic variable to deal with qualitative assessments. It has proven its usefulness in many problems, e.g., in decision making, quality evaluation, information retrieval models, etc[12, 23, 24, 33, 34, 39, 40]. Ordinal fuzzy linguistic modelling [15] is a very useful kind of fuzzy linguistic approach proposed as an alternative tool to the traditional fuzzy linguistic modelling which simplifies the computing with words process as well as linguistic aspects of problems. It is defined by considering a finite and totally ordered label set $S = \{s_i\}, i \in \{0, ..., g\}$ in the usual sense, i.e., $s_i \geq s_j$ if $i \geq j$, and with odd cardinality (usually 7 or 9 labels). The mid term represents an assessment of "approximately 0.5", and the rest of the terms are placed symmetrically around it. The semantics of the label set is established from the ordered structure of the label set by considering that each label for the pair (s_i, s_{g-i}) is equally informative [3]. For example, we can use the following set of seven labels to represent the linguistic information:

$S = \{$ N=*Null*, VL=*Very Low*, L=*Low*, M=*Medium*, H=*High*, VH=*Very High*, P=*Perfect*$\}$.

In any linguistic model we also need some management operators for linguistic information. An advantage of the ordinal fuzzy linguistic modeling is the simplicity and speed of its computational model. It is based on the symbolic computational model [15] and acts by direct computation on labels by taking into account the order of such linguistic assessments in the ordered structure of labels. Usually, the ordinal fuzzy linguistic model for computing with words is defined by establishing i) a negation operator, ii) comparison operators based on the ordered structure of linguistic terms, and iii) adequate aggregation operators of ordinal fuzzy linguistic information. In most ordinal fuzzy linguistic approaches the negation operator is defined from the semantics associated to the linguistic terms as

$$NEG(s_i) = s_j \mid j = (g - i)$$

and there are defined two comparison operators of linguistic terms:

1. *Maximization operator: MAX*$(s_i, s_j) = s_i$ if $s_i \geq s_j$; and
2. *Minimization operator: MIN*$(s_i, s_j) = s_i$ if $s_i \leq s_j$.

Using these operators it is possible to define automatic and symbolic aggregation operators of linguistic information, as for example the LOWA operator [18]:

Definition 1. Let $A = \{a_1, ..., a_m\}$ be a set of labels to be aggregated, then the LOWA operator, ϕ, is defined as:

$$\phi(a_1,\ldots,a_m) = W \cdot B^T = \mathscr{C}^m\{w_k,b_k,k=1,\ldots,m\}$$
$$= w_1 \odot b_1 \oplus (1-w_1) \odot \mathscr{C}^{m-1}\{\beta_h,b_h,h=2,\ldots,m\},$$

where $W = [w_1,\ldots,w_m]$ is a weighting vector, such that, $w_i \in [0,1]$ and $\Sigma_i w_i = 1$. $\beta_h = w_h/\Sigma_2^m w_k$, and $B = \{b_1,\ldots,b_m\}$ is a vector associated to A, such that, $B = \sigma(A) = \{a_{\sigma(1)},\ldots,a_{\sigma(m)}\}$, where, $a_{\sigma(j)} \le a_{\sigma(i)} \ \forall\, i \le j$, with σ being a permutation over the set of labels A. $\mathscr{C}^m$ is the convex combination operator of m labels and if $m=2$, then it is defined as:

$$\mathscr{C}^2\{w_i,b_i,i=1,2\} = w_1 \odot s_j \oplus (1-w_1) \odot s_i = s_k,$$

such that, $k = \min\{g,i + round(w_1 \cdot (j-i))\}$, s_j, $s_i \in S$, $(j \ge i)$, where "*round*" is the usual round operation, and $b_1 = s_j$, $b_2 = s_i$. If $w_j = 1$ and $w_i = 0$, with $i \ne j\, \forall i$, then the convex combination is defined as: $\mathscr{C}^m\{w_i,b_i,i=1,\ldots,m\} = b_j$.

An important question of the LOWA operator is the determination of the weighting vector W. In [48], it was defined an expression to obtain W that allows to represent the concept of fuzzy majority [26] by means of a fuzzy linguistic nondecreasing quantifier Q:

$$w_i = Q(i/n) - Q((i-1)/n), \ i = 1,\ldots,n.$$

When a fuzzy linguistic quantifier Q is used to compute the weights of LOWA operator ϕ, it is symbolized by ϕ_Q.

2.3   Mobile Technologies Usage in GDM Problems

During the last decade, organizations have moved from face-to-face group environments to virtual group environments using communication technology. More and more workers use mobile devices to coordinate and share information with other people. The main objective is that the members of the group could work in an ideal way where they are, having all the necessary information to take the right decisions [25, 31, 43, 44].

To support the new generation of decision makers and to add real-time process in the GDM problem field, many authors have proposed to develop decision support systems based on mobile technologies [9, 41]. Similarly, we propose to incorporate mobile technologies in a DSS obtaining a Mobile DSS (MDSS). Using such a technology should enable a user to maximize the advantages and minimize the drawbacks of DSSs.

The need of a face-to-face meeting disappears with the use of this model because the computer system acts as moderator and experts can communicate with the system directly using their mobile device from any place in the world and at any time. Hereby, a continuous information flow among the system and each member of the group is produced, which can help to reach the consensus between the experts in a faster way and to obtain better decisions.

In addition, MDSS can help to reduce the time constraint in the decision process. Thus, the time saved by using the MDSS can be used to do an exhaustive analysis of the problem and obtain a better problem definition. This time also could be used to identify more feasible alternative solutions to the problem, and thus, the evaluation of a large set of alternatives would increase the possibility of finding a better solution. The MDSS helps to the resolution of GDM problems providing a propitious environment for the communication, increasing the satisfaction of the user and, in this way, improving the final decisions [38].

3   A New Mobile Decision Support System

In this section, we present the implemented prototype to deal with dynamic decision making situations, explaining the architecture and the work flow that summarizes the functions of this system. We show how the consensual and selection processes are controlled.

A DSS can be built in several ways, and the used technology determines how a DSS has to be developed [11, 36]. The most used architecture for mobile devices is the "Client/Server" architecture, where the client is a mobile device. The client/server paradigm is founded on the concept that clients (such as personal computers, or mobile devices) and servers (computers) are both connected by a network enabling servers to provide different services for the clients. When a client sends a request to a server, this server processes the request and sends a response back to client.

We have chosen a *thick-client* model for our implementation. This allows us to use the software in all the mobile devices without taking into account the kind of browser. Furthermore, the technologies that we have used to implement the prototype comprise Java and Java Midlets for the client software, PHP for the server functions and MySQL for the database management.

So, the prototype allows user to send his/her preferences by means of a mobile device, and the system returns to the experts the final solution or recommendations to increase the consensus levels, depending on the status of the decision process. An important aspect is that the user-system interaction can be done anytime and anywhere which facilitates expert's participation and the resolution of the decision process. In what follows, we describe the client and server of the prototype in detail.

3.1   Client Side

The client software shows the next seven interfaces to the experts:

- *Authentication:* The device asks a user and a password to access the system.
- *Connection:* The device must be connected to the network to send/receive information to the server.
- *Problem description:* When a decision process is started, the device shows to the experts a brief description of the problem and the set of alternatives.

- *Insertion of preferences:* The device will have a specific interface to insert the linguistic preferences using a set of labels. To introduce or change the preferences using the interface, the user has to use the keys of the device.
- *Swap of Alternatives:* When a new alternative appears in the environment of the problem because some dynamic external factors have changed and this alternative deserves to be a member of the discussion subset or when an alternative have a low dominance degree to the current temporary solution of consensus, the system asks the experts if they want to modify the discussion subset by swapping these alternatives. The experts can assess if they agree to swap the alternatives sending their answer to the question received. The user can select the chosen degree by using the cursor keys of the device.
- *Feedback:* When opinions should be modified, the device shows experts the recommendations and allows experts to send their new preferences.
- *Output:* At the end of the decision process, the device will show the set of solution alternatives as an ordered set of alternatives.

On the technical side of the development of the client part, it is worth noting that the client application complies with the MIDP 2.0 specifications [1] and that the J2ME Wireless Toolkit 2.2 [2] provided by SUN was used in the development phase. This wireless toolkit is a set of tools that provide J2ME developers with some emulation environments, documentation, and examples to develop MIDP-compliant applications. The application was later tested using a JAVA-enabled mobile phone on a GSM network using a GPRS-enabled SIM card. The MIDP application is packaged inside a JAVA archive (JAR) file, which contains the applications classes and resource files. This JAR file is the one that actually is downloaded to the physical device (mobile phone) along with the JAVA application descriptor file when an expert wants to use our prototype.

3.2   Server Side

The server is the main side of the prototype. It implements the main modules and the database that stores the problem data as well as problem parameters and the information generated during the decision process. The communication with the client to receive/send information from/to the experts is supported by mobile Internet (M-Internet) technologies (see Figure 2). Concretely, the three modules of the server are:

3.2.1   Decision Module

In a GDM problem the experts can present their opinions using different types of preference representation (preference orderings, utility functions or preference relations) [6, 47, 46], but in this contribution, we assume that the experts give their preferences using fuzzy linguistic preference relations.

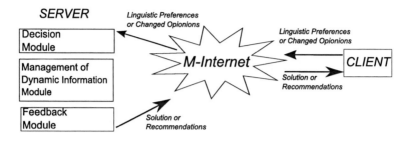

Fig. 2 *Server modules*

Definition 2. A Fuzzy linguistic Preference Relation (FLPR) P^i given by an expert e_i is a fuzzy set defined on the product set $X \times X$, that is characterized by a linguistic membership function

$$\mu_{pi} : X \times X \longrightarrow S$$

where the value $\mu_{pi}(x_l, x_k) = p_{lk}^i$ is interpreted as the linguistic preference degree of the alternative x_l over x_k for the expert e_i.

Once experts have sent their preferences, the server starts the decision module to obtain a temporary solution of the problem. In this module the consensus measures are also calculated. This module has two different processes: 1) *selection process* and 2) *consensus process*.

1. *Selection Process:* This process has two different phases [17]:

   a. Aggregation phase:
   This phase defines a collective preference relation, $P^c = (p_{lk}^c)$, obtained by means of the aggregation of all individual linguistic preference relations $\{P^1, P^2, \ldots, P^m\}$. It indicates the global preference between every pair of alternatives according to the majority of experts' opinions. The aggregation is carried out by means of a LOWA operator ϕ_Q guided by a fuzzy linguistic non-decreasing quantifier Q [18]:

   $$p_{lk}^c = \phi_Q(p_{lk}^1, \ldots, p_{lk}^m)$$

   b. Exploitation phase:
   This phase transforms the global information about the alternatives into a global ranking of them, from which the set of solution alternatives is obtained. The global ranking is obtained applying these two choice degrees of alternatives on the collective preference relation [14]:
   • $QGDD_l$: This quantifier guided dominance degree quantifies the dominance that one alternative x_l has over all the others in a fuzzy majority sense:

   $$QGDD_l = \phi_Q(p_{l1}^c, p_{l2}^c, \ldots, p_{l(l-1)}^c, p_{l(l+1)}^c, \ldots, p_{ln}^c)$$

This measure allows us to define the set of non-dominated alternatives with maximum linguistic dominance degree:

$$X^{QGDD} = \{x_l \in X \mid QGDD_l = sup_{x_k \in X} QGDD_k\}$$

- $QGNDD_l$: This quantifier guided non-dominance degree gives the degree in which each alternative x_l is not dominated by a fuzzy majority of the remaining alternatives:

$$QGNDD_l = \phi_Q(NEG(p_{1l}^s), NEG(p_{2l}^s), \ldots,$$

$$NEG(p_{(l-1)l}^s), NEG(p_{(l+1)l}^s), \ldots, NEG(p_{nl}^s))$$

where

$$p_{lk}^s = \begin{cases} s_0 & \text{if } p_{lk}^c < p_{kl}^c \\ s_{I(p_{lk}^c) - I(p_{kl}^c)} & \text{if } p_{lk}^c \geq p_{kl}^c \end{cases}$$

being $I : S \to \{0, \ldots, g\} \mid I(s_p) = p \; \forall s_p \in S$.

represents the degree in which x_l is strictly dominated by x_k. The set of of non-dominated alternatives with maximum linguistic non-dominance degree is

$$X^{QGNDD} = \{x_l \in X \mid QGNDD_l = sup_{x_k \in X} QGNDD_k\}$$

2. *Consensus Process:*
   We assume that the consensus is a measurable parameter whose highest value corresponds to unanimity and lowest one to complete disagreement. We use some consensus degrees to measure the current level of consensus in the decision process. They are given at three different levels [19, 20, 35]: pairs of alternatives, alternatives and relations. The computation of the consensus degrees is carried out as follows:

   a. For each pair of experts, e_i, e_j $(i < j)$, a similarity matrix, $SM^{ij} = (sm_{lk}^{ij})$, is defined where

   $$sm_{lk}^{ij} = 1 - \frac{|I(p_{lk}^i) - I(p_{lk}^j)|}{g}.$$

   b. A consensus matrix, CM, is calculated by aggregating all the similarity matrices using the arithmetic mean as the aggregation function $\bar{x}$:

   $$cm_{lk} = \bar{x}(sm_{lk}^{ij}; \; i = 1, \ldots, m-1, \; j = i+1, \ldots, m).$$

   c. Once the consensus matrix, CM, is computed, we proceed to calculate the consensus degrees:
      i. *Consensus degree on pairs of alternatives, cp_{lk}.* It measures the agreement on the pair of alternatives (x_l, x_k) amongst all the experts.

      $$cp_{lk} = cm_{lk}.$$

ii. *Consensus degree on alternatives, ca_l.* It measures the agreement on an alternative x_l amongst all the experts.

$$ca_l = \frac{\sum_{k=1}^{n} cp_{lk}}{n}.$$

iii. *Consensus degree on the relation, cr.* It measures the global consensus degree amongst the experts' opinions.

$$cr = \frac{\sum_{l=1}^{n} ca_l}{n}.$$

Initially, in this consensus model we consider that in any nontrivial GDM problem the experts disagree in their opinions so that decision making has to be viewed as an iterative process. This means that agreement is obtained only after some rounds of consultation. In each round, we calculate the consensus measures and check the current agreement existing among experts using cr.

3.2.2 Management of Dynamic Information Module

Classical GDM models are defined in static frameworks. In order to make the decision making process more realistic, this module is able to deal with dynamic parameters in decision making. The main parameter that could vary through the decision making process is the set of alternatives of the problem because it could depend on dynamical external factors like the traffic [8, 32], or the meteorological conditions [7], and so on. In such a way, we can solve dynamic decision problems in which, at every stage of the process, the discussion is centered on different alternatives.

This tool allows to introduce new alternatives in the discussion subset, but this change has to be approved by the experts. To do so, the mechanism has two phases. At the first one, the system identifies the new alternative to include in the set of discussion alternatives (discussion subset) and the worst alternative of the current discussion subset. The second one is to ask experts about if they agree with the replacement and updating the discussion subset [38].

3.2.3 Feedback Module

To guide the change of the experts' opinions, the DSS simulates a group discussion session in which a feedback mechanism is applied to quickly obtain a high consensus level. This mechanism is able to substitute the moderator's actions in the consensus reaching process. The main problem for the feedback mechanism is how to find a way of making individual positions converge and, therefore, how to support the experts in obtaining and agreeing with a particular solution [22]. To do that, we compute others additional consensus measures, called proximity measures [19].

These measures evaluate the agreement between the individual experts' opinions and the group opinion. To compute them for each expert, we need to use the collective FLPR, $P^c = (p_{lk}^c)$, calculated previously.

1. For each expert, e_i, a proximity matrix, $PM^i = (pm^i_{lk})$, is obtained where

$$pm^i_{lk} = 1 - \frac{|I(p^i_{lk}) - I(p^c_{lk})|}{g}.$$

2. Computation of proximity measures at three different levels:

   a. *Proximity measure on pairs of alternatives, pp^i_{lk}*. It measures the proximity between the preferences on each pair of alternatives of the expert e_i and the group.

   $$pp^i_{lk} = pm^i_{lk}.$$

   b. *Proximity measure on alternatives, pa^i_l*. It measures the proximity between the preferences on each alternative x_l of the expert e_i and the group.

   $$pa^i_l = \frac{\sum_{k=1}^n pp^i_{lk}}{n}.$$

   c. *Proximity measure on the relation, pr^i*. It measures the global proximity between the preferences of each expert e_i and the group.

   $$pr^i = \frac{\sum_{l=1}^n pa^i_l}{n}.$$

These measures allow us to build a feedback mechanism so that experts change their opinions and narrow their positions [21, 35]. In section 4, we show the use of the mechanism in a practical case of use.

3.3 *Communication and Work Flow of the Prototype*

Between client and server some communication functions are developed. In what follows, we present how the modules are connected together with the database, and the order in which each of them is executed.

0. **Initialization:** An initial step is to insert in the database all the initial parameters of the linguistic GDM problem.
1. **Verify user messages and store the main information:** When an expert wants to access the system, he has to send a message through M-Internet using his/her mobile device. The user can send two kinds of messages:
   i) A preferences message: It is composed by authentication information (login and password) and his/her preferences about the problem, using a set of labels to represent a FLPR.
   ii) A change of alternatives message: It is composed by authentication information (login and password) and his/her linguistic level of agreement with the proposed change of alternatives.
   These messages are verified by the server, checking the login and password in the database. If the authentication process is correct, the rest of the information

of the message is stored in the database and the server decides if the consensus stage should start (if all experts have provided their preferences) or, if the managing module of dynamic information can be finished (if enough experts provide their agreement degrees on the proposed change of alternatives).

2. **Calculate the set of solution alternatives and the consensus measures:** The decision module returns the solution set of alternatives in each stage of the decision process. All the information about the temporary solution is saved in the database.

3. **Control the consensus state:** In this step, the server determines if the required agreement degree has been reached (and thus, the decision process can be finished) or if we must begin a new round of consensus using the feedback mechanism that generates recommendations to change the experts' preferences.

4. **Management of new alternatives:** When the minimum consensus level has not been reached, the system checks if some new good alternatives appear in the problem environment or an old alternative deserves to be removed.

5. **Generate the recommendations:** In this step, the server calculates the proximity measures and generates the recommendations to change the FLPRs. It sends a message to the experts advising that they can use the software again for reading the recommendations and in such a way to start a new consensus stage. In order to avoid that the collective solution does not converge after several discussion rounds, the prototype stops if the number of rounds surpass MAX-CYCLES. These recommendations are saved in the database and sent to the experts through M-Internet.

In the following section we present a practical example on the use of the prototype to provide more detail about its operation.

4 Case of Use: Medical Diagnosis

Medical diagnosis is a GDM scenario that presents all the characteristics to take the advantages of our system. There is a patient who presents some symptoms, but all of them are common to several diseases. These diseases shape the set of alternatives of the problem. In addition, there are some doctors considered specialists in differential diagnosis. They form the set of experts of the problem and they have to jointly diagnose which is the disease that the patient has contracted. The experts work in different hospitals of different countries and they can not have a meeting to discuss and reach the consensual solution. Moreover, this environment is dynamic in the sense that the patient is now moved to the hospital and, at any moment, he could present new symptoms or he could set better due to the medication, and thus, any change of state of the patient might be taken into account by the doctors. So, the experts might decide to use our system because they can use the mobile communication technologies to reach the consensus, and they can change some possible diseases in the discussion set of alternatives according with the current patient's state.

The first step to solve a problem using our prototype is to insert all the initial parameters of the problem (experts, alternatives, thresholds, timing...) in the database. We assume a set of three experts (doctors), $\{e_1, e_2, e_3\}$, and a set of four alternatives (possible diseases) $\{x_1 = Cold,\ x_2 = Swine\ Flu,\ x_3 = Cancer,\ x_4 = Lupus\}$. The remaining parameters (see table 1) are used by the system to obtain the necessary consensus degree among the experts.

Table 1 Initial parameters of the problem

Name	Value	Description
Ndiseases	4	Number of diseases in the discussion subset
Nexperts	3	Number of experts (doctors)
minConsDegree	0.75	Minimum consensus level required by the problem
minProxDegree	0.75	Minimum proximity level required for the experts to be noted to change
MAXCYCLES	4	Maximum number of iterations of the consensus process
maxTime	12 (hours)	Maximum time of waiting for the experts opinions to change
minQGDD	L	Minimum dominance level that an alternative has to reach to avoid to be changed

When the initial parameters of the problem are defined, the decision making process starts.

4.1 First Round

The three experts send their FLPRs using their mobile devices and the following set of seven labels (see Figure 3): $S = \{s_0 = N, s_1 = VL, s_2 = L, s_3 = M, s_4 = H, s_5 = VH, s_6 = P\}$, where N=*Null*, VL=*Very Low*, L=*Low*, M=*Medium*, H=*High*, VH=*Very High* and P=*Perfect*.

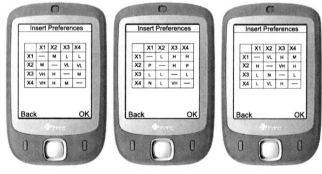

(a) Expert 1 (b) Expert 2 (c) Expert 3

Fig. 3 Expert Preferences

4.1.1 Decision Module

1. Selection Process:

In this phase we obtain the collective temporary solution by aggregating the experts' preferences.

1. Aggregation:
   We aggregate he FPLRs by means of the LOWA operator. We use the linguistic quantifier *most of* defined as $Q(r) = r^{1/2}$. Then, we obtain the following collective FLPR:

$$P^c = \begin{pmatrix} - & L & H & M \\ VH & - & H & VH \\ H & M & - & M \\ M & M & H & - \end{pmatrix}$$

2. Exploitation:
   Using again the same linguistic quantifier *most of*, we obtain $QGDD_i$ and $QGNDD_i \; \forall x_i \in X$:

Table 2 Choice degrees

	x_1	x_2	x_3	x_4
$QGDD_i$	M	VH	H	H
$QGNDD_i$	VH	P	P	P

and, the maximal sets are:

$$X^{QGDD} = \{x_2\} \text{ and } X^{QGNDD} = \{x_2, x_3, x_4\}.$$

2. Consensus Process:

In this phase the system calculates the consensus measures.

1. *Similarity matrices:*

$$SM_{12} = \begin{pmatrix} - & 0.83 & 0.66 & 0.66 \\ 0.50 & - & 0.50 & 0.16 \\ 0.50 & 0.66 & - & 0.83 \\ 0.16 & 0.66 & 0.66 & - \end{pmatrix}$$

$$SM_{13} = \begin{pmatrix} - & 0.66 & 0.66 & 0.83 \\ 0.83 & - & 0.33 & 0.50 \\ 0.50 & 0.33 & - & 0.83 \\ 0.50 & 0.50 & 0.83 & - \end{pmatrix}$$

$$SM_{23} = \begin{pmatrix} - & 0.83 & 1.00 & 0.83 \\ 0.66 & - & 0.83 & 0.66 \\ 1.00 & 0.66 & - & 1.00 \\ 0.66 & 0.83 & 0.83 & - \end{pmatrix}$$

2. *Consensus matrix:*

$$CM = \begin{pmatrix} - & 0.77 & 0.77 & 0.77 \\ 0.66 & - & 0.55 & 0.44 \\ 0.66 & 0.55 & - & 0.88 \\ 0.44 & 0.66 & 0.77 & - \end{pmatrix}$$

3. *Consensus degrees on pairs of alternatives.* The element (l,k) of CM represents the consensus degrees on the pair of alternatives (x_l, x_k).
4. *Consensus on alternatives:*

$$ca^1 = 0.77 \quad ca^2 = 0.55 \quad ca^3 = 0.69 \quad ca^4 = 0.62$$

5. *Consensus on the relation:*

$$cr = 0.66$$

As $cr < minConsDegree = 0.75$ is satisfied, then it is concluded that there is no consensus amongst the experts, and consequently, the system should continue by executing the next two processes: managing process of dynamic information to replace some alternatives in the discussion subset and feedback process to support the experts' changes in their preferences in order to increase cr.

4.1.2 Management of Dynamic Information Module

As soon as the system has verified that the minimum consensus level among the experts has not been reached and before beginning a new round of consensus, it is necessary to update all the information of the problem that could be changed during the process.

In this case, the patient, due to the medication, has started to show a new symptom that is typical of a disease that was not included in the initial discussion subset of the problem and should be included now ($x_5 = Allergy$). This new situation does not pose any problem because the system manages the dynamic information. We identify those alternatives with low choice degrees ($x_1 = Cold$) and ask the experts if they agree to replace those identified alternatives by the new suitable alternative (See Figure 4a).

The experts' answers were the following: *(Agree, Nor Agree/Nor Disagree and Completely Agree)*. The system applies the LOWA operator to aggregate these opinions and obtain a collective agreement degree. In this case we obtain ,(Agree), what represents an affirmative position to introduce the changes of alternatives. Therefore, the change of *Cold* by *Allergy* is done. The experts will be informed about it and then they are urged to refill their preferences by considering in this occasion the new alternative.

4.1.3 Feedback Module

- **Computation of proximity measures:**

  1. *Proximity matrices:*

$$PM_1 = \begin{pmatrix} - & 0.83 & 0.66 & 0.83 \\ 0.66 & - & 0.50 & 0.33 \\ 0.83 & 0.83 & - & 1.00 \\ 0.66 & 0.83 & 0.83 & - \end{pmatrix}$$

$$PM_2 = \begin{pmatrix} - & 1.00 & 1.00 & 0.83 \\ 0.83 & - & 1.00 & 0.83 \\ 0.66 & 0.83 & - & 0.83 \\ 0.50 & 0.83 & 0.83 & - \end{pmatrix}$$

$$PM_3 = \begin{pmatrix} - & 0.83 & 1.00 & 1.00 \\ 0.83 & - & 0.83 & 0.83 \\ 0.66 & 0.50 & - & 0.83 \\ 0.83 & 0.66 & 1.00 & - \end{pmatrix}$$

  2. *Proximity on pairs of alternatives:* $PP_i = PM_i$.
  3. *Proximity on alternatives (See Table 3):*

Table 3 Proximity measures on alternatives

x_1	x_2	x_3	x_4
$pa_1^1 = 0.77$	$pa_1^2 = 0.50$	$pa_1^3 = 0.88$	$pa_1^4 = 0.77$
$pa_2^1 = 0.94$	$pa_2^2 = 0.88$	$pa_2^3 = 0.77$	$pa_2^4 = 0.72$
$pa_3^1 = 0.94$	$pa_3^2 = 0.83$	$pa_3^3 = 0.66$	$pa_3^4 = 0.83$

  4. *Proximity on the relation:*

$$pr_1 = 0.73 \quad pr_2 = 0.83 \quad pr_3 = 0.81$$

- **Production of advice:**

  1. *Identification phase:*
     a. Identification of experts:

$$EXPCH = \{e_i \mid pr_i < minProxDegree\} = \{e_1\}$$

     b. Identification of alternatives:

$$ALT_1 = \{x_l \in X \mid pa_l^i < minProxDegree \wedge e_i \in EXPCH\} = \{x_2\}$$

     c. Identification of pairs of alternatives to generate recomendations:

$$PALT_1 = \{(x_2, x_1), (x_2, x_3), (x_2, x_4)\}$$

2. *Recommendation phase:*

In this phase, we have to take into account that alternative x_1 has been replaced in the previous process by x_5. So, x_1 does not need rules to be modified and there is a new alternative in the discussion subset, x_5, that needs new preference values. The recommendations interface for the expert e_1 is shown in Figure 4b.

  a. Rules to change the opinions:

    – Because x_1 has been replaced, p_{21}^1 does not need to be modified.

    – Because $p_{23}^1 < p_{23}^c$, expert e_1 is advised to increase the assessment of this preference value.

    – Because $p_{24}^1 < p_{24}^c$, expert e_1 is advised to increase the assessment of this preference value.

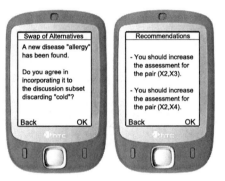

(a) Swap of alterna- (b) Recommendations
tives

Fig. 4

4.2 Second Round

The experts send their preferences about the new discussion subset to start the second round (see Figure 5).

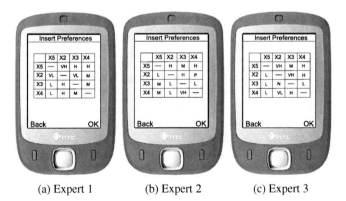

    (a) Expert 1   (b) Expert 2   (c) Expert 3

Fig. 5 Expert Preferences

4.2.1 Decision Module

1. Selection Process:

1. Aggregation:
   The collective FLPR is:

   $$P^c = \begin{pmatrix} - & VH & H & H \\ L & - & H & VH \\ M & M & - & M \\ M & M & H & - \end{pmatrix}$$

2. Exploitation:
   Using again the same linguistic quantifier "most of", we obtain the following choice degrees:

Table 4 Choice degrees in 2nd round

	x_5	x_2	x_3	x_4
$QGDD_i$	VH	H	M	H
$QGNDD_i$	P	VH	VH	VH

Clearly, the maximal sets are:

$$X^{QGDD} = \{x_5\} \text{ and } X^{QGNDD} = \{x_5\}.$$

2. Consensus Process:

Consensus on the relation:

$$cr = 0.79$$

Because $cr > minConsDegree$, then it is concluded that there is the required consensus amongst the experts, and consequently, the current solution is the final solution, that is stored and sent to the experts (see Figure 6).

Fig. 6 Final solution

According to these results, doctors agree that the most suitable disease, taking into account all the dynamic symptoms, is allergy. In such a way, the patient can receive the most appropriate treatment.

5   Conclusions

We have presented a new prototype of DSS based on dynamic information and mobile technologies which provides consensual and selection support to deal with dynamic decision making situations. There are a large number of scenarios in which the deployment of DSSs on mobile devices is desirable. So, it is specifically designed to deal with GDM problems based on dynamic sets of alternatives, which uses the advantages of mobile Internet technologies to improve the user-system interaction through decision process. In this prototype we allow the experts to use linguistic preference relations to express their preferences. In short, with this new mobile decision support system we shall be able to deal with linguistic GDM problems in which experts could interact anywhere and anytime, quickly, in a flexible way and under dynamic frameworks.

Acknowledgements. This work has been developed with the financing of FEDER funds in FUZZYLING Project TIN200761079, FUZZYLING-II Project TIN201017876, PETRI Project PET20070460, Andalusian Excellence Project TIC-05299, and project of Ministry of Public Works 90/07.

References

1. http://java.sun.com/products/midp/
2. http://java.sun.com/products/sjwtoolkit/
3. Alonso, S., Herrera-Viedma, E., Chiclana, F., Herrera, F.: Individual and social strategies to deal with ignorance situations in multi-person decision making. International Journal of Information Technology & Decision Making 8(2), 313–333 (2009)
4. Ben-Arieh, D., Chen, Z.: Linguistic-labels aggregation and consensus measure for autocratic decision making using group recommendations. IEEE Transactions on Systems, Man, and Cybernetics. Part A: Systems and Humans 36(3), 558–568 (2006)
5. Cabrerizo, F.J., Alonso, S., Herrera-Viedma, E.: A consensus model for group decision making problems with unbalanced fuzzy linguistic information. International Journal of Information Technology & Decision Making 8(1), 109–131 (2009)
6. Chiclana, F., Herrera, F., Herrera-Viedma, E.: Integrating three representation models in fuzzy multipurpose decision making based on fuzzy preference relations. Fuzzy Sets and Systems 97(1), 33–48 (1998)
7. Clarke, H.: Classical decision rules and adaptation to climate change. The Australian Journal of Agricultural and Resource Economics 52, 487–504 (2008)
8. Dia, H.: An agent-based approach to modelling driver route choice behaviour under the influence of real-time information. Transportation Research Part C 10, 331–349 (2002)
9. Eren, A., Subasi, A., Coskun, O.: A decision support system for telemedicine through the mobile telecommunications platform. Journal of Medical Systems 32(1) (2008)

10. Fodors, J., Roubens, M.: Fuzzy preference modelling and multicriteria decision support. Kluwer Academic Publishers, Dordrecht (1994)
11. French, S., Turoff, M.: Decision support systems. Communications of the ACM 50(3), 39–40 (2007)
12. García-Lapresta, J.L., Meneses, L.C.: Modelling rationality in a linguistic framework. Fuzzy Sets and Systems 160, 3211–3223 (2009)
13. Herrera, F., Alonso, S., Chiclana, F., Herrera-Viedma, E.: Computing with words in decision making: Foundations, trends and prospects. Fuzzy Optimization and Decision Making 8(4), 337–364 (2009)
14. Herrera, F., Herrera-Viedma, E.: Choice functions and mechanisms for linguistic preference relations. European Journal of Operational Research 120, 144–161 (2000)
15. Herrera, F., Herrera-Viedma, E.: Linguistic decision analysis: steps for solving decision problems under linguistic information. Fuzzy Set and Systems 115, 67–82 (2000)
16. Herrera, F., Herrera-Viedma, E., Martínez, L.: A fuzzy linguistic methodology to deal with unbalanced linguistic term sets. IEEE Transactions on Fuzzy Systems 16(2), 354–370 (2008)
17. Herrera, F., Herrera-Viedma, E., Verdegay, J.L.: A sequential selection process in group decision making with a linguistic assessment approach. Information Sciences 85(4), 223–239 (1995)
18. Herrera, F., Herrera-Viedma, E., Verdegay, J.L.: Direct approach processes in group decision making using linguistic owa operators. Fuzzy Sets and Systems 79, 175–190 (1996)
19. Herrera, F., Herrera-Viedma, E., Verdegay, J.L.: A model of consensus in group decision making under linguistic assessments. Fuzzy Sets and Systems 78(1), 73–87 (1996)
20. Herrera, F., Herrera-Viedma, E., Verdegay, J.L.: Linguistic measures based on fuzzy coincidence for reaching consensus in group decision making. International Journal of Approximate Reasoning 16, 309–334 (1997)
21. Herrera-Viedma, E., Alonso, S., Chiclana, F., Herrera, F.: A consensus model for group decision making with incomplete fuzzy preference relations. IEEE Transactions on Fuzzy Systems 15(5), 863–877 (2007)
22. Herrera-Viedma, E., Herrera, F., Chiclana, F.: A consensus model for multiperson decision making with different preference structures. IEEE Transactions on Systems, Man, and Cybernetics. Part A: Systems and Humans 32(3), 394–402 (2002)
23. Herrera-Viedma, E., Pasi, G., López-Herrera, A.G., Porcel, C.: Evaluating the information quality of web sites: A methodology based on fuzzy. Journal of the American Society for Information Science and Technology 57(4), 538–549 (2006)
24. Herrera-Viedma, E., Peis, E.: Evaluating the informative quality of documents in SGML-format using fuzzy linguistic techniques based on computing with words. Information Processing & Management 39(2), 195–213 (2003)
25. Imielinski, T., Badrinath, B.R.: Mobile wireless computing: challenges in data management. Communications of the ACM 37(10), 18–28 (1994)
26. Kacprzyk, J.: Group decision making with a fuzzy linguistic majority. Fuzzy Sets and Systems 18, 105–118 (1986)
27. Kacprzyk, J., Fedrizzi, M.: A soft measure of consensus in the setting of partial (fuzzy) preferences. Eur. J. Oper. Res. 34, 316–323 (1988)
28. Kacprzyk, J., Fedrizzi, M.: Multiperson decision making models using fuzzy sets and possibility theory. Kluwer Academic Publishers, Dordrecht (1990)
29. Kacprzyk, J., Fedrizzi, M., Nurmi, H.: Group decision making and consensus under fuzzy preferences and fuzzy majority. Fuzzy Sets and Systems 49, 21–31 (1992)
30. Kacprzyk, J., Nurmi, H., Fedrizzi, M.: Consensus under fuzziness (1997)
31. Katz, J.: Handbook of Mobile Communication Studies. MIT Press, Cambridge (2008)

32. Kim, N., Seok, H., Joo, K., Young, J.: Context-aware mobile service for routing the fastest subway path. Expert Systems with Applications 36, 3319–3326 (2009)
33. Kraft, D., Bordogna, G., Pasi, G.: An extended fuzzy linguistic approach to generalize boolean information retrieval. Journa of Information Science-Applications 2(3) (1995)
34. Lopez-Herrera, A.G., Herrera-Viedma, E., Herrera, F.: Applying multi-objective evolutionary algorithms to the automatic learning of extended boolean queries in fuzzy ordinal information retrieval systems. Fuzzy Sets and Systems 160, 2192–2205 (2009)
35. Mata, F., Martínez, L., Herrera-Viedma, E.: An adaptive consensus support model for group decision making problems in a multi-granular fuzzy linguistic context. IEEE Transactions on Fuzzy Systems 17(2), 279–290 (2009)
36. Muntermann, J.: Event-Driven mobile Finantial Information Services: Mobile notification and Decision Support for private investors. DUV (2008)
37. Nurmi, H.: Fuzzy social choice: a selective retrospect. Soft Computing 12, 281–288 (2008)
38. Pérez, I.J., Cabrerizo, F.J., Herrera-Viedma, E.: A Mobile Decision Support System for Dynamic Group Decision Making Problems. IEEE Transactions on Systems, Man and Cybernetics - Part A: Systems and Humans 40(6), 1244–1256 (2010)
39. Porcel, C., Lopez-Herrera, A.G., Herrera-Viedma, E.: A recommender system for research resources based on fuzzy linguistic modeling. Expert Systems with Applications 36(3), 5173–5183 (2009)
40. Porcel, C., Moreno, J.M., Herrera-Viedma, E.: A multi-disciplinar recommender system to advice research resources in university digital libraries. Expert Systems with Applications 36(10), 12520–12528 (2009)
41. Ricci, F., Nguyen, Q.N.: Acquiring and revising preferences in a critique-based mobile recommender system. IEEE Intelligent Systems 22(3) (2007)
42. Roubens, M.: Fuzzy sets and decision analysis. Fuzzy Sets and Systems 90(2), 199–206 (1997)
43. Schiller, J.: Mobile Communications, 2nd edn. Addison-Wesley, Reading (2003)
44. Wen, W., Chen, Y.H., Pao, H.H.: A mobile knowledge management decision support system for automatically conducting an electronic business. Knowledge-Based Systems 21(7) (2008)
45. Xu, Z.S.: Deviation measures of linguistic preference relations in group decision making. Omega 33(3), 249–254 (2005)
46. Xu, Z.S.: Group decision making based on multiple types of linguistic preference relations. Information Sciences 178(2), 452–467 (2008)
47. Xu, Z.S., Chen, J.: Magdm linear-programming models with distinct uncertain preference structures. IEEE Transactions on Systems, Man and Cybernetics part B-Cybernetics 38(5), 1356–1370 (2008)
48. Yager, R.R.: On ordered weighted averaging aggregation operators in multicriteria decision making. IEEE Transactions on Systems, Man, and Cybernetics 18(1), 183–190 (1988)
49. Zadeh, L.A.: The concept of a linguistic variable and its applications to approximate reasoning. Information Sciences, Part I, II, III 8,8,9, 199–249 (1975)
50. Zhang, Z., Chu, X.: Fuzzy group decision-making for multi-format and multi-granularity linguistic judgments in quality function deployment. Expert Systems with Applications 36(5), 9150–9158 (2009)

Building Consensus in On-Line Distributed Decision Making: Interaction, Aggregation and the Construction of Shared Knowledge

Luca Iandoli

Abstract. In this chapter I discuss the possibility of exploiting large-scale knowledge sharing and mass interaction taking place on the Internet to build decision support systems based on distributed collective intelligence. Pros and cons of currently available collaborative technologies are reviewed with respect to their ability to favor knowledge accumulation, filtering, aggregation and consensus formation. In particular, I focus on a special kind of collaborative technologies, online collaborative mapping, whose characteristics can overcome some limitations of more popular collaborative tools, in particular thanks to their capacity to support collective sense-making and the construction of shared knowledge objects. By reviewing some of the work in the field, I argue that the combination of online mapping and computational techniques for beliefs aggregation can provide an interesting basis to support the construction of systems for distributed decision-making.

1 Introduction: Aggregating Intelligence through the Internet

By exploiting the vast amount of collaboration taking place on the Internet it is now feasible to draw together large groups of knowledgeable and interested individuals and support large-scale discussions and information accumulation virtually on any topic. The effective combination of intelligent collective behavior and Internet capabilities has created a number of successful cases like the Open source communities and Wikipedia. These successes in turn have induced many scholars to think that it is possible to harness such a combination to enable forms of "collective intelligence" over the Internet for a number of diverse tasks like collective prediction (Sunstein, 2006), collective deliberation (Klein, Cioffi and Malone,

Luca Iandoli
University of Naples Federico II (Italy), Dept. of Business and
Managerial Engineering
e-mail: iandoli@unina.it

E. Herrera-Viedma et al. (Eds.): Consensual Processes, STUDFUZZ 267, pp. 339–355.
springerlink.com   © Springer-Verlag Berlin Heidelberg 2011

2007) or to support distributed problem solving and creativity in non-profit as well as business communities (Gloor, 2006; McAfee, 2006; Raymond, 2001; Tapscott and Williams, 2006, von Krogh and von Hippel, 2006).

Reframing the issue in computational terms, we can say that complex problems can be represented on a very large, unexplored and partially unknown solution space. Through the contributions of a large number (up to many thousands) of knowledgeable users, a virtual community can enable unprecedented breadth of exploration of the solution space and convergence on high-quality solutions in less time compared to small, co-located groups.

In practice, however, there is no guarantee that people will converge on a shared, good quality conclusion and create consensus in an online discussion. In particular, the most popular collaborative technologies, such as forums, wikis and blogs, while successfully enabling information sharing and accumulation through mass interaction, do not help to evaluate and organize the accumulated knowledge for later re-use. Also they are not designed to help people to converge towards mutual understanding or the identification of a best option among a set of alternatives; in fact, online discussions can evolve in any possible direction and quite easily allow for the emergence of well-known group thinking pitfalls as debate polarization or errors propagation.

In this paper I provide an evaluation of current online technologies that could be considered as possible candidates to enable distributed decision-making and introduce a possible approach to support collective intelligence through on-line collaborative mapping tools. Collaborative Mapping tools (CMT) have been proposed to support the collaborative construction of knowledge maps by geographically-dispersed groups to systematically explore a debate, evaluate competing points of view, and possibly come to a shared decision. Unlike conversational technologies such as forums, wikies and the like, collective maps, by providing a spatial rather than a time-based representation of a debate, are expected to facilitate sense-making and the construction of shared, explicit and re-usable knowledge objects.

In this paper I analyze to which extent and how CMTs can support knowledge aggregation and consensus formation. Drawing from recent studies, I will report some evidence and analysis to show that the combination of such tools with beliefs aggregation and consensus operators can significantly improve online mapping tools' potential to support large scale, distributed decision making.

The paper is structured as follows. In the next section I outline the main limitations of using currently available collaborative technologies for decision-making purposes, including web 2.0 technologies and Information Aggregation Markets. In the second section a quick introduction to online collaborative mapping tools is provided, along with an analysis of their points of weakness. Finally the results offered by some recent studies on online decision making & deliberation are reported to show how the combination of computational engines for beliefs aggregation and mapping tools can represent both a promising research field to build technologies able to harness collective intelligence as well as an interesting application field for scholars in the consensus and information aggregation areas.

2 Limitations of Current Collaborative Technologies to Support Online, Distributed Decision Making

2.1 WEB 2.0 Technologies: Sharing and Accumulating Knowledge

Though different definitions have been provided after the initial one proposed by O'Reilly (2005), the expression web 2.0 is generally referred to a sort of evolution of the Internet toward a new paradigm in which online applications based on an intense interaction between users are dominant (blogs, forums, micro-blogging, social networks, information sharing services like YouTube, MySpace and the like). But how well can current popular Internet web 2.0 technologies support aggregation of information for decision-making purposes?

While current, popular, online collaborative tools like wikies and forums have not been specifically designed to support decision making, there is a widespread expectation, as well as some empirical findings, that suggests that large groups, possibly facilitated by Internet technologies, can leverage their collective intelligence to outperform skilled but isolated individuals or small groups for such tasks as problem solving (Page, 2007), collective prediction (Surowiecki, 2004), and deliberation on complex systemic problems (Malone et al., 2009), *assuming* groups made up of a sufficiently large number of capable and cognitively diverse members. There is, however, also a substantial literature about factors that can prevent groups from being smarter than their average member, factors that appear to be present in unstructured large-scale Internet-based conversations such as online forums (Sunstein, 2006).

By far the most commonly used collaborative technologies, including wikis, blogs, and discussion forums, are *sharing tools* (de Moor and Aahkus, 2006). Sharing tools perform very well on facilitating interaction, harvesting and sharing of information among a large number of users. Instead they are very ineffective in aggregating information, generating consensus and building reusable knowledge representations.

Thus, while such tools have been remarkably successful at enabling a global explosion of ideas and knowledge sharing, they face serious shortcomings when applied for distributed decision-making purposes. In particular, these tools do not provide enough filtering and aggregation of information. The content is typically of highly variable quality, since online conversation tools do not inherently encourage or enforce any quality standard.

In his book Infotopia, Sunstein (2006) outlines several causes that can induce deliberating groups to fail in making accurate, truthful and reliable decisions as well as some conditions under which group deliberation can work. He points out that deliberating group typically suffer from three major problems:

- they do not elicit all the relevant information that their members have (low *information disclosure,* a problem also known as hidden profiles in group decision making studies);
- they are subject to *cascade effects*: sequential information propagation in the group may produce errors amplification and premature convergence;

- they show a tendency toward group *polarization*: often deliberating groups may assume a position on an issue which is even more extreme than the average opinion, in particular when they are very homogeneous and when the issue is related to values and social identity.

The failures outlined above have been detected in experiments in which groups were required to deliberate about an issue and reach a collective decision. It is important to remark that this literature is largely concerned with small-scale, closed, physically co-located groups of individuals involved in direct interaction in typical social situations, such as political and management committees, juries, assemblies, focus groups, meetings, etc. While to our knowledge no systematic evidence is available to compare the quality of online VS face-to-face conversations for deliberation and decision making purposes, there are no reasons to believe that unconstrained, conversational interaction on an online medium would not reproduce the same undesirable social dynamics usually leading to group decision making failure in face-to-face situations.

In addition to the quality issues above outlined, there is also a scalability problem. Forum conversations do not scale well because it is virtually impossible for latecomers to make sense of and effectively join a conversation that has already been started by other participants. While wikis are able to cope with some of the above limitations, they tend to fare poorly when applied to controversial topics, often leading to such phenomena as "edit wars". Wikies are able to support some kind of consensus formation that ultimately leads to the creation of shared documents whose quality can be surprising. This capability is due to two main characteristics:

1. wikies are additive: diverse perspectives can be added up in a text, even when they express different points of view on the same topic;
2. zero revert costs: the costs of repairing the damages created by bad contributions (off-topic, vandalism, low quality, etc.) is virtually zero thanks to the possibility of reverting to the last version of the document.

While wikies can support the creation of shared texts whose contents can be used in principle to inform decision-making, they can not be considered Decision Support Systems for obvious reasons and primarily because a wiki does not help to create a computable model of the problem under examination.

2.2 Funneling Technologies: Filtering, Weighting and Aggregating Information

In *Funneling* technologies (de Mooor and Aakhus, 2006) I include information aggregation markets (IAMs), e-voting and various forms of collaborative filtering. E-voting systems can be employed for such tasks as ideas competitions and online polls, but chances that they will produce random or highly biased outcomes can be high. With more structure and sophistication e-voting can be used to build on line trust and reputation systems (Jøsang et al., 2007) in which users assign scores expressing their confidence about how useful/relevant/trustable etc. a certain source of information or a given service provider is. With additional design and computational effort, trust and reputation systems can be used as a basis to design collaborative filtering tools able to make recommendations by collecting information on collective preferences from many users. The main limitation of e-voting, trust & reputation or collaborative filtering tools is that they just aggregate individual preferences while the knowledge and the cognitive models that are used to generate those same preferences remain not coded or implicit.

Good evidence for "wisdom of the crowd" has appeared in studies on on-line IAMs, especially when used for forecasting purposes and then called prediction markets (Wolfers and Zitzewitz, 2004). From many respects prediction markets can be considered the leading technology among the ones built on the idea of collective intelligence.

Prediction markets are implemented as online trading software platforms through which participants trade in contracts whose payoff depends on unknown future events. The market exchange of contracts determines their price: in general, the higher the price of a contract, the higher the confidence of the market in the future occurrence of the associated event.

IAMs work by virtue of the efficient market hypothesis, but, like ordinary markets, they can display deviations from perfect rationality that appear in other real markets as the emergence of speculative bubbles. In addition to the possibility of irrational behavior the main limitation of a market is that it does not give any insight on the knowledge traders use to make their decisions. In other words, a prediction market only aggregates information about the probability of occurrence of an event in the future, but it will not tell *why* the event is going to happen. While markets capture knowledge into prices, they do not capture the reasoning behind market choices.

In conclusion, while funneling technologies can be effective at aggregating information provided by large groups made up by diverse and independent individuals, they i) do not help groups develop a shared understanding and consensus, ii) provide no visibility on the process through which members retrieve and create knowledge; iii) are not able to create reusable knowledge, for instance in the forms of collective models or other explicit knowledge representations.

3 Mapping Tools: Building Shared Knowledge Representation

Online Mapping Tools help groups collaboratively define networks of issues and concepts in the form of visual maps. Such maps can be used to support diverse cognitive tasks like online collaborative learning (Novak and Canas, 2006; Suthers et al. 2008), sense-making (Kirschner et al., 2003) and deliberation (Iandoli, Klein and Zollo, 2009).

A special class of mapping tools is on-line argument mapping. Arguments maps contain a question to be answered, possible alternative solutions, and chains of pros and cons arguments to support or attack competing positions. In fig. 1 an example of an argument map realized through the software tool COMPENDIUM using the IBIS formalism (Issue Based Information systems) is reported. The question to be answered is what will be the price trend in the European Real Estate market after the recent economic crisis.

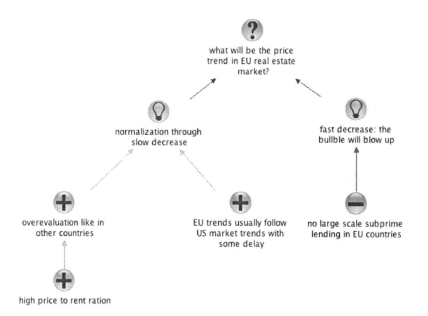

Fig. 1 an example of an on-line argument map

Two competing ideas are proposed as possible answers ("normalization" VS "bubble will blow up"). Each idea can be linked to chains of arguments (pros and cons, indicated by plus or minus signs) supporting or attacking existing positions or other arguments.

Arguments mapping is expected to encourage critical thinking, by implicitly requiring that users express the evidence and logic in favor of the options they prefer. The results are captured in a compact, visual form that helps users understand and reflect on what has been discussed to-date. While there is evidence that online collaborative mapping has proven to be effective for education and learning purposes, mapping tools have shown several limitations for large-scale, online distributed decision-making. Most of these systems have been originally developed to support physically co-located team meetings where a single facilitator captures free-form discussions in the form of a commonly viewable map (Conklin, 2006) while only a few have been used to enable non-facilitated deliberations, with physically distributed participants over the Internet (Shum et al., 2006).

In the only experiment, to our knowledge, that has involved a considerable number of users (about two hundreds) in online deliberation supported by argument mapping, my colleagues and I (Gurkan et al., 2009) have provided evidence to show that a large group of users need considerable learning effort, guidance (in the form of training and debate moderation), and strong incentives to build usable shared knowledge representations in the form of an argument map. Even when the debate map is properly constructed, it becomes soon very hard to navigate and manage as its size increase thanks to the contribution provided by many users at the same time.

Finally, mapping tools are not natively computable. For instance, most of the well-known argument or concept mapping tools currently available do not support computation of argument weights, positions strength or concept relevance. Nevertheless, since maps can be described with little effort through formal and explicit mathematical objects (e.g. graphs or trees), they provide a basis for the development of computational tools able to aggregate individual preferences to evaluate and rank alternative solutions. Thus, the combination of maps and belief aggregation algorithms can be used in principle to build online, group decision support systems.

In the following section I show a proposal of how such a combination can be implemented and discuss the advantages that belief aggregation can bring to collaborative, online mapping for distributed decision-making applications. I will focus specifically on argument mapping for the following reasons: 1) argument maps have been historically developed to support deliberation, mainly in the legal field; so they are aimed at supporting decision making through the comparison between competing alternatives backed by plausible rationales; 2) unlike other kind of online mapping tools, there is a number of attempts in the literature in which computation and argument mapping have been combined to implement decision support tools[1].

4 Empowering Mapping Tools through Beliefs Aggregation

Online argumentation (also known as Computer-Supported Argument Visualization – CSAV) has received considerable attention by scholars, particularly in the field on Computer Supported Cooperative Work (Kirschner at al., 2003, Shum et al., 2006). The idea is to use argument mapping to display users' contributions to an online discussion in the form of visual maps made up of positions and associated chains of pros and cons. Usually such maps come in the form of an argument tree.

Work on online argumentation has focused mainly on the development of effective formalisms for argument representation and their implementation into usable tools; the issue of providing users with systems to evaluate the quality of ideas and arguments has been largely neglected.

Most attempts to fill this gap have emerged from the field of Artificial Intelligence (e.g. Dung, 1995, Pollock, 2001), but such approaches are not easily exportable to human-centric collaborative environments, for several reasons.

First, they impose excessive formal constraints on the interaction among the participants, thus creating barriers to participation. For instance in Dung's the argumentation process is formalized as a cooperative n-persons game. Second, they are usually complex and hard to use and understand for human participants. For example, Pollock's framework requires users to declare tacit argumentation components, in the form of premise-warrant-conclusions chains, which can increase users cognitive burden in the absence of clearly perceived benefits. Third, since AI approaches have often been developed for rational agents like machines, they can

---

[1] While aggregation algorithms are available also for causal maps (Kosko, 1992), to our knowledge causal maps have not been used for online collaborative mapping.

be easily misused or even deliberately gamed by human participants. For instance, the "The one who has the last word laughs best" principle in Dung and the idea of partial defeat of a strong argument when under attack by a weak one in Pollock could be instrumentally used by malicious users to game the system, and have to be dealt with through additional countermeasures and constraints.

A belief aggregation system based on Dempster-Shafer (D-S) theory and reducing to a reasonable extent many of the above criticalities has been developed and applied by Das (2005) and Introne (2008, 2009). In this framework, arguments are chained and visually represented through an argument tree. The representation also incorporates subjective probabilities offered by the decision makers representing the argument strength in the form of degree of belief. D-S theory is used to compute degrees of belief for decision options by aggregating the probabilistic arguments for and against the decision options along the tree branches. The proposed decision making framework has been successfully applied in a variety of domains ranging from automated tasks as theater missile defense and army fire support to expert decision making tasks like medical diagnosis. D-S aggregation is a generalization of Bayes theorem obtained by relaxing some assumptions, in particular the necessity to know in advance all the a priori conditional probabilities needed for the calculation in Bayes' formula.

For the analytical details we suggest readers to refer to the original work (Shafer, 1976), while in the following we report an example of application of the D-S aggregation rule applied to an argument tree from Das (2005).

In this example, an answer to a question takes the form of an option followed by a subjective argument along with a subjective probability abut how much the proposed argument actually supports the option, for example, according to a decision maker, the probability that the status of a game is "delayed" is 0.6 given a "heavy rain condition".

The set of the mutually exclusive positions is called *frame of discernment*. In the game example the frame could be $\Omega = \{$on, delayed, cancelled$\}$, while "heavy rain" can be an argument to support, to some extent, one of the available position, e.g. that they game will be cancelled. A basic probability assignment (BPA) is a function:

m: $\Pi(\Omega) \rightarrow [0,1]$

quantifying the amount of belief committed to a given subset A of Ω, also called focal element. The value $m(A)$ is also called belief mass. The total belief committed to A can be obtained by computing the belief function Bel(A) through the summation of all its belief masses:

$$Bel(A) = \sum_{B \subseteq A} m(B)$$

To combine two independent belief assignments m_1 and m_2 Dempster and Shafer proposes the following aggregation formula:

$$m_{12} = \begin{cases} \dfrac{\sum\limits_{B \cap C = A} m_1(B)m_2(C)}{1 - \sum\limits_{B \cap C = \Phi} m_1(B)m_2(C)} & A \neq \Phi \\ \\ 0 & A = \Phi \end{cases}$$

To illustrate how D-S aggregation works we report the game example from Das (2005). Suppose the following arguments are provided to support alternative positions:

1. "heavy-rain" supports the game is "not on" with a belief 0.7;
2. "club_financial_crisis" supports the game is "not cancelled" with belief 0.6.

In order for the D-S aggregation to work properly all the evidences to be combined must be independent. The first condition states that the decision maker has a 0.7 subjective probability for the game not being on (i.e. cancelled or delayed) given rainy weather. If not specified otherwise a zero degree of belief (not 0.3) that the game is on has to be assumed. In other words $m_1(\{\text{Cancelled,Delayed}\}) = 0.7$. Since nothing is said about the remaining probability, this is allocated to the whole frame of discernment as $m_1(\{\text{On, Cancelled, Delayed}\}) = m_1(\{\Omega\}) = 0.3$.

The decision maker also believes that the current financial situation of the club is bad, resulting in a 0.6 subjective probability that the game will not be cancelled in this situation. The new evidence provides a belief $m_2(\{\text{On, Delayed}\}) = 0.6$ while the remaining probability is allocated to the whole frame of discernment as $m_2(\{\text{On, Cancelled, Delayed}\}) = 0.4$.

We can combine m_1 and m_2 through the D-S aggregation rule stated above. Since the focal elements overlap, the denominator in the D-S- rule is equal 1 and only the products m_{12} have to be calculated, as reported in table 1.

Table 1 an example of calculation of joint beliefs using D-S aggregation rule with consistent evidence

	$m_2(\{\text{On,Delayed}\}) = 0.6$	$m_2(\{\Omega\}) = 0.4$
$m_1(\{\text{Cancelled,Delayed}\}) = 0.7$	$m_{12}(\{\text{Delayed}\}) = 0.42$	$m_{12}(\{\text{Cancelled,Delayed}\}) = 0.28$
$m_1(\{\Omega\}) = 0.3$	$m_{12}(\{\text{On, Delayed}\}) = 0.18$	$m_{12}(\{\Omega\}) = 0.12$

Finally, summing up the mass beliefs for all the relevant events we get:

$\text{Bel}(\{\text{Delayed}\}) = 0.42$
$\text{Bel}(\{\text{On,Delayed}\}) = 0.60$
$\text{Bel}(\{\text{Cancelled,Delayed}\}) = 0.70$
$\text{Bel}(\{\Omega\}) = 1$

Table 2 Table 1: an example of calculation of joint beliefs using D-S aggregation rule with consistent and inconsistent evidence

	$m_3(\{Cancelled\}) = 0.8$	$m_3(\{\Omega\}) = 0.2$
$m_{12}(\{Delayed\}) = 0.42$	k = 0.336	$m(\{Delayed\}) = 0.084$
$m_{12}(\{On,Delayed\}) = 0.18$	k = 0.144	$m(\{On, Delayed\}) = 0.036$
$m_{12}(\{Cancelled,Delayed\}) = 0.28$	$m(\{Cancelled\}) = 0.224$	$m(\{Cancelled,Delayed\}) = 0.056$
$m_{12}(\{\Omega\}) = 0.12$	$m(\{Cancelled\}) = 0.096$	$m(\{\Omega\}) = 0.024$

We could now get further evidence, e.g. that the game has been cancelled due to a player's injury with degree $m_3(Cancelled) = 0.8$. To the light of the new evidence the decision makers may want to revise their overall beliefs about the final outcome. The D-S rule of combination applies as before, but with one modification. When the evidences are inconsistent, i.e. focal elements intersection is empty, a single measure of inconsistency is computed, say k, corresponding to the term $\Sigma m_1(B)m_2(C)$ with B and C such that $B \cap C = \varnothing$, in the denominator of the D-S rule. The total mass of evidence assigned to inconsistency k is 0.336+0.144 = 0.48. The normalizing factor is 1-k =0.52. The computations are reported in table 2.

Finally, summing up the consistent evidences and discounting for the inconsistent ones we get our revised beliefs:

Bel ($\{Cancelled\}$) = 0.62
Bel ($\{Delayed\}$) = 0.16
Bel ($\{On,Delayed\}$) = 0.23
Bel ($\{Cancelled,Delayed\}$) = 0.89

Introne (2008) has developed an online collaborative argumentation platform that combines argument visualization with a decision model based on D-S theory to mediate collaborators' decision making. His empirical findings demonstrate that the platform addresses a known deficiency in information pooling known as the problem of common knowledge or hidden profiles. Common knowledge is a known group decision-making pitfall in which group members do not disclose all the information they have but only the one they share. In such cases, the group will make decisions using the available information in a suboptimal way. The empirical findings indicate that the combination of argument visualization and belief aggregation helps people to adhere to a rational model when combining information that is raised during their deliberation more effectively than argument visualization alone.

Das and Introne studies show that by empowering online mapping with beliefs aggregation, at least in the specific case of argument mapping, it is possible to improve collaborative tools from many respects:

1. mapping tools become computable objects and can be used to build collaborative group decision support systems;
2. the quality of the collaborative decision making process improves thanks to the attenuation of some group deliberation pitfalls as the common knowledge problem.

In addition to the above advantages it is possible to speculate that the availability of a belief aggregation engine embedded into online mapping tools could create a rich basis of data to compute other useful meta-aggregates. Such additional indicators could be helpful during the decision process to support sense-making and better understanding of the debate by participants: for instance, users could be interested to know on which aspects of the problem there is more conflict. The availability of useful meta-information is expected to improve both the quality of the decision process as well as to provide an immediate payoff to users and debate managers.

An example is to use consensus operators to measure the level of agreement within the group. Stephanou and Lu (1988) have developed a consensus operator based on the D-S framework. The proposed consensus operator is based on the concept of generalized entropy, intended as a measure of the uncertainty in a knowledge source. Stephanou and Lu prove that the pooling of evidence by D-S rule of combination decreases the total amount of generalized entropy in the knowledge sources and that the decrease of entropy corresponds to the focusing of knowledge and can be used as a measure of consensus effectiveness.

5 Discussion

Using the Internet for more intelligent tasks than sharing and searching for data requires dealing with a number of challenges. In this paper I have focused on the idea of building collective decision tools to leverage the huge amount of information and intelligence provided by thousands of users on the Internet.

I have reported some evidence that the combination of online collaborative mapping tools and beliefs aggregation is a promising base to build a technology able to effectively exploit the collective intelligence of many Internet users. Much work however needs to be done to achieve this result. In the following I will discuss some of the main challenges to deal with. I have classified them in two areas: technical design and organizational design.

5.1 Issues in the Technical Design

The inclusion of a rating and aggregation procedures in on-line mapping tools could provide users with several benefits like:

1. Decision making support: users preferences could be aggregated to identify most supported ideas;
2. Meta-information and visualization: aggregated preferences could be somehow linked to the way information is displayed, e.g.: emphasize most rated posts, highlight debate areas characterized by high conflict, etc.
3. Incentives: rates could be associated to authors' reputation scores, which could be used as incentives for most active users.

A further benefit can emerge from the empirical evidence reported above from which it appears that when beliefs aggregation is available people use argument platforms in a more rational way, for instance improving on information pooling. Additional research is needed to verify empirically if the availability of belief aggregation also helps with other pitfalls in the group decision-making process. The design of a rating and aggregation tools requires to deal with several issues, namely:

1. argumentation format: the way information is represented and structured will constrain the way it is rated. For instance, it could be possible to make different choices about the argumentation framework to be used among the many available in the literature;
2. preference representation: which semantics do we assign to users preferences? What do we want to measure?
3. preferences aggregation and use of aggregated ratings: a consistent preference aggregation method has to be defined. This issue is interlaced with some organizational issues that will be described in the next section: definition of a voting procedures (steps, scheduling, etc.), identification of voters (who can vote what), conflict management, etc.

The above issues are interdependent, for instance a choice made on preference representation can constrain the way preferences are aggregated.

As far as the *argument format* is concerned, there is an intrinsic ambiguity about rating an argument since an argument is made up by many elements: for instance, in the Toulmin's framework (Toulmin, 1959) there are grounds, warrants, conclusions, etc. In general we can distinguish at least three parts: a premise, a conclusion and an inferential scheme, so that the conclusion is produced by applying the inferential scheme to one or more premises. So, what should we ask users to evaluate?

We may ask voters to understand the difference between the different parts of an argument and ask them to vote accordingly, but unbundling argument structures would probably favor possible misconceptions and increase users' cognitive burden.

In terms of *preference representation* there are many alternative ways of expressing users' preferences, like for instance traditional probabilities in a Bayesian framework, subjective probabilities, or linguistic decision making (Herrera et al., 2002). Additional research is needed here to find out if the way preferences are represented has any significant impact.

As far as *preferences aggregation* is concerned, the choice of a suitable method will ultimately depend on the argument and references representation format. The aggregation of preferences over maps poses however a major problem. To be consistent, every aggregation procedure requires that we structure the decision problems in terms of the available evidence and decision options. For instance, the Dempster-Shafer method requires we are able to define a set of mutually exclusive propositions, connect all available and *independent* evidence to the relevant positions and express belief masses. However, the open-ended ongoing nature of the discussion happening through an online medium could not ensure this level of structure. For instance, it is very common the case in which one can find several ideas under a same issue which are neither exhaustive nor mutually exclusive. A second problem has to do with scale: when the map is really big, information about a same alternative can be scattered in several "corners" of the map. This can be due both to a dysfunction (redundancy) as well as to the different role a same piece of information may play in different local sub-debates. For the Dempster Shafer to work correctly (but this applies also to other approaches) one has to recollect pertinent evidence and restructure the decision problem properly.

There are two possible options to solve the problem. A first one is that moderators are able to reconfigure the map in real time so that the requirements for the chosen aggregation procedure are met (the argument map is reorganized so to match the information processing requirements of the aggregation method). This requirement can be too strict since moderators are only very good users but are not under control and might not be able to re-aggregate contents when the map becomes too large. Real time moderation is also very time consuming and requires a permanent, fully dedicated team of skilled meta-users.

A second option is to see the deliberation process as made by two separated steps: the exploration phase, in which people debate and discuss, and the exploitation or convergence step, when people are involved in a convergence task, as identifying the best option among a set of alternatives.

5.2   Issue in the Organizational Design

In the following I will briefly discuss four challenges related to the design of the decision-making community that in my opinion are highly critical for the development of an Internet-based DSS and that correspond to research areas in which further advancements would be greatly welcome:

1. Attract & retain participants;
2. Create the conditions for collective intelligence;
3. Develop effective mediating technologies;
4. Establish community governance.

Attract & retain participants. The success stories of other web 2.0 communities show that participation is entirely bottom-up. Participants elect a virtual community as a personal point of reference and act mostly under the pressure of intrinsic incentives. Two motivational drivers are really powerful: to feel part of a joint, high-impact project; and to access information and other knowledge resources. On top of that other important mechanisms based on status and reputation can be built, but those come later, when the seeds for a users' base have already been planted.

Create the conditions for collective intelligence. The conditions under which collectives can produce better decisions than those made by single decision makers or small groups have not been largely investigated. This is a wide debate and there is no room here to address it. There are still many controversial aspects, but some results are clear enough. The three basic ingredients are motivation, scale and diversity. Page (2007) has showed that a complex problem can be thought of as a rugged landscape where peaks correspond to local solutions. If we model a decision problem as a search problem in the space of the possible solutions, we expect that having a large number of diverse agents will help to explore this space more effectively by larger and more homogeneous sampling.

This conceptual representation has direct implications for the design of a collective DSS. The tool has to support both exploration and exploitation behavior. Here timing is a crucial issue: how much time has to be assigned to exploration and how much to exploration? Do we need iterations? If so, how many? Is it better to have synchronous or asynchronous collaborative work sessions?

Develop effective mediating technologies. As I have already pointed out elsewhere (Iandoli and Quinto, 2010), designing a new knowledge representation does not mean we have created a good mediating technology. Mediation is a complex issue. It has to do with adoption of any new technology. A big difference between human beings and other creatures is that we mediate interaction with our environment with an increasingly complex bundle of technological artifacts.

Since our experience is so deeply entrenched with the current set of available, conventional tools, we need a strong incentive to switch to new technologies.

In this case the challenge is to introduce a new communication format able to favor the passage from unstructured conversational interaction to the construction of collective, computable knowledge representations to be used as a base for decision making. This may prefigure a shift toward new online communication codes and patterns needed to spark Internet-based, large-scale collaborative modeling initiatives. Since we are actually asking users to deeply revise their conventional communication practices, resistance to adoption can be enormous. As designers we have really to give them a compelling reason for switching.

Establish community governance. Online communities are often said to be characterized by high decentralization and limited structure, but this does not mean that they are unregulated. The main difference between more traditional organizational models and online communities is that online communities tend to have fuzzy boundaries and are self-policing. They may give rise to quite sophisticated governance structure like those found in open source projects or even in less technical venues like the popular website slashdot.org.

Tasks and workload allocation are achievable in online communities because task choice and effort are self-determined by each participant and hierarchies are defined on the base of reputation. For decision making, however, it is definitely more critical than for other applications to manage conflicts of interests and contrast attempts to manipulate decision outcomes. Internet-based DSS have to be robust with respect to various kind of malicious activities like attacks and manipulation, which could be frequent especially when the topic is hot or the decision outcome may have some influence on real world decision makers like politicians or managers. Unfortunately the novelty of collective decision tools is such that we do not have any record of significant experience on this matter.

6 Conclusions

In a society that becomes more complex every day and which faces impressive challenges like climate change or global financial crises, humanity has to develop new ways and new technologies to address and solve complex problems. What seems to be missing is the availability of a collective memory that, as individual memory, could accumulate and store collective experience and refine successful models of action, but that unlike individual memory could be updated and improved by a huge number of intelligent contributors in a matter of seconds, just as Wikipedians fix flawed definitions or repair vandalism. The new participative Internet seems to be growing in that direction, but we are still looking for a locus and viable technologies.

In this paper I hope I have been able to provide some reasons to encourage research on the development of intelligent collaborative mapping tools and I believe that the combination of visual maps and computational techniques could provide a good basis to build more effective distributed decision making tools.

However, the construction of new virtual agoras is not just a matter of technical development, but will require a major cultural shift toward decentralization, openness, and increasing level of democracy. Although such characteristics are favored on the Internet, they are not, to the same extent at least, when the new tools are used within more traditional and hierarchical organizations, such as companies or public bodies. Both markets and centralized governance of elites (even when democratically elected) seem to be unable to address effectively the complexity of our hyper-connected world, as the financial and the environmental crises we are facing clearly show. The time has come, perhaps, to rely more on a different, diffused form of participative intelligence to achieve the breakthroughs we need to solve systemic problems and, at the same time, promote transparency and traceability of decision making.

Acknowledgements

This paper reports some of the results obtained through the research work I developed as a visiting researcher at the Center for Collective Intelligence of the Massachusetts Institute of Technology. I wish to acknowledge the support I received by the Italian Fulbright Program who sponsored my visiting with a research grant. I am also grateful to the CCI Director prof. Thomas Malone and to Josh Introne, Robert Laubacher and Mark Klein, members of the MIT Collaboratorium research team, for the ideas and insights they offered me in the conversations and the research meetings I attended during my visiting.

References

Cowgill, B., Wolfers, J., Wharton, U.P., Zitzewitz, E.: Using Prediction Markets to Track Information Flows: Evidence from Google. working paper (2008),
http://bocowgill.com/GooglePredictionMarketPaper.pdf

Conklin, J.: Dialogue Mapping: Building Shared Understanding of Wicked Problems. Wiley, Chichester (2006)

Das, S.: Symbolic argumentation for decision making under uncertainty. In: 8th International Conference on Information Fusion (2005)

de Moor, A., Aakhus, A.M.: Argumentation support from technologies to tools. Communication of the ACM 49, 93–98 (2006)

Dung, P.M.: On the acceptability of arguments and its fundamental role in nonmonotonic reasoning, logic programming and n-person games. Artificial intelligence 77, 321–357 (1995)

Gloor, P.: - Swarm Creativity Competitive advantage through collaborative networks. Oxford University Press, New York (2006)

Gurkan, A., Iandoli, L., Klein, M., Zollo, G.: Mediating debate through on-line large-scale argumentation: evidence from the field. Accepted for publication on the Journal fo Information Sciences (forthcoming)

Herrera, F., Herrera-Viedma, E., Martinez, L.Mesiar,R.: Representation models for aggregating linguistic information: issues and analysis. In: Di, T., Calvo, G., Mayor, R. (eds.) Aggregation Operators: new trends and applications, Springer, Heidelberg

Iandoli, L., Klein, M., Zollo, G.: Enabling On-Line Deliberation and Collective Decision-Making through Large-Scale Argumentation: A New Approach to the Design of an Internet-Based Mass Collaboration Platform. International Journal of Decision Support System Technology 1, 69–92 (2009)

Introne, J.: Adaptive Mediation in Groupware. PhD Dissertation, Brandeis University (2008)

Introne, J.: Supporting group decisions by mediating deliberation to improve information pooling. In: Proceedings of the ACM, international conference on Supporting group work, pp. 188–198 (2009)

Jøsang, A., Ismail, R., Boyd, C.: A survey of trusts and reputation systems for online service provision, Decision Support Systems 43, 618–644 (2007)

Kirschner, P.A., Shum, S.B., Carr, C.S.(eds.): Visualizing Argumentation. Springer, Heidelberg (2003)

Klein, M., Cioffi, M., Malone, T.: Achieving Collective Intelligence via Large Scale Online Argumentation. Working paper, MIT Center for Collective Intelligence, Cambridge (2007)

Kosko, B.: Neural networks and fuzzy systems. Prentice-Hall, Englewood Cliffs (1992)

Malone, T.W., Laubacher, R., Introne, J., Klein, M., Abelson, H., Sterman, J. Olson, G.: The Climate Collaboratorium: Project Overview, MIT Center for Collective Intelligence Working Paper No. 2009-03,
http://cci.mit.edu/publications/CCIwp2009-03.pdf

McAfee, A.: Enterprise 2.0: The Dawn of Emergent Collaboration. MIT Sloan Management Review 47(3), 21–28 (2006)

Novak, J.D., Canas, A.J.: The Theory Underlying Concept Maps and How to Construct and Use Them. Technical Report IHMC CmapTools, -01 Rev. 2006-01 (2006),
http://www.cmap.org

Page, S.E.: The Difference: How the Power of Diversity Creates Better Groups, Firms, Schools, and Societies. Princeton University Press, NJ (2007)

Pollock, J.L.: Defeasible reasoning with variable degrees of justification. Artifical Intelligence 133, 233–282 (2001)

Raymond, E.S.: The cathedral and the bazaar. O'Reilly, Sebastopol (2001)

Shafer, G.: A mathematical theory of evidence. Princeton University Press, Princeton (1976)

O'Reilly, T.: What Is Web 2.0. Design Patterns and Business Models for the Next Generation of Software (2005),
http://oreilly.com/web2/archive/what-is-web-20.html

Shum, S.J.B., Selvin, A.M., Sierhuis, M., Conklin, J., Haley, C.B., Nuseibeh, B.: Hypermedia Support for Argumentation-Based Rationale: 15 Years on from gIBIS and QOC. In: Dutoit, A.H., McCall, R., Mistrik, I., Paech, B. (eds.) Rationale Management in Software Engineering. Springer, Berlin (2006)

Stephanou, H.E., Lu, S.Y.: Measuring Consensus Effectiveness by a Generalized Entropy Criterion. IEEE Transactions on Pattern Analysis and Machine Intelligence 10(4) (1988)

Sunstein, C.R.: Infotopia. Oxford University Press, New York (2006)

Surowiecki, J.: The Wisdom of Crowds: Why the Many Are Smarter Than the Few and How Collective Wisdom Shapes Business, Economies. Societies and Nations, Societies and Nations, Doubleday, New York (2004)

Suthers, D.D., Vatrapu, R., Medina, R., Joseph, S., Dwyer, N.: Beyond threaded discussion: Representational guidance in asynchronous collaborative learning environments. Computers & Education 50(4), 1103–1127 (2008)

Tapscott, D., Williams, A.D.: Wikinomics. Penguin Book, New York (2006)

Toulmin, S.: The Uses of Arguments. Cambridge University Press, Cambridge (1959)

von Krogh, G., von Hippel, E.: The Promise of Research on Open Source Software. Management Science 52, 975–983 (2006)

Wolfers, J., Zitzewitz, E.: Prediction Markets. The Journal of Economic Perspectives 18, 107–126 (2004)

A Web-Based Consensus Support System Dealing with Heterogeneous Information

Francisco Mata, Juan Carlos Martínez, and Rosa Rodríguez

Abstract. The study of the consensus is an important research field in Decision Making. Several authors have addressed the analysis of the consensus processes from different points of view (techniques, models, information domains, etc.). In this contribution we show a novel web application about a consensus support system to carry out consensus reaching processes with heterogenous information, i.e., the decision makers may use different information domains (particulary, numerical, interval-valued and linguistic assessments) to express their opinions. The software application has the following main characteristic: i) it automatizes virtual consensus reaching processes in which experts may be situated in different places, ii) experts may use information domains near their work areas to provide their preferences and, iii) it is able to run on any computer and operating system. This application may be seen as a practical development of a theoretical research on the consensus modelling. It could be used by any organization to carry out virtual consensus reaching processes.

1 Introduction

Group decision making (GDM) problems may be defined as decision situations where several people (commonly called decision makers or experts) try to reach

Francisco Mata
Dept. of Computer Science, University of Jaén
e-mail: fmata@ujaen.es

Juan Carlos Martínez
Dept. of Computer Science, University of Jaén
e-mail: jcmartinez@loscerros.org

Rosa Rodríguez
Dept. of Computer Science, University of Jaén
e-mail: rmrodrig@ujaen.es

E. Herrera-Viedma et al. (Eds.): Consensual Processes, STUDFUZZ 267, pp. 357–381.
springerlink.com   © Springer-Verlag Berlin Heidelberg 2011

a common solution to a problem from their opinions or preferences. So, given a set of alternatives, experts try together to find the best alternative to solve the problem.

In the literature we can see several attempts to solve decision problems where experts use the same domain of information to express their preferences [11, 18, 21]. However, in many cases it may be advisable that experts express their points of view through domains more consistent with either the nature of the alternatives or their area of expertise. For example, experts of different departments of a company (marketing, accounting, psychology, ...) may prefer to express their opinions using a domain of information closer to their knowledge fields. Moreover, in decision problems we can deal with alternatives whose nature is quantitative and others whose nature is qualitative. The first ones can be assessed by means of precise values. However, when alternatives are related to qualitative aspects, it may be difficult to qualify them using precise values. In such cases, where the uncertainty is present, other types of assessments as interval-valued [23, 30] or linguistic values [11, 32] could be more suitable. In such circumstances, we can consider the decision problems are defined into a heterogenous context.

Usually GDM problems have been resolved through selection processes where the experts get the best set of alternatives from the preferences expressed by themselves [9, 26]. However it may happen that some experts consider that their preferences were not taken into account to obtain the solution and therefore may disagree with that solution. To avoid this situation, it is advisable to carry out a consensus process (see Fig. 1) where experts discuss and change their preferences to reach enough agreement before making the selection process [7, 12, 15, 19].

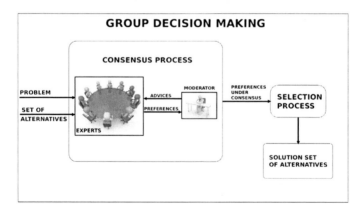

Fig. 1 Group decision making process

The consensus is an interesting research area in decision making that has been approached from different points of view [1, 2, 4, 5, 8, 10, 15, 19, 28, 31, 33]. It can be defined as a mutual-agreement state among the members of a group where all opinions have been expressed and listened to the satisfaction of the group [27]. The consensus-building process is a dynamic and iterative process consisting of several rounds, where experts express and discuss their points of view in order to achieve a common solution. Traditionally this process is coordinated by a human moderator, which calculates the consensus among experts in each round using different consensus measures [17, 22]. If the agreement is not enough, the moderator encourages experts to change their preferences further from the group's opinion in an effort to make them closer in the next consensus round [5, 34].

Roughly speaking, the consensus processes of the real world have two main characteristics:

a) A human moderator, who may become a controversial figure because experts may have complaints about his/her lack of objectivity. Moreover, in heterogeneous contexts, he/she may have problems to understand all the different information domains and scales in a proper way.
b) The physical presence of the experts in the same place. It is true that many consensus sessions are carried out by using the services given by new technologies as for example the videoconference, but the on-line presence of the expert is required. This is a problem since it is very difficult to get together experts or the cost is too high.

Several authors have proposed different models to approach the consensus reaching processes [4, 6, 7, 12, 16, 24, 28] but none of them has finally been implemented as a software solution to carry out virtual consensus processes. Therefore, taking into account the lack of software developments to tackle the consensus problems and the characteristics of the consensus reaching processes introduced previously, the aim of this chapter is to present a novel web application about a web consensus support system (WCSS). This application is supported on the last web technologies to accomplish consensus processes with heterogenous information and without limitations of the human moderator and the physical presence of the experts. This web application allows that experts, who are in different places, can participate in virtual consensus sessions using a simple browser and internet connection.

The chapter is structured as follows. In the Section 2, GDM problems defined in a heterogeneous setting as well as the process to unify heterogeneous information are introduced. In the Section 3, we present the consensus reaching model implemented in the web application. Finally, Section 4 shows the features of the WCSS.

2   Preliminaries

In this section, firstly we shall introduce the GDM problems with heterogeneous information. In the following, we shall briefly review the heterogeneous preferences unification process used by the WCSS to unify the preferences given by the experts.

2.1   Group Decision Making Problems in Heterogenous Settings

GDM problems are decision situations in which several individuals or experts, $E = \{e_1, e_2, \ldots, e_m\}$ $(m \geq 2)$, provide their preferences on a set of alternatives,

$X = \{x_1, x_2, \ldots, x_n\}$ $(n \geq 2)$, to derive a solution (an alternative or set of alternatives). Depending on different factors as the knowledge degree on the alternatives and/or the features of the problem, experts may use different structures to express their preferences.

In fuzzy contexts, a popular way to provide the experts' preferences are the fuzzy preference relations [18, 29]. A preference relation may be defined as a matrix $\mathbf{P_i} \subset X \times X$,

$$\mathbf{P_i} = \begin{pmatrix} p_i^{11} & \cdots & p_i^{1n} \\ \vdots & \ddots & \vdots \\ p_i^{n1} & \cdots & p_i^{nn} \end{pmatrix}$$

where the value $\mu_{P_{e_i}}(x_l, x_k) = p_i^{lk}$ is meant as the preference degree of the alternative x_l over x_k given by the expert e_i.

Assuming $p_i^{lj} \in [0,1]$, then:

1. $p_i^{lj} = 1$ indicates the maximum degree of preference of x_l over x_j
2. $0.5 \leq p_i^{lj} \leq 1$ indicates a definitive preference of x_l over x_j
3. $p_i^{lj} = 0.5$ indicates the indifference between x_l and x_j

The fuzzy preference relations may satisfy some of the following properties:

- Reciprocity: $p_i^{lj} + p_i^{jl} = 1, \forall l, j$
- Completeness: $p_i^{lj} + p_i^{jl} \geq 1, \forall l, j$
- Max-Min Transitivity: $p_i^{lk} \geq Min(p_i^{lj}, p_i^{jk}), \forall l, j, k$
- Max-Max Transitivity: $p_i^{lk} \geq Max(p_i^{lj}, p_i^{jk}), \forall l, j, k$
- Restricted Max-Min Transitivity: $p_i^{lj} \geq 0.5, p_i^{lk} \geq 0.5 \Rightarrow p_i^{lk} \geq Min(p_i^{lj}, p_i^{jk})$
- Restricted Max-Max Transitivity: $p_i^{lj} \geq 0.5, p_i^{lk} \geq 0.5 \Rightarrow p_i^{lk} \geq Max(p_i^{lj}, p_i^{jk})$
- Additive Transitivity: $p_i^{lj} + p_i^{jk} - 0.5 = p_i^{lk}, \forall l, j, k$

The ideal situation in a GDM problem is that all experts have a wide knowledge about the alternatives and provide their opinions in a numerical precise scale. However, in many cases, experts belong to distinct research areas and/or may have different levels of knowledge about the alternatives. In these situations, experts may prefer to express their preferences by means of different information domains and we can consider the problem is defined in a heterogeneous context.

In this chapter we deal with heterogenous GDM problems where experts provide their preferences by means of fuzzy preference relations with numerical, interval-valued or linguistic assessments. Under these circumstances, a process to unify these assessments on an unique information domain is necessary.

2.2 Unification of Heterogeneous Information

In GDM problems with heterogeneous information, there are no standard operators to operate directly with the preferences expressed in different information domains.

Herrera et al. proposed in [14] to unify all experts' preferences into a common utility space called basic linguistic term set (BLTS), $S_T = \{s_0, ..., s_g\}$ (see Fig. 2). They defined different transformation functions to transform each numerical, interval-valued and linguistic preference value into a fuzzy set on the BLTS, $F(S_T)$:

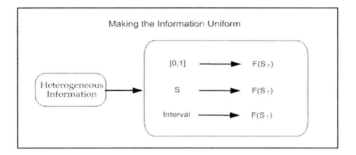

Fig. 2 Unification process of heterogeneous information

Assuming that the WCSS is supported on these transformation functions to unify the experts' preferences, we shall briefly review these functions. More details and examples can be consulted in [14].

2.2.1 Transforming Numerical Values into Fuzzy Sets

To transform a numerical value in $[0,1]$ into a fuzzy set on S_T, we use the following function. Let ϑ be a numerical value, $\vartheta \in [0,1]$, and $S_T = \{s_0, ..., s_g\}$ the BLTS. The function τ_{NS_T} that transforms a numerical value ϑ into a fuzzy set on S_T is defined as [13]:

$$\tau_{NS_T} : [0,1] \rightarrow F(S_T)$$

$$\tau_{NS_T}(\vartheta) = \{(s_0, \alpha_0), ..., (s_g, \alpha_g)\}, s_i \in S_T \text{ and } \alpha_i \in [0,1]$$

$$\alpha_i = \mu_{s_i}(\vartheta) = \begin{cases} 0, & \text{if } \vartheta \notin Support(\mu_{s_i}(x)) \\ \frac{\vartheta - a_i}{b_i - a_i}, & \text{if } a_i \leq \vartheta \leq b_i \\ 1, & \text{if } b_i \leq \vartheta \leq d_i \\ \frac{c_i - \vartheta}{c_i - d_i}, & \text{if } d_i \leq \vartheta \leq c_i \end{cases}$$

Remark 1: We consider membership functions, $\mu_{s_i}(\cdot)$, for linguistic labels, $s_i \in S_T$, are represented by a parametric function (a_i, b_i, d_i, c_i). As a particular case, triangular membership functions with $b_i = d_i$.

2.2.2 Transforming Interval-Valued into Fuzzy Sets

To transform an interval-valued into a fuzzy set on S_T, we use the following function. Let $I = [\underline{i}, \overline{i}]$ be an interval valued in $[0,1]$ and $S_T = \{s_0, ..., s_g\}$ the BLTS. Then, the function τ_{IS_T} that transforms the interval-valued I into a fuzzy set on S_T is defined as:

$$\tau_{IS_T} : I \rightarrow F(S_T)$$
$$\tau_{IS_T}(I) = \{(s_k, \alpha_k^i) / k \in \{0, ..., g\}\},$$
$$\alpha_k^i = \max_y \min\{\mu_I(y), \mu_{s_k}(y)\}$$

where $F(S_T)$ is the set of fuzzy sets defined in S_T, and $\mu_I(\cdot)$ and $\mu_{s_k}(\cdot)$ are the membership functions associated with the interval-valued I and terms s_k, respectively.

2.2.3   Transforming Linguistic Terms into Fuzzy Sets

To transform a linguistic value in S into a fuzzy set on S_T, we use the following function. Let $S = \{l_0, ..., l_p\}$ be a linguistic term set and $S_T = \{s_0, ..., s_g\}$ the BLTS, such that, $g \geq p$. Then, the function τ_{SS_T} that transforms $l_i \in S$ into a fuzzy set on S_T is defined as:

$$\tau_{SS_T} : S \rightarrow F(S_T)$$
$$\tau_{SS_T}(l_i) = \{(s_k, \alpha_k^i) / k \in \{0, ..., g\}\}, \forall l_i \in S$$
$$\alpha_k^i = \max_y \min\{\mu_{l_i}(y), \mu_{s_k}(y)\}$$

where $F(S_T)$ is the set of fuzzy sets defined in S_T, and $\mu_{l_i}(\cdot)$ and $\mu_{s_k}(\cdot)$ are the membership functions of the fuzzy sets associated with the terms l_i and s_k, respectively.

Remark 2: If S is chosen as S_T, then the fuzzy set that represents a linguistic term will be **0** except the value of the ordinal of the linguistic label that will be **1**.

3   A Consensus Support System Model for GDM Problems with Heterogeneous Information

Several theoretical models have been proposed in the literature to approach consensus processes [6, 7, 16, 20, 24, 28]. In this section we present the comprehensive consensus support system model implemented in the web application. The aim of this model is to automate consensus processes without the direct intervention of the human moderator. The model has the following main features:

1. It is able to carry out consensus processes in heterogeneous GDM problems with numerical, interval-valued and linguistic assessments. Experts provide their opinions by means of preference relations.
2. It uses two types of measurements,*consensus degrees* to check the level of agreement reached among the experts in each round and *proximity measures* to measure the distance among the experts' individual preferences and the collective preference.
3. It has a guided advice generator to suggest the changes of preferences to experts in order to increase the level of agreement in each new consensus round. A set of advice rules based on both measurements is used to guide the direction of the changes of the experts' preferences.

The CSS model consists of the following phases depicted in Fig. 3:

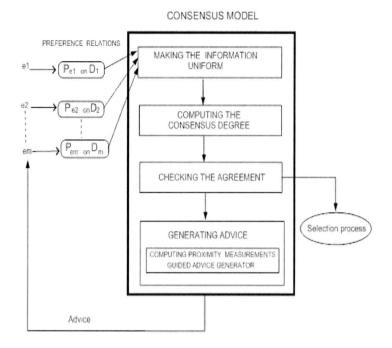

Fig. 3 A consensus support system model with heterogeneous information

A brief explanation about every phase is presented in the following subsections.

3.1 Making the Information Uniform

As we have said before, there are no standard operators to manage heterogeneous information directly, so we have to transform all preferences into an unique domain in order to operate with this information. By using the transformation functions introduced in Sect. 2.2, the model transforms all experts' preferences into fuzzy sets. So, assuming each fuzzy set, $\tilde{p}_i^{lk} = (s_0, \alpha_{i0}^{lk}), \ldots, (s_g, \alpha_{ig}^{lk})$, is represented by means of its respective membership degrees, $(\alpha_{i0}^{lk}, \ldots, \alpha_{ig}^{lk})$, the result of this phase is a matrix of fuzzy sets, $\tilde{\mathbf{P}}_i$:

$$\tilde{\mathbf{P}}_i = \begin{pmatrix} \tilde{p}_i^{11} = (\alpha_{i0}^{11}, \ldots, \alpha_{ig}^{11}) & \cdots & \tilde{p}_i^{1n} = (\alpha_{i0}^{1n}, \ldots, \alpha_{ig}^{1n}) \\ \vdots & \ddots & \vdots \\ \tilde{p}_i^{n1} = (\alpha_{i0}^{n1}, \ldots, \alpha_{ig}^{n1}) & \cdots & \tilde{p}_i^{nn} = (\alpha_{i0}^{nn}, \ldots, \alpha_{ig}^{nn}) \end{pmatrix}$$

3.2   Computing the Consensus Degree

The consensus degree evaluates the level of agreement achieved among the experts in each consensus round. So, if experts' preferences are similar, the consensus degree will be high, otherwise, if preferences are very different, the consensus degree will be low. To compute the level of agreement, a consensus matrix is obtained by aggregating the values that represent the similarity or distance among the experts' preferences, comparing each other.

The distance between two preferences $\tilde{p}_i^{lk}$ and $\tilde{p}_j^{lk}$ is computed by means of the similarity function $s(\tilde{p}_i^{lk}, \tilde{p}_j^{lk})$ measured in the unit interval $[0,1]$ [16]:

$$s(\tilde{p}_i^{lk}, \tilde{p}_j^{lk}) = 1 - \left| \frac{cv_i^{lk} - cv_j^{lk}}{g} \right| \tag{1}$$

The cv_i^{lk} is the central value of the fuzzy set:

$$cv_i^{lk} = \frac{\sum_{h=0}^{g} index(s_h^i) \cdot \alpha_{ih}^{lk}}{\sum_{h=0}^{g} \alpha_{ih}^{lk}}, \tag{2}$$

and represents the centre of gravity of the information contained in the fuzzy set $\tilde{p}_i^{lk} = (\alpha_{i0}^{lk}, \ldots, \alpha_{ig}^{lk})$, where $index(s_h^i) = h$ and α_{ih}^{lk} are the membership degrees. The range of this central value is the closed interval $[0,g]$.

The closer $s(\tilde{p}_i^{lk}, \tilde{p}_j^{lk})$ to 1 the more similar preferences p_i^{lk} and p_j^{lk} are, while the closer $s(\tilde{p}_i^{lk}, \tilde{p}_j^{lk})$ to 0 the more distant p_i^{lk} and p_j^{lk} are.

Once the similarity function has been introduced, the consensus degree is computed according to the following steps:

1. Central values of all fuzzy sets are calculated:

$$cv_i^{lk}; \ \forall \ i = 1, \ldots, m; \ l, k = 1, \ldots, n \wedge l \neq k. \tag{3}$$

2. For each pair of experts e_i and e_j ($i < j$), a *similarity matrix* $SM_{ij} = \left(sm_{ij}^{lk} \right)$ is calculated, where,

$$sm_{ij}^{lk} = s(\tilde{p}_i^{lk}, \tilde{p}_j^{lk}). \tag{4}$$

3. A *consensus matrix* CM is obtained by aggregating all the similarity matrices at the level of pairs of alternatives,

$$CM = \begin{pmatrix} cm^{11} & \cdots & cm^{1n} \\ \vdots & \ddots & \vdots \\ cm^{n1} & \cdots & cm^{nn} \end{pmatrix}$$

where,

$$cm^{lk} = \phi(sm_{12}^{lk}, sm_{13}^{lk}, \ldots, sm_{1m}^{lk}, sm_{23}^{lk}, \ldots, sm_{2m}^{lk}, \ldots, sm_{(m-1)m}^{lk})$$

for $l,k \in \{1, \ldots, n\}$.

As we can deduce from (1), if two preferences $\tilde{p}_i^{lk}$ and $\tilde{p}_j^{lk}$ are very similar, the similarity value between them, $s(\tilde{p}_i^{lk}, \tilde{p}_j^{lk})$, is high and therefore we can consider the agreement is high too. Taking into account this assumption, in order to evaluate the level of agreement among experts, the model computes the similarity values between each expert's preference with respect to the rest of experts' preferences and it aggregates them. So, the more similar the preferences are the more high aggregated values are and therefore the agreement will be higher. In this model, we use the arithmetic mean as the aggregation function ϕ, although, other aggregation operators could be used according to other points of view (majority concept, penalization techniques, etc.) [25].

4. The consensus degree is computed in three different levels: pairs of alternatives, alternatives and relations. In this way, we can know in a precise way the level of agreement in each pair of alternatives and so to identify the pairs as well as the alternatives in which there exists greater disagreement.

Level 1.   Consensus on pairs of alternatives, cp^{lk}, represents the agreement on the pair of alternatives (x_l, x_k). They are obtained from the consensus matrix CM,

$$cp^{lk} = cm^{lk}, \quad \forall l, k = 1, \ldots, n \ \wedge \ l \neq k.$$

Values of cp^{lk} close to 1 mean a greater agreement. This measurement allows to identify those pairs with a poor level of agreement.

Level 2.   Consensus on alternatives, ca^l, represents the agreement on the alternative x_l,

$$ca^l = \frac{\sum_{k=1, l \neq k}^n (cp^{lk} + cp^{kl})}{2(n-1)}. \tag{5}$$

Level 3.   Consensus on relations or global consensus, cr, represents the global agreement among all experts' preferences,

$$cr = \frac{\sum_{l=1}^n ca^l}{n}. \tag{6}$$

The model uses this value to check the level of agreement achieved in each round. If cr is high enough then the system stops the consensus process.

Assuming the computing of the consensus degree is a key phase of the model, an example is presented to facilitate the understanding.

Example 1. . *Let e_1, e_2 and e_3 be three experts who use numerical, linguistic and interval-valued assessments respectively. Their preferences are the following* [1]:

---

[1] Notice that the properties of the preference relations defined in Sect. 2.1 have been smoothed in order to make more flexible the entrance of experts' preferences.

$$
\mathbf{P_1} = \begin{pmatrix} - & .5 & .8 & .4 \\ .3 & - & .9 & .3 \\ .3 & .2 & - & .4 \\ .9 & .8 & .5 & - \end{pmatrix} ; \ \mathbf{P_2} = \begin{pmatrix} - & H & VH & M \\ L & - & H & VH \\ VL & N & - & VH \\ L & VL & N & - \end{pmatrix}
$$

$$
\mathbf{P_3} = \begin{pmatrix} - & [.7,.8] & [.65,.7] & [.8,.9] \\ [.3,.35] & - & [.6,.7] & [.8,.85] \\ [.3,.35] & [.3,.4] & - & [.7,.9] \\ [.1,.2] & [.2,.4] & [.1,.3] & - \end{pmatrix}
$$

After carrying out the unification process, experts' preferences have been transformed into matrices of fuzzy sets:

$$
\tilde{\mathbf{P}}_1 = \begin{pmatrix} - & (0,0,0,1,0,0,0) & (0,0,0,0,.19,.81,0) & (0,0,.59,.41,0,0,0) \\ (0,.19,.81,0,0,0,0) & - & (0,0,0,0,0,.59,.41) & (0,.19,.81,0,0,0,0) \\ (0,.19,.81,0,0,0,0) & (0,.81,.19,0,0,0,0) & - & (0,0,.59,.41,0,0,0) \\ (0,0,0,0,0,.59,.41) & (0,0,0,0,.19,.81,0) & (0,0,0,1,0,0,0) & - \end{pmatrix}
$$

$$
\tilde{\mathbf{P}}_2 = \begin{pmatrix} - & (0,0,0,0,1,0,0) & (0,0,0,0,0,1,0) & (0,0,0,1,0,0,0) \\ (0,0,1,0,0,0,0) & - & (0,0,0,0,1,0,0) & (0,0,0,0,0,1,0) \\ (0,1,0,0,0,0,0) & (1,0,0,0,0,0,0) & - & (0,0,0,0,0,1,0) \\ (0,0,1,0,0,0,0) & (0,1,0,0,0,0,0) & (1,0,0,0,0,0,0) & - \end{pmatrix} ;
$$

$$
\tilde{\mathbf{P}}_3 = \begin{pmatrix} - & (0,0,0,0,.81,.81,0) & (0,0,0,.12,1,.19,0) & (0,0,0,0,.19,1,.41) \\ (0,.19,1,.12,0,0,0) & - & (0,0,0,.41,1,.19,0) & (0,0,0,0,.19,1,.12) \\ (0,.19,1,.12,0,0,0) & (0,.19,1,.41,0,0,0) & - & (0,0,0,0,.81,1,.41) \\ (.41,1,.19,0,0,0,0) & (0,.81,1,.41,0,0,0) & (.41,1,.81,0,0,0,0) & - \end{pmatrix}
$$

From these matrices, the model computes the global consensus degree cr applying the following steps:

1. *Central values.* Applying (2), the model computes the central values of the fuzzy sets:

$$
cv_1 = \begin{pmatrix} - & 3 & 4.81 & 2.41 \\ 1.81 & - & 5.41 & 1.81 \\ 1.81 & 1.19 & - & 2.41 \\ 5.41 & 4.81 & 3 & - \end{pmatrix} ; \ cv_2 = \begin{pmatrix} - & 4 & 5 & 3 \\ 2 & - & 4 & 5 \\ 1 & 0 & - & 5 \\ 2 & 1 & 0 & - \end{pmatrix}
$$

$$
cv_3 = \begin{pmatrix} - & 4.5 & 4 & 5.13 \\ 1.94 & - & 3.86 & 4.94 \\ 1.94 & 2.13 & - & 4.81 \\ 0.86 & 1.81 & 1.18 & - \end{pmatrix}
$$

2. *Similarity matrices. The model computes a similarity matrix between each pair of experts by using the distance function (1):*

$$SM_{12} = \begin{pmatrix} - & 0.83 & 0.96 & 0.9 \\ 0.96 & - & 0.76 & 0.46 \\ 0.86 & 0.8 & - & 0.56 \\ 0.43 & 0.36 & 0.5 & - \end{pmatrix}; \quad SM_{13} = \begin{pmatrix} - & 0.75 & 0.87 & 0.54 \\ 0.97 & - & 0.74 & 0.47 \\ 0.97 & 0.84 & - & 0.59 \\ 0.24 & 0.5 & 0.69 & - \end{pmatrix}$$

$$SM_{23} = \begin{pmatrix} - & 0.91 & 0.84 & 0.64 \\ 0.99 & - & 0.97 & 0.99 \\ 0.84 & 0.64 & - & 0.97 \\ 0.81 & 0.86 & 0.8 & - \end{pmatrix}$$

3. *Consensus matrix. The model calculates the consensus matrix by aggregating the similarity matrices:*

$$CM = \begin{pmatrix} - & 0.83 & 0.89 & 0.69 \\ 0.97 & - & 0.82 & 0.64 \\ 0.89 & 0.76 & - & 0.71 \\ 0.49 & 0.57 & 0.66 & - \end{pmatrix}$$

4. *Consensus degrees. The model computes the consensus degree at different levels:*

Level 1.   *Consensus on pairs of alternatives. The element (l,k) of CM represents the consensus degree on the pair of alternatives (x_l, x_k).*
Level 2.   *Consensus on alternatives.*

$$ca^1 = 0.79, \ ca^2 = 0.80, \ ca^3 = 0.78, \ ca^4 = 0.63$$

Level 3.   *Consensus on the relations or global consensus. Finally the consensus degree among experts is,*
$$cr = 0.73$$

3.3   Checking the Agreement

In this phase the model controls the level of agreement achieved in the current consensus round. Before starting the process, a minimum consensus threshold, $\beta \in [0,1]$, is fixed, which will depends on the particular problem we are dealing with. When the consequences of the decision are of a transcendent importance, the minimum level of consensus required to make that decision should be logically high, for example $\beta = 0.8$ or higher. At the other extreme, when the consequences are not so transcendental (but are still important) and it is urgent to obtain a solution, a fewer consensus threshold near to 0.5 could be required.

In any case, independently of the value β, when the global consensus cr reaches β, the consensus process stops, and the selection process is applied to obtain the solution. However, there is the possibility that the global consensus will not converge to consensus threshold and the process will get block. In order to avoid this circumstance, the model incorporates a parameter, *Maxcycles*, to limit the number of consensus rounds to carry out. The performance of this phase is shown in Fig. 4.

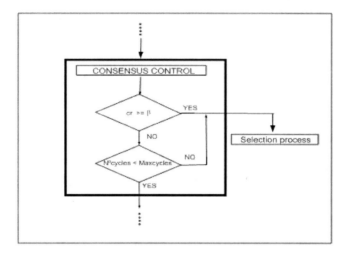

Fig. 4 Consensus control

3.4 Generating Advice

When the agreement is not big enough, i.e. $cr < \beta$, experts should modify their preferences to make them closer and to increase the level of agreement. In order to identify the furthest experts' preferences from the collective opinion, the model computes proximity measurements among experts'preferences and the collective opinion. Once the furthest preferences have been identified, a guided advance generator suggests the changes on these preferences with the objective of increasing the agreement in the next consensus round. Both processes are presented in detail below.

3.4.1 Computing Collective Preference and Proximity Measurements

The proximity measurements evaluate the distance between the individual experts' preferences and the collective preference. To calculate them, firstly the model obtains a collective preference $\tilde{\mathbf{P}}_c$,

$$
\tilde{\mathbf{P}}_c = \begin{pmatrix} \tilde{p}_c^{11} & \cdots & \tilde{p}_c^{1n} \\ \vdots & \ddots & \vdots \\ \tilde{p}_c^{n1} & \cdots & \tilde{p}_c^{nn} \end{pmatrix}
$$

which represents the group's opinion. $\bar{P}_c$ is calculated by aggregating the set of individual preference relations $\{\bar{P}_{e_1}, \ldots, \bar{P}_{e_m}\}$,

$$\tilde{p}_c^{lk} = \psi(\tilde{p}_1^{lk}, \ldots, \tilde{p}_m^{lk}) = (\alpha_{c0}^{lk}, \ldots, \alpha_{cg}^{lk})$$

where

$$\alpha_{cj}^{lk} = \psi(\alpha_{1j}^{lk}, \ldots, \alpha_{mj}^{lk})$$

and ψ is the average of the membership degrees.

Once the model has obtained the collective preference relation, it computes a proximity matrix, PM_i, for each expert e_i,

$$PM_i = \begin{pmatrix} pm_i^{11} & \cdots & pm_i^{1n} \\ \vdots & \ddots & \vdots \\ pm_i^{n1} & \cdots & pm_i^{nn} \end{pmatrix}$$

by using the similarity function defined in expression (1),

$$pm_i^{lk} = s(\tilde{p}_i^{lk}, \tilde{p}_c^{lk}).$$

These matrices contain the necessary information to know the distance of the preferences of each expert with respect to the collective opinion.

The experts' proximities are computed at level of pairs of alternatives, alternatives and relations.

Level 1. Proximity on pairs of alternatives, pp_i^{lk}, represents the proximity between the expert's preference and the collective one on the pair of alternatives (x_l, x_k),

$$pp_i^{lk} = pm_i^{lk}, \quad \forall l, k = 1, \ldots, n \; \wedge \; l \neq k.$$

Level 2. Proximity on alternatives, called pa_i^l, represents the proximity between the expert's preferences and the collective one on the alternative x_l,

$$pa_i^l = \frac{\sum_{k=1, k \neq l}^n (pp_i^{lk} + pp_i^{kl})}{2(n-1)}. \tag{7}$$

Level 3. Proximity on the relation, pr_i, represents the global proximity between the expert's preferences and the collective one,

$$pr_i = \frac{\sum_{l=1}^n pa_i^l}{n}. \tag{8}$$

3.4.2 Guided Advice Generator

The purpose of the guided advice generator is to identify the furthest experts' preferences and to suggest how to change them in order to increase the consensus. The model will only suggest increasing or decreasing the current assessments and each expert will decide the ratio of the changes according his/her both experience and the preferences expression domain.

To attain this objective, the guided advice generator uses two types of advice rules: *identification rules* and *direction rules*.

a) *Identification rules (IR)*. These rules identify what experts, alternatives and pairs of alternatives should be changed. In this way, the model only focuses on the preferences in disagreement and will not recommend to change those which the agreement is enough. The model uses three types of rules:

a.1) An identification rule of experts. Experts whose proximity, pr_i, is smaller than the average of proximities have to change some of their preferences.

  IR.1. $\forall e_i \in E \cap EXPCH$, then e_i must change his/her preferences, where

$$EXPCH = \{e_i \mid pr_i < \overline{pr}\}$$

 and

$$\overline{pr} = \frac{\sum_{i=1}^{m} pr_i}{m}.$$

a.2) An identification rule of alternatives. Alternatives whose consensus degree, ca^l, is smaller than the global consensus, cr, should change some of its pairs.

  IR.2. $\forall e_i \in EXPCH$, e_i should change some assessments associated to the pairs that belong to the alternative x_l, such that, $x_l \in ALT$ and $ALT = \{x_l \in X \mid ca^l < cr\}$.

a.3) An identification rule of pairs of alternatives. Pairs of alternatives whose proximity value at level of pairs, pp_i^{lk}, is smaller that the average.

  IR.3. $\forall (x_l \in ALT \wedge e_i \in EXPCH)$, if $(x_l, x_k) \in PALT_i$, then e_i should change the pair p_i^{lk}, where

$$PALT_i = \{(x_l, x_k) \mid x_l \in ALT \wedge e_i \in EXPCH \wedge pp_i^{lk} < \overline{pp}^{lk}\}.$$

 and

$$\overline{pp}^{lk} = \frac{\sum_{i=1}^{m} pp_i^{lk}}{m}.$$

b) *Direction rules (DR)*. Once the model has identified the pairs of alternatives to be changed, it uses a set of direction rules to suggest the directions of the changes. For each preference value to be changed, the model will suggest increasing or decreasing the current assessment. Now each expert increases or decreases the his/her currents assessments by using his/her respective information domain. The following directions rules are used taking into account the central values of the fuzzy sets of both experts' preferences and collective ones.

  DR.1. If $(cv(\tilde{p}_i^{lk}) - cv(\tilde{p}_c^{lk})) < 0$ then the expert e_i should increase the assessment associated to the pair of alternatives (x_l, x_k).

  DR.2. If $(cv(\tilde{p}_i^{lk}) - cv(\tilde{p}_c^{lk})) > 0$ then the expert e_i should decrease the assessment associated to the pair of alternatives (x_l, x_k).

  DR.3. If $(cv(\tilde{p}_i^{lk}) - cv(\tilde{p}_c^{lk})) = 0$ then the expert e_i should not modify the assessment associated to the pair of alternatives (x_l, x_k).

4 WCSS Features

In this section we present the features of the web consensus support system to carry out virtual consensus processes with heterogeneous information. This web application has been developed by the research group Intelligent Systems Based on Fuzzy Decision Analysis ($Sinbad^2$) of the University of Jaén. As we said in the Sect. 1, the consensus may be seen as a technique to achieve good results in any GDM problem. If we take as example consensus processes of the real world, the physic presence of the experts (in the same place or by videoconference) is required. In many cases this is a problem since it is difficult to get together experts (different cities or ever countries, engagement books, ...) or the cost is too high for the organization. In these cases, carrying out virtual consensus sessions supported on web technologies with experts participating from different places might be very interesting. Different theoretic consensus models have been proposed in the literature, but very few of them have been implemented and none of them considers the possibility for accomplishing virtual consensus processes.

The characteristics of the web application are the following:

a) It automatizes virtual consensus reaching processes where the moderator's tasks are assumed by the system and the experts may be situated in different places.
b) Ability to manage different information domains (numerical, linguistic and interval-valued assessments).
c) It has been implemented using the last technologies related to the developments of web applications.
d) Multi-user and multi-platform system.
e) Storage of information in databases.
f) Smooth handling and ease of use.

This system may be considered as an example of research transfer from the theoretical model proposed in Sect. 3 to a real system to solve consensus reaching problems.

This section has been structured in two main parts: system architecture and system functionality. Details of the implementation code have been obviated because we have considered these are out of the scope of this book.

4.1 System Architecture

In this section we briefly describe the main elements of the model along with technologies used to design and to implement the system. As it is depicted in the Fig. 5, the system has been developed following the typical client-server architecture and the object-oriented paradigm. So, the system is hosted in a computer which makes the role of internet server [2]. Remote clients can connect with this server through a web browser. Users send requests that are parsed on the server and consistent answers are sent back to client. This architecture frees the end user of the task of

---

[2] In this version of the application, Apache Tomcat 6.0 web server has been used since it is open source software, very popular among programmers and it is used as web server in many organizations.

installing the application on his computer. Moreover, it is highly scalable and extensible to add new clients and servers. Note that a database system is also used to store all the information related with the process, that is, information about problems, experts, preferences given in every round, etc.

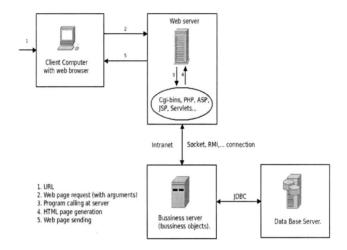

Fig. 5 System architecture

Regarding web technologies and programming languages, the application has been implemented with the Java language, particulary JavaServer Pages (JSP) because allow to generate dynamic web pages by using HTML and XML documents. JSP generate pages that are compiled and executed on the server to deliver an HTML or XML document. The compiled pages and dependent Java libraries use Java bytecode rather than a native software format. In this way the application can be executed within the Java Virtual Machine on any computer and operating system.

Although it is not our intention to explain the implementation details, we think it is interesting to describe the package used to manage the heterogenous information. The design of this package may guide other researchers to define other similar packets in other contexts.

4.1.1 Package for Managing Heterogeneous Information

The Fig. 6 depicts the class diagram to deal with heterogeneous information used by our system. This diagram has been designed taking into account the nomenclature proposed by Unified Modeling Language (UML) [3].

The class diagram is clair, we only focus on the class *Valuation* because it is interesting from of point of view of the heterogeneous information. This class is an abstract class of "Numeric", "Interval" and "Linguistic" classes, which inherit the unification() operation. This method transforms all experts' preferences into fuzzy sets according to the expression domain (see Sect. 2.2). As we can see all classes

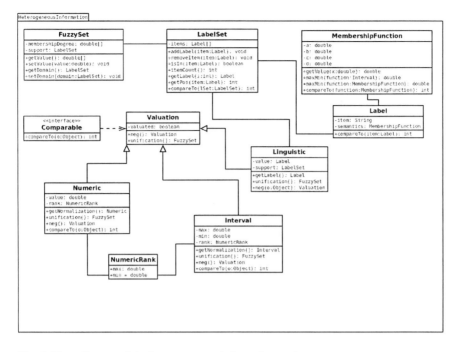

Fig. 6 Class diagram of the heterogeneous information package

have their own attributes (e.g. the attributes of the class Interval are the extremes values of the interval and the range of the possible values) and their methods.

4.1.2 Database

As it seems logic, any consensus process generates a lot information that has to be stored. For example, about the experts we manage personal details, about the problems, we have parameters (consensus threshold, number of rounds), and about the consensus rounds we store the consensus degrees achieved, number of changes, etc. All this information have to be organized properly and should be accessible by the system, administrator and experts, therefore a database [3] is needed. In the Fig. 7 we show the Entity- Relationship diagram used by the system.

It is clearly viewed that admin, expert, alternative, problem, round, etc., are entities in our conceptual schema. Regarding relationships, we describe some examples: in a problem several experts take part and an expert takes part in one or several problems (R1); an expert uses an information type and an information type can be used by several experts (R2); a problem is set on several alternatives and an alternative is included in only one problem (R3); etc.

---

[3] MySQL 5.0 database system has been used due to it is easy to use, reliable and quite fast.

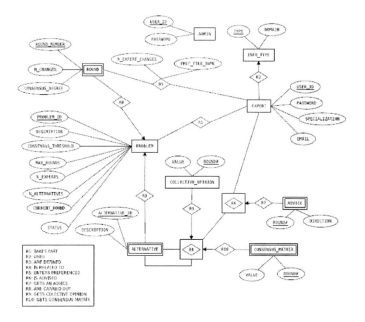

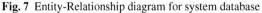

Fig. 7 Entity-Relationship diagram for system database

4.2   System Functionality

In this subsection the system functionality is described by following a structure similar to the consensus model depicted in the Fig. 3. To improve the understanding of the system, we use an example which allows us to see the system performance during a consensus reaching process. So, a food company wants to launch a new type of juice but it has doubts about the best juice taste (lemon, apple, orange and peach taste). To solve this decision, the company decides to consult several experts from different areas (sports, medicine, marketing, ...) who live in different cities and countries. The company management wants to carry out a consensus process among the experts in order to attain a high level of agreement about the preferred taste before deciding the juice to be produced. Due to experts live in different cities, the company decides to use our web application to reproduce a virtual consensus reaching process.

4.2.1   Users Identification

Two different user profiles have been considered in this application from the point of view of the functionality:

- Experts, persons who participate in the consensus process by providing their opinions about a problem.
- Administrators, persons who are in charge of the system performance. They manage all the information concerning the problems, experts, and they check that the consensus process is developed properly.

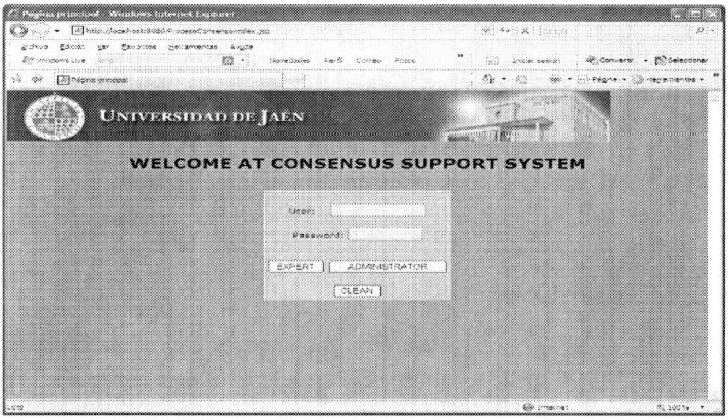

Fig. 8 Access to the system

The system identifies the user profile by means of the typical identification mechanism based on an user's name and password (see Fig. 8.) In the case of experts, the access codes are assigned by the administrator when experts' accounts are created.

4.2.2   Gathering Information

The information about the problems (description, set of alternatives, consensus threshold, ...) is exclusively inserted and/or updated by the administrator such as it is depicted in the Fig. 9. This information is requested by the system when the administrator creates the problem. Note that the information may be changed easily

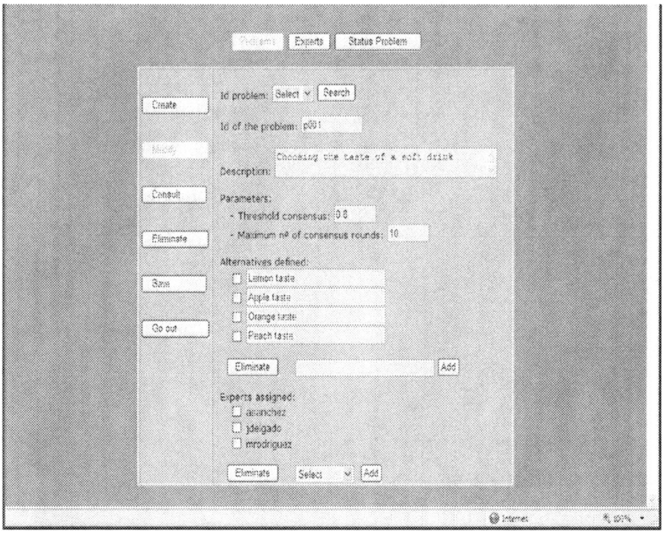

Fig. 9 Information and characteristic about a problem

because all parameters are edited. For example, the administrator can add new alternatives (button Add below list of alternatives) or change the consensus threshold. In this way, the system is more flexible and able to adapt to new situations without having to introduce all the information again.

From the point of view of the experts' opinions, once the expert has been identified by the system, this shows a empty preference relation to be filled out with the expert's preferences.

4.2.3 Computing the Consensus Degree

Once the preferences have been inserted by the experts, the administrator can compute the consensus degree. The result of this action is shown in the Fig. 10 and Fig. 11:

As we can see in the Fig. 10, the system shows the following information:

- General information on the problem.
- Status of the problem. Two status are possible: *progress*, when the consensus process is open or *finished* when the process is over.
- Consensus Matrix. We can see the consensus values on each pair of preferences.
- Consensus degree achieved in each consensus round.
- List of experts and number of changes of preferences (see Fig. 11).
- Preferences given by each expert and round.

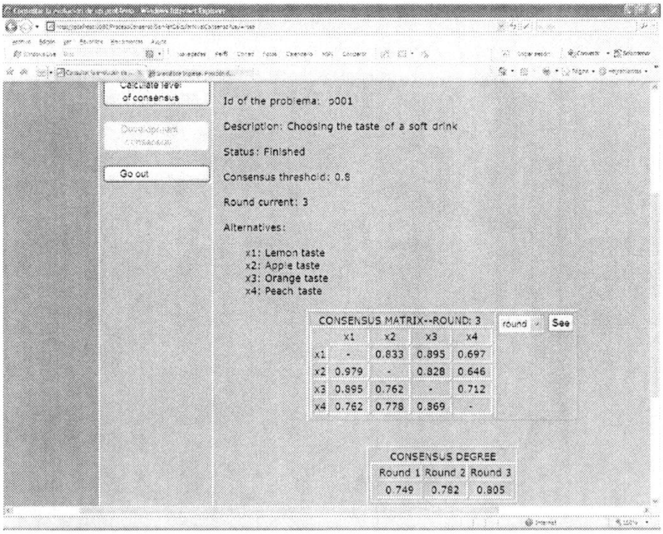

Fig. 10 Status of the consensus process

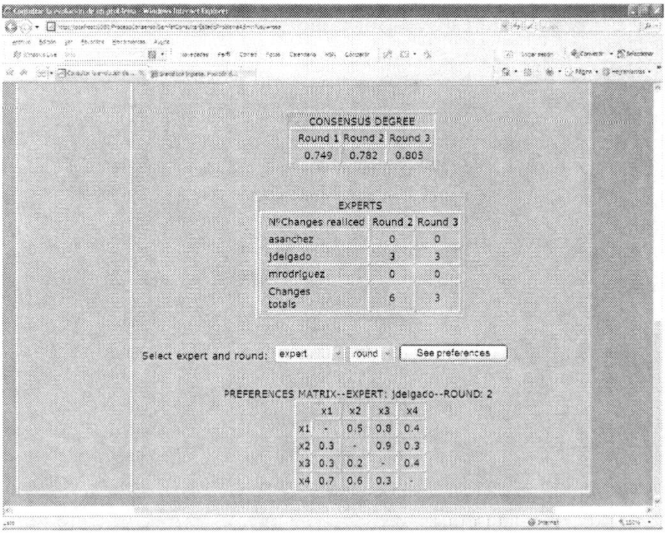

Fig. 11 Status of the consensus process: Expert's preferences

This information allows administrator to know with accuracy the evolution of the consensus reaching process and to accomplish a sensibility analysis in order to extract conclusions or relationship among the system parameters, preferences and changes. The information is stored in the database and can be consulted in the future to study the details of each problem. Note that all mathematic operations to unify the heterogenous information and to compute the consensus degree are transparent for experts and administrator.

4.2.4 Checking the Agreement

If the consensus process progresses in the right way, the level of agreement is achieved and the system shows a message reporting about the success of the process (see Fig. 12). Otherwise, the system shows other message reporting about the failure of the consensus process because the maximum number of consensus rounds has been exceeded.

4.2.5 Generating Advice

If the level of agreement is not high enough, each expert may know the changes suggested on his/her preferences in order to increase the agreement in the next consensus round. Before applying the changes, the expert may consult the current state of the problem (see Fig. 13). This window shows information about the number of rounds, current consensus degree, preferences given in each round, etc.

To help expert about the right direction of the changes, the system shows the window of the Fig. 14. As we can see, on the one hand we have the expert's preferences given in the last consensus round and by means of a color code the system

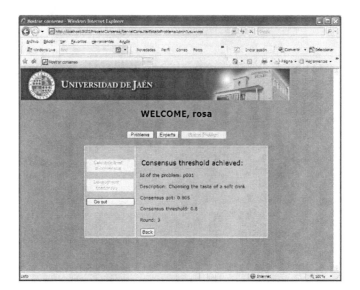

Fig. 12 Consensus has been achieved

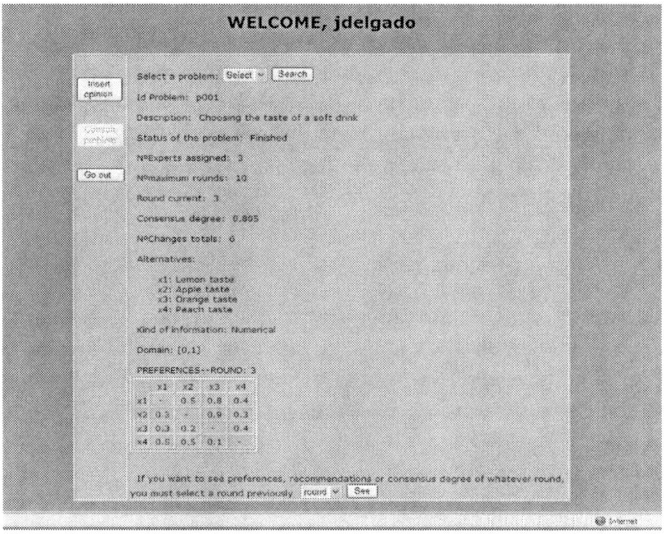

Fig. 13 Information about the problem

indicates the direction of the changes. So, values in red color have to be decreased while values in green color have to be increased. The rest of preferences have not to be modified. On the other hand, at the bottom of the window, the system shows a preference relation where the preferences to be changed are empty. In this way the system allows expert to insert new assessments which will be taken into account for the next consensus round.

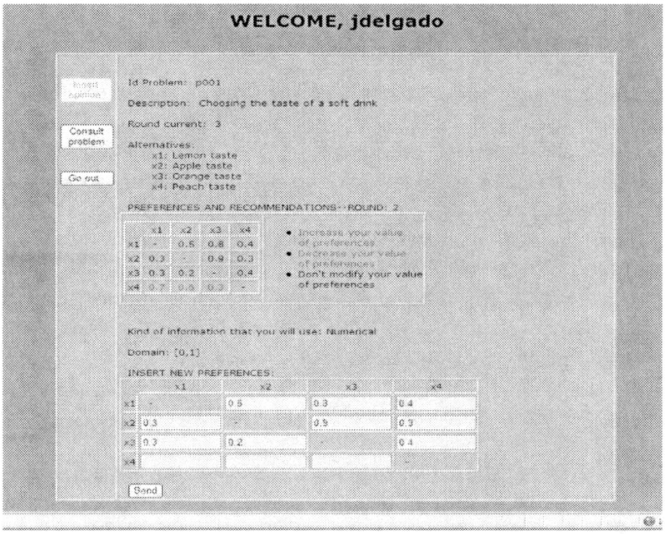

Fig. 14 Changes suggested by the system

Finally, note that the system has a wide repository of warning messages about possible errors (identifiers already used, empty fields of the forms) which makes easy the use of the application.

5 Conclusions

In this chapter we have shown a web consensus support system to automate the consensus processes in GDM problems with heterogeneous information. This WCSS uses a methodology based on transformation functions proposed by Herrera et al. [14] to unify the heterogeneous information. Different approaches to consensus process have been proposed in the literature but very few of them have been implemented as real software applications. Here we present a novel software to carry out virtual consensus reaching processes in which the experts may be situated in different places. Three main features may be highlighted of the system: i) it automatizes virtual consensus reaching process without the direct intervention of human moderator, ii) experts may use numerical, interval-valued and linguistic assessments and iii) it is a multi-platform system able to run on any computer and operating system. The system has been developed following the object-oriented paradigm and uses open source software tools. This system may be seen as a transfer of knowledge from the theoretical research about the consensus process in GDM to a practical development.

Acknowledgements. This work has been supported by the University of Jaén, research projects UJA2009/12/25.

References

1. Alonso, S., Cabrerizo, F., Chiclana, F., Herrera, F., Herrera-Viedma, E.: Group decision making with incomplete fuzzy linguistic preference relations. International Journal Intelligent System 24(2), 201–222 (2009)
2. Ballester, M., Garcia-Lapresta, J.: Sequential consensus for selecting qualified individuals of a group. International Journal of Uncertainty Fuzziness and Knowledge-Based System 16, 57–68 (2008)
3. Booch, G., Rumbaugh, J.: The Unified Modeling Language User Guide. Addison-Wesley, Reading (2005)
4. Bordogna, G., Fedrizzi, M., Pasi, G.: A linguistic modeling of consensus in group decision making based on OWA operators. IEEE Transactions on Systems, Man and Cybernetics, Part A: Systems and Humans 27, 126–132 (1997)
5. Bryson, N.: Group decision-making and the analytic hierarchy process: Exploring the consensus-relevant information content. Computers and Operational Research 1(23), 27–35 (1996)
6. Cabrerizo, F., Alonso, S., Herrera-Viedma, E.: A consensus model for group decision making problems with unbalanced fuzzy linguistic information. International Journal of Information Technology and Decision Making 8(1), 109–131 (2009)
7. Carlsson, C., Ehrenberg, D., Eklund, P., Fedrizzi, M., Gustafsson, P., Lindholm, P., Merkuryeva, G., Riissanen, T., Ventre, A.: Consensus in distributed soft environments. European Journal of Operational Research 61, 165–185 (1992)
8. Fan, Z., Chen, X.: Consensus measures and adjusting inconsistency of linguistic preference relations in group decision making. In: Wang, L., Jin, Y. (eds.) FSKD 2005. LNCS (LNAI), vol. 3613, pp. 130–139. Springer, Heidelberg (2005)
9. Fodor, J., Roubens, M.: Fuzzy Preference Modelling and Multicriteria Decision Support. Kluwer Academic Publishers, Dordrecht (1994)
10. Garcia-Lapresta, J.: Favoring consensus and penalizing disagreement in group decision making. Journal of Advanced Computational Intelligence and Intelligent Informatics 12(5), 416–421 (2008)
11. Herrera, F., Herrera-Viedma, E.: Linguistic decision analysis: Steps for solving decision problems under linguistic information. Fuzzy Sets and Systems 115, 67–82 (2000)
12. Herrera, F., Herrera-Viedma, E., Verdegay, J.: A model of consensus in group decision making under linguistic assessments. Fuzzy Sets and Systems 79, 73–87 (1996)
13. Herrera, F., Martínez, L.: An approach for combining linguistic and numerical information based on 2-tuple fuzzy representation model in decision-making. International Journal of Uncertainty, Fuzziness and Knowledge-Based Systems 8(5), 539–562 (2000)
14. Herrera, F., Martínez, L., Sánchez, P.: Managing non-homogeneous information in group decision making. European Journal of Operational Research 166(1), 115–132 (2005)
15. Herrera-Viedma, E., Herrera, F., Chiclana, F.: A consensus model for multiperson decision making with different preference structures. IEEE Transactions on Systems, Man and Cybernetics-Part A: Systems and Humans 32(3), 394–402 (2002)
16. Herrera-Viedma, E., Martínez, L., Mata, F., Chiclana, F.: A consensus support system model for group decision-making problems with multi-granular linguistic preference relations. IEEE Transactions on Fuzzy Systems 13(5), 644–658 (2005)
17. Herrera-Viedma, E., Mata, F.S., Martínez, L., Chiclana, F., Pérez, L.G.: Measurements of consensus in multi-granular linguistic group decision-making. In: Torra, V., Narukawa, Y. (eds.) MDAI 2004. LNCS (LNAI), vol. 3131, pp. 194–204. Springer, Heidelberg (2004)

18. Kacprzyk, J.: Group decision making with a fuzzy linguistic majority. Fuzzy Sets and Systems 18, 105–118 (1986)
19. Kacprzyk, J., Fedrizzi, M., Nurmi, H.: Consensus Under Fuzziness. Kluwer Academic Publishers, Dordrecht (1997)
20. Kacprzyk, J., Zadrozny, S.: An internet-based group decision support system. Management VII 28, 2–10 (2003)
21. Kim, S., Choi, S., Kim, J.: An interactive procedure for multiple attribute group decision making with incomplete information: Range-based approach. European Journal of Operational Research 118, 139–152 (1999)
22. Kuncheva, L.: Five measures of consensus in group decision making using fuzzy sets. International Conference on Fuzzy Sets and Applications IFSA 95, 141–144 (1991)
23. Kundu, S.: Min-transitivity of fuzzy leftness relationship and its application to decision making. Fuzzy Sets and Systems 86, 357–367 (1997)
24. Mata, F., Martínez, L., Herrera-Viedma, E.: An adaptive consensus support system model for group decision-making problems in a multigranular fuzzy linguistic context. IEEE Transactions on Fuzzy Systems 17(2), 279–290 (2009)
25. Mata, F., Martínez, L., Martínez, J.: A preliminary study of the effects of different aggregation operators on consensus processes. In: Ninth International Conference on Intelligent Systems Design and Applications, Pisa, Italy, pp. 821–826 (2009)
26. Roubens, M.: Fuzzy sets and decision analysis. Fuzzy Sets and Systems 90, 199–206 (1997)
27. Saint, S., Lawson, J.R.: Rules for Reaching Consensus. A Modern Approach to Decision Making, Jossey-Bass (1994)
28. Szmidt, E., Kacprzyk, J.: A consensus reaching process under intuitionistic fuzzy preference relations. International Journal of Intelligent System 18(7), 837–852 (2003)
29. Tanino, T.: On Group Decision Making Under Fuzzy Preferences. In: Multiperson Decision Making Using Fuzzy Sets and Possibility Theory, pp. 172–185. Kluwer Academic Publishers, Dordrecht (1990)
30. Téno, J.L., Mareschal, B.: An interval version of PROMETHEE for the comparison of building products' design with ill-defined data on environmental quality. European Journal of Operational Research 109, 522–529 (1998)
31. Xu, Z.: An automatic approach to reaching consensus in multiple attribute group decision making. Computers and Industrial Engineering 56, 1369–1374 (2009)
32. Yager, R.: An approach to ordinal decision making. International Journal of Approximate Reasoning 12, 237–261 (1995)
33. Yager, R.: Protocol for Negotiations among Multiple Intelligent Agents. In: Consensus Under Fuzziness, pp. 165–174. Kluwer Academic Publishers, Dordrecht (1997)
34. Zadrozny, S.: An approach to the consensus reaching support in fuzzy environment. In: Consensus under fuzziness, pp. 83–109. Kluwer Academic Publishers, Dordrecht (1997)

A Fuzzy Hierarchical Multiple Criteria Group Decision Support System – Decider – and Its Applications

Jun Ma, Guangquan Zhang, and Jie Lu

Abstract. Decider is a Fuzzy Hierarchical Multiple Criteria Group Decision Support System (FHMC-GDSS) designed for dealing with subjective, in particular linguistic, information and objective information simultaneously to support group decision making particularly on evaluation. In this chapter, the fuzzy aggregation decision model, functions and structure of Decider are introduced. The ideas to resolve decision and evaluation problems we have faced in the development and application of Decider are presented. Two real applications of the Decider system are briefly illustrated. Finally, we discuss our further research in this area.

1 Introduction

Decision making is complex. An appropriate decision is often made by a group person in terms of several evaluation criteria. On the one hand, the rapidly increasing amount of data (information) provides necessary decision support, but at the same time, brings difficulties to appropriate decision making due to the reduced quality. Except for this reason, a decision maker's personal experience and knowledge in related fields are restricted. Hence, individual decision making often deviates from

Jun Ma
Centre of QCIS, Faculty of Engineering and Information Technology,
University of Technology, Sydney (UTS)
e-mail: junm@it.uts.edu.au

Guangquan Zhang
Centre of QCIS, Faculty of Engineering and Information Technology,
University of Technology, Sydney (UTS)
e-mail: zhangg@it.uts.edu.au

Jie Lu
Centre of QCIS, Faculty of Engineering and Information Technology,
University of Technology, Sydney (UTS)
e-mail: jielu@it.uts.edu.au

E. Herrera-Viedma et al. (Eds.): Consensual Processes, STUDFUZZ 267, pp. 383–403.
springerlink.com   © Springer-Verlag Berlin Heidelberg 2011

the appropriate one. Group decision making can redeem this deviation to some extent. On the other hand, an appropriate decision should be a result after deliberately synthesizing many related aspects of a decision problem. A decision just focusing a single aspect of a problem is dangerous in real application. Multi-criteria decision making (MCDM) can reduce the danger through consideration of a set, usually conflicting, of criteria simultaneously.

Multi-criteria group decision-making (MCGDM), which combines MCDM and GDM methods, has been proved to be a very effective technique to increase the degree of overall satisfaction for the final decision across the group [7], and is particularly suitable in problems such as quality evaluation, policy selection, employee nomination, and designing assessment [4, 19]. These problems have some common features. For example, evaluation criteria are often in a multiple-level hierarchy; evaluators are from different departments; assessments are expressed in various forms. Traditional decision support systems can efficiently help decision makers resolve some of those problems. However, because they are mainly data-centred, they have obvious limitation to deal with subjective data which is a primary representation of assessments from evaluators. Therefore, how to efficiently deal with subjective information becomes an crucial issue in developing a real application of an MCGDM decision support system [6].

In practice, subjective data is often expressed by natural or artificial language, such as linguistic terms. Since Fuzzy sets technique is proved in practices that it is a powerful tool to handle subjective information, that combining fuzzy sets technique with MCGDM technique and studying FMCGDM technique is necessary and possible. In our opinion, FMCGDM technique is an important basis for developing people-centred intelligent systems including decision support. Based on aforementioned analysis and our related work, we developed the Fuzzy Hierarchical Multiple Criteria Group Decision Support System (FHMC-GDSS), named Decider, as a platform to test developed FMCGDM process algorithms. We also successfully applied this system to resolve subjective information process problems in industry applications. In this chapter, we will briefly introduce the main modules and the main functions of this system, and two of its applications.

The remaining sections in this chapter are organized as follows: Section 2 gives a simple overview of related research works on MCGDM and FMCGDM techniques and, then, lists used concepts and notions in the following sections; In Section 3 we will introduce the structure of the Decider system and its functions and implementation; Next, Section 4 illustrates two applications of the Decider system and gives a short analysis; Finally, we conclude the chapter and presents our further research.

2   Related Works

Multi-criteria group decision making (MCGDM) is widely-used in various fields including managements [19], industry [20], social sciences [3], highway infrastructure management [23], spatial data processing [8], and urban water supply [10]. Techniques such as the Analytic Hierarchy Process (AHP) and evolutionary complutation have been applied in MCGDM [11, 22]. In practice, MCGDM is conducted in complicated context with heterogeneous information sources [18]. The collected information is often in two primary forms: subjective and objective. Effective process models and methods for integrating heterogeneous information are required.

Subjective information is often expressed by linguistic terms in real applications. Linguistic methods are typical techniques to integrate subjective information [2, 5, 4]. The core idea of existing linguistic methods is to develop an approximate aggregation operator to integrate linguistic information [16, 15, 24]. Because fuzzy set is the most used representation form of a linguistic term, most aggregation operators are established on fuzzy sets technique. Hundreds of aggregation operators have been developed and applied [1, 17]. However, existing linguistic methods only focus on linguistic terms process and pay little attention on objective information. In real applications, objective information is often some accurate measurements by means of devices and equipments with specific meanings and has special process requirements. Hence, it is necessary to establish information aggregation for subjective and objective information simultaneously.

In 2007, a fuzzy MCGDM decision algorithm was developed and implemented in an FMCGDM system [12] in our lab. Since then, some applications have been developed during collaboration with other researchers. Real applications indicated the great interested in such a decision support system and also presented more concrete and essential requirements. Based on the applications and their feedbacks, an expansion of that FMCGDM system is designed and named Decider, which is used as a testing and analysis platform of MCGDM algorithms and models. Since 2008, this system have been partly implemented and some planned functions of it have been readjusted based on requirements in applications. Section 3 will give more details of the Decider system.

Before introducing the Decider system, we give some basic definitions about fuzzy numbers and fuzzy algorithms which will be used in the following sections.

Definition 1 (Fuzzy set). A fuzzy set $\tilde{A}$ in a universe of discourse X is characterized by a membership function $\mu_{\tilde{A}}(x)$ which associates with each example x in X a real number in the real interval $[0, 1]$.

The function value $\mu_{\tilde{A}}(x)$ is called the membership degree of x belonging to $\tilde{A}$.

Definition 2 (Cut set). The λ-cut set of a fuzzy set $\tilde{A}$ is defined by

$$\tilde{A}_\lambda = \{x \in X | \mu_{\tilde{A}}(x) \geqslant \lambda\} \tag{1}$$

where $\lambda \in [0, 1]$ is a real number.

If $\tilde{A}_\lambda$ is a non-empty bounded closed interval in X, then it can be denoted by $\tilde{A}_\lambda = [\tilde{A}_\lambda^L, \tilde{A}_\lambda^R]$, where $\tilde{A}_\lambda^L$ and $\tilde{A}_\lambda^R$ are the lower and upper end points of the closed interval.

Definition 3 (Fuzzy number). [9] A fuzzy set $\tilde{a}$ on $\mathbb{R}$ is called a fuzzy number, if $\tilde{a}$ satisfies:

(1) $\tilde{a}$ is a normal fuzzy set, i.e. $\tilde{a}_1$ is not empty;
(2) $\tilde{a}_\lambda$ is a closed interval for any $\lambda \in (0,1]$;
(3) the support of $\tilde{a}$, $\tilde{a}_{0+}$ is bounded.

In the following, the set of fuzzy numbers on X is denoted by $\mathscr{F}(X)$.

Definition 4 (Basic Algorithms). For any $\tilde{a}, \tilde{b} \in \mathscr{F}(\mathbb{R}^+)$ and $\alpha \in \mathbb{R}$, let

$$\tilde{a} \oplus \tilde{b} = \bigcup_{\lambda \in (0,1]} \lambda \left[\tilde{a}_\lambda^L + \tilde{b}_\lambda^L, \tilde{a}_\lambda^R + \tilde{b}_\lambda^R \right],$$

$$\alpha \tilde{a} = \bigcup_{\lambda \in (0,1]} \lambda \left[\alpha \tilde{a}_\lambda^L, \alpha \tilde{a}_\lambda^R \right],$$

$$\tilde{a} \otimes \tilde{b} = \bigcup_{\lambda \in (0,1]} \lambda \left[\tilde{a}_\lambda^L \times \tilde{b}_\lambda^L, \tilde{a}_\lambda^R \times \tilde{b}_\lambda^R \right].$$

In Definition 4, we use $\oplus$ and $\otimes$ to replace $+$ and $\times$ in conventional definitions in order to emphasize that the algorithm is applied to fuzzy numbers.

Definition 5 (Triangular fuzzy number). A triangular fuzzy number $\tilde{a}$ is defined by a triplet (a_0^L, a, a_0^R) and the membership function $\mu_{\tilde{a}}(x)$ is given

$$\mu_{\tilde{a}}(x) = \begin{cases} 0, & x < a_0^L \\ \dfrac{x - a_0^L}{a - a_0^L}, & a_0^L \leqslant x < a \\ \dfrac{a_0^R - x}{a_0^R - a}, & a \leqslant x \leqslant a_0^R \\ 0, & a_0^R < x \end{cases} \tag{2}$$

where $a = a_1^R = a_1^L$.

Definition 6 (Normalized positive fuzzy number). A fuzzy number $\tilde{a}$ is called a normalized positive fuzzy number if $0 < a_\lambda^L \leqslant a_\lambda^R \leqslant 1$ for any $\lambda \in (0,1]$.

Definition 7 (Quasi-distance). Let $\tilde{a}, \tilde{b} \in \mathscr{F}(\mathbb{R})$ be two normalized positive fuzzy numbers, the quasi-distance of $\tilde{a}$ and $\tilde{b}$ is

$$d(\tilde{a}, \tilde{b}) = \left(\int_0^1 \frac{1}{2} \left[\left(\tilde{a}_\lambda^L - \tilde{b}_\lambda^L \right)^2 + \left(\tilde{a}_\lambda^R - \tilde{b}_\lambda^R \right)^2 \right] d\lambda \right)^{1/2}. \tag{3}$$

3   Decider: A Fuzzy Hierarchical Multiple Criteria Decision Support System

This section introduces the design and implementation of the Decider system. As an expansion of our previous FMCGDM system, Decider also implemented the fuzzy

MCGDM algorithm in that system. Besides, it also implemented other functions. At the beginning of its redesign, Decider is a testing and analysis platform of different MCGDM algorithms and models. However, with the progress in collaborative application developments, new requirements were concerned and new functions were added to this system.

Decider has some features to meet the demands of real applications.

(1) Decider is a cross-platform system. In applications, we noticed that some processes are not conducted on Windows operating system. Hence, when we redesigned the Decider system, we selected the Java programming language as developing tool and developed Decider on both MS Windows and Linux operating systems.

(2) Decider extents the hierarchies for criteria and evaluators. In our previous FMCGDM system, the level of criteria is restricted due to the limitation of used data structure, and only one level of evaluators was permitted. Considering the application requirements, we extended the hierarchies of criteria and evaluators in tree-like structures. Moreover, we have designed an information source level in order to represent network structure in criteria.

(3) Decider deals with subjective and objective simultaneously. It uses fuzzy numbers to represent subjective information such as linguistic terms, and applies a fuzzfication algorithm to convert objective information to subjective information.

The main components of Decider are shown in Fig. 1. Decider includes four basic modules, i.e., problem input, decision (MCGDM) process, decision display, and analysis/comparison. Users set the impacts and relationships/organizations of criteria, evaluators, alternatives, and other decision related information through "problem input" module. This information is then sent to the "decision process" module to generate decision result. The decision result is shown to users through the "decision play" module. Further, users can adjust decision parameters and process models by means of "analysis/comparison" module to check the change. Information between users and the Decider system forms two process circles called the basic process and the analysis process respectively.

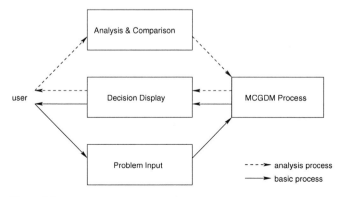

Fig. 1 Decider architecture

3.1   Decision Information Input

In the "problem input" module, users mainly set two kinds of decision information. The first kind of decision information is called basic information which is directly collected for a decision problem. Basic information includes the relationships among criteria and their impacts (or weights, support degrees) to a decision problem; the organizations of evaluators and their impacts (or weights, reliabilities) to the decision problem; the assessments for each of alternatives (decision); and the information aggregation (fusion) strategy for the decision problem. The second kind of information is called conversion information which is taken from knowledge related to a decision problem. This information includes the relative distribution of assessments; the corresponding relations between subjective and objective assessments; and the cost/profit feature of criteria. The Decider system partially implements process for above information.

The most important basic information is the structure of criteria and their impacts. In the Decider system, the criteria related to a decision problem are organized in a multi-level hierarchy named criteria tree. The criteria tree is established through a cause-and-effect problem analysis from the general decision problem down to the detailed indicators. In the criteria tree, nodes except the root node are called criteria. In particular, leaf nodes of the criteria tree are called indicators; the children nodes of the root nodes are called aspects; and the rest criteria are called factors. The root node of the criteria tree represents the decision problem or decision goal/target derived from it; the aspects are general considerations which support the final decision; the indicators are detailed considerations on which assessments about alternatives are collected directly; and the factors illustrate the knowledge from indicators to final decision. Fig. 2 shows an example of a typical criteria tree in Decider.

Similar to the criteria tree, the Decider system uses an evaluator tree to represent organization of evaluators. There are two kinds of evaluators, i.e., the real evaluators and the virtual evaluators. A real evaluator refers to a person (expert) or a device, which provides assessments on alternatives directly. A virtual evaluator represents an evaluator group, a set of devices, or combination of human evaluators or devices. It is corresponding to a department or an assembly line in real applications.

Another kind of basic information is the impacts of criteria and evaluators on the decision problem and the assessments on alternatives. The Decider system provides two kinds of representations for above information, i.e., the subjective linguistic terms and the objective numeric values. Users can assign linguistic weights such as "Important" or numeric grade such as "4" to a criterion (or an evaluator) to describe its impact on the decision problem. Users can also input linguistic assessments such as "Very high" or numeric value such as "35.2" as assessment about an alternative. The used subjective and objective representations are listed in Table 1.

The conversion information is related to the processing method and derived from knowledge of the decision problem. Users can determine the corresponding relation between different information representation forms and select a process model from provided process models in the Decider system. As this information is closely related to decision process, details of it will be introduced in Section 3.2.

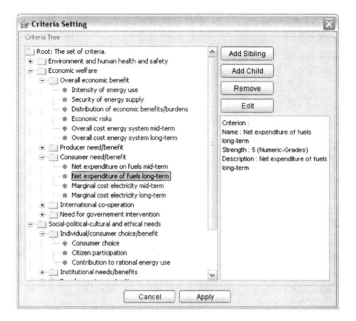

Fig. 2 An example of criteria tree.

Table 1 Information representation terms/values.

Types	Named expression	Terms/values	Applied to
Subjective	Standard Score (SS)	0, 1, …, 100	criteria evaluators assessments
	Linguistic Weights (LW)	Absolutely unimportant (AU) Unimportant (U) Less important (LI) Important (I) More important (MI) Strongly important (SI) Absolutely important (AI)	criteria evaluators
	Linguistic Scores (LS)	Lowest (LE) Very low (VL) Low (L), Medium (M) High (H), Very high (VH) highest (HE)	assessments criteria evaluators
	Numeric Grade (NG)	1, 2, 3, 4, 5, 6, 7	criteria evaluators assessments
Objective	Range (R)	User defined interval of real numbers	assessments
Boolean		True (T), False (F)	assessments

3.2 Decision Process

Based on user selected process model and information about criteria, evaluators, and assessments, a basic process circle can be implemented. The main decision procedure is conducted in the "decision process" module. Decision process mainly implements the information conversion and the selected process model.

3.2.1 Information Conversion

Information conversion conducts two kinds of information transformation. First, it determines the corresponding between different kinds of information representation forms. Second, it determines relative distributions of those terms or values. Concretely, suppose S and T are two sets of terms/values used in a decision process. The first transformation determines mapping between S and T. Hence, for any $s_i \in S$, a term t_j in T is given as its corresponding and vice versa. Considering two representation forms may have different number of terms/values, a common reference is used in Decider. This common reference is the real interval [0, 1]. All six information representation forms in Table 1 are mapped into this interval. The second transformation determines the relative distribution of terms in a specific representation form. For instance, "Linguistic Weights (LW)" is composed of seven terms. Adjusting the mapping from LW to the interval [0, 1], the images of the seven terms determine the relative relationship among them. The two transformations are supported in the current Decider system. However, for conversion consistency purpose, i.e., one transformation does not been changed by the other, the Decider system takes different strategies to convert subjective and objective information.

For subjective representations. A natural order is defined on SS, LW, LS, and NG for subjective terms/values such that

$$SS: \qquad 0 < 1 < 2 < \cdots < 100 \tag{4}$$

$$LW: \quad AU < U < LI < I < MI < SI < AI \tag{5}$$

$$LS: LE < VL < L < M < H < VH < HE \tag{6}$$

$$NG: \qquad 1 < 2 < 3 < 4 < 5 < 6 < 7. \tag{7}$$

Then we map SS to [0, 1] such that

$$x \mapsto x/100, \quad x \in SS \tag{8}$$

and map LS, LW, NG to SS based on "equal distance" or "equal ratio" settings and some initial assignments. Take LW for example. Suppose user assigns 0 to AU, and 90 to AI, and expects the rest terms to be placed by "equal distance". Then the terms U, LI, I, MI, and SI will be assigned 15, 30, 45, 60, and 75 respectively. If the user wants the rest terms to be placed by "equal ratio", and expects the ratio γ is 3.0. Then based on this requirement and the system defined distribution function $s_{j+1} - s_j = \gamma \cdot (s_j - s_{j-1})$, where s_{j-1}, s_j, and s_{j+1} are sequential terms in LW, the terms U, LI, I, MI, and SI will be given 0, 1, 3, 9, and 27 respectively. If the user adds one more assignment 60 to I, then the terms U, LI, MI, and SI will take 20, 40, 70, and 80 under the "equal distiance" setting and 5, 18, 62, and 69 under the "equal ratio" setting with ratio $\gamma = 3.0$.

For objective representation. The Decider system converts directly the value to a fuzzy set on the real interval [0, 100]. Decider requires users to determine three parameters to convert an objective value, i.e., the lower bound (a), the upper bound (b), and the preferred value (p). Users also need to determine the interpretation of the preferred value from three interpretations: threshold value (T), medium value (M), and expected value (E). On the interpretation of the preferred value, Decider defines four kinds of orders for labeling an objective representation:

(1) the larger the better (O1);
(2) the smaller the better (O2);
(3) the near the better (O3);
(4) the farther the better (O4).

Therefore, an objective representation can be converted to a fuzzy set based on the combination of labelling order and interpretation of the preferred value. A linear transformation method is introduced in [14]. For instance, if the labelling order is O3 and the preferred value p is interpreted as expected value (E), then Decider assigns 100 to p and

$$
x \mapsto \begin{cases} \lfloor \dfrac{x-a}{p-a} \cdot 100 \rfloor, & x \in [a,p] \\ \lfloor \dfrac{b-x}{b-p} \cdot 100 \rfloor, & x \in [p,b]. \end{cases} \tag{9}
$$

For Boolean values, Decider will assign 100 to the value "True" and 0 to "False" by default. Under order reverse requirement, Decider will exchange the assignment of "True" and "False".

By the information conversion, information in different representation forms is mapped to SS. Next terms in SS will be mapped to normal positive fuzzy numbers on the interval [0, 1] in one of provided forms (triangular, quadratic, and exponential forms). For example, suppose x is the image of term t in SS, then the triangular form is given by

$$
\mu_x(y) = \begin{cases} \dfrac{y-x}{a}, & y \in [x-a,x], \\ \dfrac{y-x}{b}, & y \in [x,x+b], \end{cases} \tag{10}
$$

where a, b are two parameters to be determined by users.

3.2.2 Fuzzy Information Aggregation

The core problem in the "decision process" module is the implementation of MCGDM process models and algorithms. Developing an aggregation method is the core technique in most MCGDM process models and algorithms. The Decider system implements a fuzzy aggregation algorithm as shown below. The fuzzy aggregation algorithm is applied to the criteria tree and the evaluator tree at the same time.

Suppose c is a criterion in the criteria tree, $\tilde{v}_1$, $\tilde{v}_2$, $\tilde{v}_n$ are the assessments on its children nodes c_1, c_2, ..., c_n, and wc_1, wc_2, ..., wc_n are the impacts of its children nodes to c. Here, $\tilde{v}_j$ and wc_j $(j = 1, 2, \ldots, n)$ are normalized positive fuzzy numbers on the real interval [0, 1]. Then the assessment $\tilde{v}$ on c is obtained by

$$\tilde{v} = \sum_{j=1}^{n} \widetilde{wc}_j \otimes \tilde{v}_j \qquad (11)$$

where $\widetilde{wc}_j$ is the normalized impact for wc_j given by

$$\widetilde{wc}_j = \frac{wc_j}{\sum_{j}^{n} (wc_j)_0^R}, \quad j = 1, \ldots, n, \qquad (12)$$

where $(wc_j)_0^R$ is the right-end point of the 0-cut of wc_j.

Suppose g is a virtual evaluator in the evaluator tree and g_1, g_2, ..., g_m are its group members. Let $\tilde{u}_1$, $\tilde{u}_2$, ..., $\tilde{u}_m$ be the assessments from the m group members and we_1, we_2, ..., we_m be their impacts on the group assessment of g. Therefore, the group assessment $\tilde{u}$ is obtained by

$$\tilde{u} = \sum_{i=1}^{m} \widetilde{we}_i \otimes \tilde{u}_i \qquad (13)$$

where $\widetilde{we}_i$ is the normalized impact for we_i given by

$$\widetilde{we}_i = \frac{we_i}{\sum_{i}^{m} (we_i)_0^R}, \quad i = 1, \ldots, m, \qquad (14)$$

where $(we_i)_0^R$ is the right-end point of the 0-cut of we_i.

Notice that Eq. (11) and Eq. (13) is the same form, the order of aggregation on the criteria tree and the evaluator tree is not affecting. Hence, we can exchange the aggregation order.

The obtained assessment on the root node of the criteria tree and the evaluator tree will be used to generate the final decision. Suppose $\tilde{v}^{(i)}$ is the final assessment on the root node of the critera tree and the evaluator tree for the alternative a_i, $i = 1, 2, \ldots, k$. To generate the decision, the Decider system will compare the assessment $\tilde{v}^{(i)}$ with two predefined ideal assessments $\tilde{v}^{(-)}$ and $\tilde{v}^{(+)}$ representing the worst and the best situations. $\tilde{v}^{(-)}$ and $\tilde{v}^{(+)}$ are two special normalized positive fuzzy numbers on the real interval [0, 1] defined as follows:

$$\tilde{v}^{(-)} = \begin{cases} 1, & x = 0 \\ 0, & x \neq 0 \end{cases}, \quad \tilde{v}^{(+)} = \begin{cases} 1, & x = 1 \\ 0, & x \neq 1 \end{cases}. \qquad (15)$$

The comparison result of the assessment $\tilde{v}^{(i)}$ and these two ideal assessments is a reference distance d_i defined by

$$d_i = \frac{1}{2}\left(d(\tilde{v}^{(i)}, \tilde{v}^{(-)}) + (1 - d(\tilde{v}^{(i)}, \tilde{v}^{(+)}))\right), \tag{16}$$

where $d(\tilde{a}, \tilde{b})$ is a quasi fuzzy distance defined in Definition 7. The reference distance d_i is the final standard of genreating the decision. In most applications, for example profit related decisions, the decision is generated by the principle "the bigger the better". While in some applications, for instance cost related decisions, a decision is generated based on the reversion of above principle.

3.3 Decision Output

Users can observe the generated result through the "decision display" module. This module displays the obtained reference distances of all alternatives in colour bars. Fig. 3 displays a snapshot of an example output. In this snapshot, area 1 displays the criteria tree; area 2 displays the evaluator tree; and area 3 displays the reference distance value of each alternative. The alternative with the biggest reference distance value is displayed in red. In real decision problem, this alternative is the best choice in general. Through graphic display, users can visually observe and compare these results.

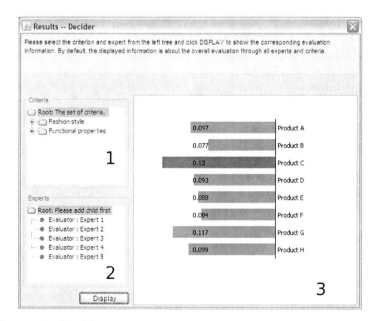

Fig. 3 Example of decision output

By default, Decider will display the final results based on the root combination of the criteria tree and the evaluator tree. Considering users' different interests in a specific criterion or group, Decider can display the results accordingly. For instance, users can observe the results of all evaluators' synthesized assessment on specific criteria by selecting that criterion (from area 1 in Fig. 3) and the root of the evaluator tree (from area 2 in Fig. 3); users can observe a group's synthesized assessment on the whole problem by selecting that group (a virtual evaluator) and the root of the criteria tree. Users, therefore, can generate a whole picture of the solution to the problem.

3.4 Analysis and Comparison

The procedure from users setting a decision problem, selecting processing model, to observing the decision results forms the first information process circle (the basic circle). The basic circle provides users with a general observation of the solution to the problem. If users need more observations, Decider provides the "analysis and comparison" module to help for it. This module is schemed but has not been completely implemented yet.

In the scheme, this module is going to implement three functions. The first one is sensitivity analysis. Decider will analyse the result change after adjusting the impacts of criteria and evaluators, and the assessments. The second one is decision models comparison. Decider will compare the result change after altering the aggregation operator and the process strategy of a used model and selecting different models. The third one is information representation comparison. Decider will compare the result change after changing the information representation forms.

Decider currently implements most of above functions in a limited way. For instance, it provides different decision models which have different requirements on information representation, aggregation operator, and decision strategy. Users can repeat the basic circle to conduct the listed analysis and comparison.

4 Examples of Applications

This section will introduce three applications we have conducted using the Decider system. Detailed case studies can be found in the references [13, 21]. Here, we just focus on the particular requirement in those applications.

4.1 Long-term Scenarios of Belgian Energy Policy

Long-term sustainable development is an issue of common concern in the whole world from developing countries to development countries. It involves problems in diversified forms. These problems include such as public health and security, environment protection and climate changes, as well as energy managements strategy. Formulating a long-term sustainable development policy is a typical multi-criteria group decision problem [19]. Since 2007, we have collaborated with researchers from Belgian Nuclear Research Centre to develop relevant knowledge and techniques for designing, evaluating, and selecting long-term energy strategy. This section gives a brief illustration of applying the Decider system to this problem.

The Long-term Belgian Energy Policy Evaluation (LBEPE) problem involves 63 criteria in 4 aspects, 16 factors and 43 indicators. Table 2 lists the relationship of the aspects, factors, and indicators. Based on Table 2, eight scenarios as alternatives are determined and named "MLCS," "MPCS," "MPLCS," "MPLCSI," "RLCS," "RPCS," "RPLCS," and "RPLCSI." [1]

According to these criteria, we collected evaluation data from 10 experts. The collected data has three features. Due to the complexity of the problem, experts' views are expressed in linguistic terms in the evaluation procedure. Due to the lack of relevant knowledge, some views are not presented. Due to the criteria nature, some views using same term are with opposite meanings.

After necessary data pre-processing, experiments applying the Decider system to this problem are conducted. The evaluation results from the expert group and an individual are shown in Fig. 4 and Fig. 5. The values displayed on bars in Fig. 4 and Fig. 5 are relevant reference distances of those alternative. The bigger the value is, the bigger the corresponding reference distance is. From these figures, the difference and similarity between group and individual evaluations is easy to observe. For instance, from the group's viewpoint, the scenario "RPCS" is the best one; while expert e1 takes the scenario "RPLCS" as the best one. At the same time, they both think that the scenario "MLCS" is the worst.

4.2   Well-being New Product Development

We have also applied the Decider system to the well-being design theme evaluation of garment new product development in an establishing digital ecosystem. Digital ecosystem was presented in the FP6 framework programme and is continued as a research hotspot in the FP7 framework programme. Under its definition, the user-centred product and service development is a primary research activity for garment enterprises in their product development procedure. These enterprises need to evaluate their product design before launch a new product to a competitive market. This evaluation procedure must combine human actions and cognition with their manufacturing.

Well-being is a design theme with both psychological and sociological features. That evaluating a new product design is appropriate or inappropriate for the

---

[1] For each scenario, the initial letter "M" and "R" represent the Market world and the Rational Perspective world; and the following letters indicate possible energy strategies: P - nuclear phase out; LCS - low carbon capture and storage; I - import of electricity. For example, the scenario "MPLCS" is read: using the Market assumptions and assuming a nuclear phase out, no import and low potential for carbon capture and storage. In such a scenario, investments on renewable and cogeneration is therefore necessary. Detailed explanation of these scenarios is referred to [21].

Table 2 Criteria of the long-term Belgian energy policy evaluation

Environment & human health and safety	
Air pollution	- Impacts of air pollution on human health: mid-term
	- Impacts of air pollution on human health: long-term
Occupational health	- Impacts on occupational health (gas + coal)
Radiological health impacts	- Radiological health impacts (nuclear)
	- Need for long-term management of HLW
Aesthetic impacts	- Visual impact on landscape
	- Noise amenity
Other environmental impacts	- Impact on natural ecosystem (air pollution): mid-term
	- Impact on natural esosystem (air pollution): long-term
	- Environmental impact from solid waste (coal)
Resource use	- Land use
	- Water use
Other energy related pressures	- Catastrophic risk: nuclear
	- Geographical distribution risk/benefits
Economic welfare	
Overall economic benefit	- Intensity of energy use
	- Security of energy supply
	- Distribution of economic benefits / burdens
	- Ecosuppressnomic risks
	- Overall cost energy system: 2010
	- Overall cost energy system: 2030
Producer need/benefit	- Overall cost energy system: 2050
	- Ability to provide specialist market
	- Marginal cost electricity: mid-term
Consumer need/benefit	- Marginal cost electricity: long-term
International cooperation	- Strategic factors for export
	- Compatibility with international R&D agenda
Need for government intervention	- Amount of direct or indirect subsidies needed
Social, political, cultural and ethical needs	
Individual/consumer choice/benefit	- Consumer choice
	- Citizen participation
	- Contribution to rational energy use
Institutional needs	- Degree of decentralisation
	- Need for intermediary storage of spent fuel
	- Control and concentration of power
	- Influence on political decision-making
	- Need for socio-political stability
	- Need for direct political intervention
	- Reversibility of technology choice
	- Knowledge specialisation
	- Need for institutional non-proliferation measures
Development opportunities	- Potential for technology transfer
	- Leaving resources for development
	- Equity (general)
Jobs	- Job opportunities
Diversification	

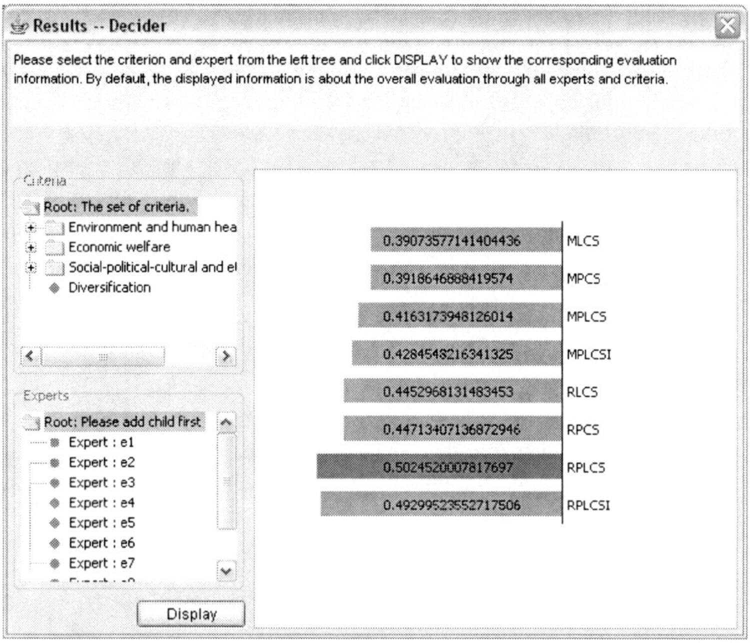

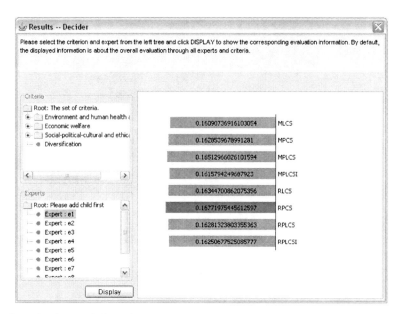

Fig. 4 Evaluation result from the expert group.

Fig. 5 Evaluation result from the expert 1.

well-being theme involves not only subjective assessments from customers and experts but also objective measurements in the manufacturing procedure simultaneously. The criteria tree is given in Table 3. Observing the criteria tree, we know that some criteria are with objective information form. Data for these criteria is generally not provided by human beings but by specific devices.

Based on the criteria, a survey was conducted among garment experts and customers. We collected assessments from these people on all subjective indicators and measurements from devices on all objective indicators. Considering that all specific devices from which the objective data is collected coexist in a whole manufacturing procedure and affect a product together, we set a virtual evaluator to represent all specific devices. Then, applying the Decider system, we evaluated eight product designs. The evaluation results are shown in Fig. 6 – Fig. 8. Comparing these results, we can observe that the product C is the one which is appropriate for the well-being theme. Because both human evaluators and the devices reach the same assessment, the manufacturing setting for product C need not to be adjusted. However, for the products B and D, the human evaluators and the devices give different assessments. Hence, the manufacturing settings for them should be adjusted.

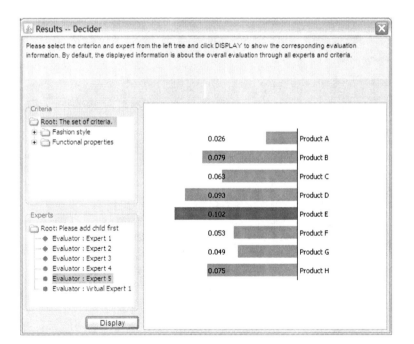

Fig. 6 Evaluation result on well-being design by an individual human evaluator.

Table 3 Criteria tree of Well-being design theme evaluation problem.

Criteria name	Information form
Fashion stype	Subjective (S)
Protection	S
Health	S
Pleasure	S
Serenity	S
Relaxation	S
Cocooning	S
Warmth	S
Health	S
Sport	S
Serenity	S
Pleasure	S
Holiday	S
Relaxation	S
Cocooning	S
Dynamism	S
Sport	S
Pleasure	S
Relaxation	S
Coolness	S
Health	S
Pleasure	S
Holiday	S
Relaxation	S
Functional properties	S
Fabric handle	S
Extensibility	Objective (O)
Density	O
Compressibility	O
Flexibility	O
Surface friction	O
Resilience	O
Surface contour	O
Thermal-Wet sensation	O
Smell	O
Sound	O
Wash and care	S
Wash requirement	S
Iron requirement	S
Storage requirement	S
Durability	O

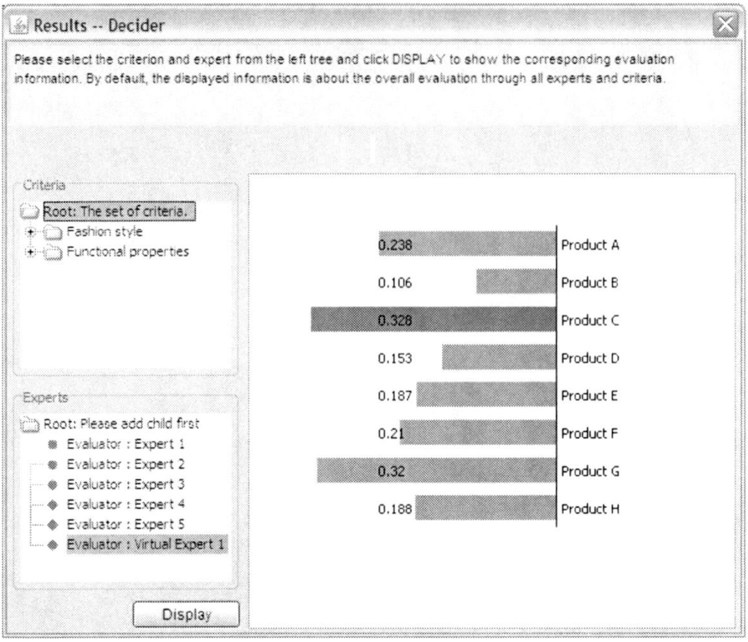

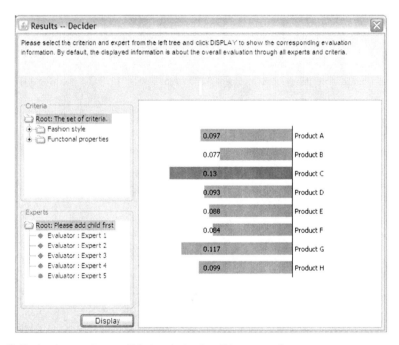

Fig. 7 Evaluation result on well-being design by devices.

Fig. 8 Evaluation result on well-being design by all human evaluators.

4.3   Further Discussion

Practices of applying the Decider system to many real applications indicate that developing such a decision support system is necessary and urgent. Different from traditional decision support system, Decider can partly handle subjective information. We have noticed during developing above applications that one of important factor which affects appropriate decision is the subjective information process. This is because that human is the centre of a decision problem and communication is a main channel of subjective information. Hence, establishing people-centred decision support system is an efficient way to improve decision quality. That is a main reason for us developing such a system.

Another important point we found in those applications is that establishing the connection between objective and subjective information is an implied way for transferring knowledge and perception. In manufacturing, experts' and customers' knowledge and perception cannot be used directly. Manufacturing is measured by devices, i.e. by objective data; while knowledge and perception is expressed by subjective information. To produce products satisfy customers' perception, transferring those perception to objective data is necessary and crucial.

Another starting point of our developing the Decider system is establishing a testing platform of decision algorithms and models. Since the requirements of real applications vary from one to another, it is possible to verify which algorithm or model is suitable to which situation and requirement. This work is very helpful for both theoretical research and industrial application. More work need to be done not only for improving the developed system but also for advancing related research.

5   Conclusions

In this chapter, we introduced a fuzzy hierarchical multiple criteria group decision support system – Decider. We overviewed the structure and function of the Decider system and its applications on two typical multiple criteria decision problems. Through developing real applications, we concluded that 1) the Decider system can partly resolve subjective information process in decision problems; but more work needs to be done to improve the developed system; 2) developing and deploying people-centred decision support systems is necessary in real application fields; 3) advancing research on human decision making with subjective information will help establish people-centred decision support. These issues are also our future works.

Acknowledgements. The work presented in this paper was supported by Australian Research Council (ARC) under Discovery Project DP0880739.

References

1. T. Calvo, G. Mayor, R. Mesiar.: Aggregation Operators: New Trends and Applications. Number 97 in Studies in Fuzziness and Soft Computing. Physica-Verlag, Springer, Heidelberg (2002)
2. Delgado, M., Herrera, F., Herrera–Viedma, E., Verdegay, J.L., Vila, M.A.: Aggregation of linguistic information based on a symbolic approach. In: Zadeh, L.A., Kacpryzk, J. (eds.) Computing with Words in Information/Intelligent Systems I. Foundations, pp. 428–440. Physica–Verlag (1999)
3. Grabisch, M., Labreuche, C.: Fuzzy Measures and Integrals in MCDA. In: Figueira, J., Greco, S., Ehrgott, M. (eds.) Multiple Criteria Decision Analysis: State of the Art Surveys. International Series in Operations Research & Management Science, vol. 78, pp. 563–604. Springer, New York (2005)
4. Herrera, F., Herrera-Viedma, E., Chiclana, F.: Multiperson decision–making based on multiplicative preference relations. European J. Operational Research 129, 372–385 (2001)
5. Herrera, F., Lopez, E., Rodriguez, M.A.: A linguistic decision model for promotion mix management solved with genetic algorithms. Fuzzy Sets and Systems 131, 47–61 (2002)
6. Herrera-Viedma, E., Chiclana, F., Herrera, F., Alonso, S.: Group decision-making model with incomplete fuzzy preference relations based on additive consistency. IEEESMCB 37(1), 176–189 (2007)
7. Huynh, V.-N., Nakamori, Y.: A satisfactory-oriented approach to multiexpert decision-making with linguistic assessments. IEEESMCB 35(2), 184–196 (2005)
8. Ascough II, J.C., Rector, H.D., Hoag, D.L., McMaster, G.S., Vandenberg, B.C., Shaffer, M.J., Weltz, M.A., Ahjua, L.R.: Multicriteria spatial decision support systems: overview, applications, and future research directions. In: Proceedings of Integrated Assessment and Decision Support (iEMSs 2002), Lugano, Switzerland, pp. 175–180 (June 2002)
9. Klir, G.J., Yuan, B.: Fuzzy Sets and Fuzzy Logic: Theory and Applications. Prentice-Hall Inc., NJ (1995)
10. Prashanthi Nirmala Kodikara. Multi-Objective Optimal Operation of Urban Water Supply Systems. PhD thesis, School of Architectural, Civil and Mechanical Engineering (Februvary 2008)
11. Laumann, M., Zitzler, E., Thiele, L.: Multiple criteria decision support by evolutionary computation. In: Hilty, L.M., Gilgen, P.W. (eds.) Sustainability in the Information Society, 15th International Symposium Informatics for Environmental Protection, Zurich, Metropolis Verlag, Marburg (2001)
12. Lu, J., Zhang, G., Ruan, D., Wu, F.: Multi-Objective Group Decision Making – Methods, Software and Applications with Fuzzy Set Technology. Imperial College Press, London (2007)
13. Lu, J., Zhu, Y., Zeng, X., Koehl, L., Ma, J., Zhang, G.: A fuzzy decision support system for garment new product development. In: Wobcke, W., Zhang, M. (eds.) AI 2008. LNCS (LNAI), vol. 5360, pp. 532–543. Springer, Heidelberg (2008)
14. Ma, J., Lu, J., Zhang, G.: Decider: A fuzzy multi-criteria group decision support system. Knowledge-Based Systems 23(1), 23–31 (2010)
15. Marichal, J.L.: An axiomatic approach of the discrete Choquet integral as a tool to aggregate interacting criteria. IEEE Transactions on Fuzzy Systems 8(6), 800–807 (2000)
16. Marichal, J.-L.: An axiomatic approach of the discrete Sugeno integral as a tool to aggregate interacting criteria in a qualitative framework. IEEETFS 9(1), 164–172 (2001)

17. Mesiar, R., Kolesárová, A., Calvo, T., Komorníková, M.: A review of aggregation functions. In: Bustince, H., Herrera, F., Montero, J. (eds.) Fuzzy Sets and Their Extensions: Representation, Aggregation and Models. Studies in Fuzziness and Soft Computing, vol. 220, pp. 121–144. Springer, Heidelberg (2008)
18. Motro, A., Anokhin, P.: Fusionplex: resolution of data inconsistencies in the integration of heterogeneous information sources. Information Fusion 7(2), 176–196 (2006)
19. Munda, G.: A conflict analysis approach for illuminating distributional issues in sustainability policy. EJOR 194(1), 307–322 (2009)
20. Carchon, R.: Linguistic assessment approach for managing nuclear safeguards indicator information. Logistic Information Management 16(6), 401–419 (2003)
21. Laes, E., Zhang, G., Ma, J.: Multi-criteria group decision support with linguistic variables in long-term scenarios for belgian energy policy. Journal of Universal Computer Science 15(1), 103–120 (2010)
22. Vihakapirom, P., Li, R.K.-Y.: A framework for distributed group multi-criteria decision support systems. In: Proceedings of the Ninth Australia World Wide Web Conference, Hyatt Sanctuary Cove, Gold Coast (2003)
23. Šelih, J., Kne, A., Srdić, A., Žura, M.: Multiple-criteria decision support system in highway infrastructure management. Transport 23(4), 299–305 (2008)
24. Yager, R.R.: Families of OWA operators. FSS 59, 125–148 (1993)

Product Design Compromise Using Consensus Models

David Ben-Arieh and Todd Easton

Abstract. Obtaining a group consensus is a critical step in making effective business decisions. In this chapter the consensus process is defined as a dynamic and interactive group decision process, which is coordinated by a moderator, who helps the experts to gradually move their opinions until a consensus is reached. This paper describes the importance of group consensus and the need to minimize the cost of this process. Moreover, this work focuses on product design compromise and discusses how group consensus can be used in this process. The paper demonstrates the importance of the consensus process to the product design compromise process and presents several models that can be used to obtain such a compromise.

The paper discusses the costs associated with decision making using group consensus, and then describes three methods of reaching a minimum cost consensus assuming quadratic costs for a single criterion decision problem. The first method finds the group opinion (consensus) that yields the minimum cost of reaching throughout the group. The second method finds the opinion with the minimum cost of the consensus providing that all experts must be within a given threshold of the group opinion. The last method finds the maximum number of experts that can fit within the consensus, given a specified budget constraint.

Keywords: Consensus, Multi-Person Decision Making, Product Design Compromise.

1 Introduction

Group decision making is a key process to many of the crucial decisions that businesses have to face. Susskind, et al (1999) states that applications of group decision making range from nebulous decision such as public health spending or air pollution effects, to detailed decisions such as product selection, or advertising

David Ben-Arieh · Todd Easton
Dept. of Industrial and Manufacturing Systems Engineering
Kansas State University
216 Durland Hall Manhattan, KS 66506, United States

E. Herrera-Viedma et al. (Eds.): Consensual Processes, STUDFUZZ 267, pp. 405–423.
springerlink.com   © Springer-Verlag Berlin Heidelberg 2011

budgeting. Group decisions can be used in business decisions ranging from setting strategic goals for the year to temporary groups trying to accomplish a common goal. There are several advantages and disadvantages to group decision making (Griffin, 2005), as shown in Table 1.

Table 1 Advantages and Disadvantages to Group Decision Making

Advantages	Disadvantages
1. More information and knowledge are available. 2. More alternatives are likely to be generated. 3. More acceptance of the final decision is likely. 4. Enhanced communication of the decision may result. 5. Better decisions generally emerge.	1. The process takes longer than individual decision making, so it is costlier. 2. Compromise decisions resulting from indecisiveness may emerge. 3. One person may dominate the group. 4. Group dynamics, politics and other irrelevant issues may interfere with the decision.

One aspect of group decision making is the concept of group consensus, representing a single outcome that the entire group endorses. A prevalent opinion is that a group consensus is a natural outcome of a group decision process in which a full and unanimous agreement can be found between all the experts involved. A more reasonable definition was presented by Ness and Hoffman (1998), in which consensus is defined as "a decision that has been reached when most members of the team agree on a clear option and the few who oppose it think they have had a reasonable opportunity to influence that choice. All team members agree to support the decision." Clearly, a consensus process must be used in order to aggregate the opinions of the experts. This opinion is supported by Shanteau (2001) and Shanteau and Weiss (2002), who explain that experts by the virtue of their expertise can never agree to the same opinion, and thus some moderator supporting a consensus process is required.

In this paper the consensus process is defined as a dynamic and interactive group decision process, which is coordinated by a moderator, who helps the experts to gradually move their opinions closer to that of the group opinion (Herrera-Viedma et al, 2002). It is impractical to believe that a group of experts will all share the same opinion, and therefore, many times the opinions of individual experts must be influenced in order for them to consent to an overall group opinion.

As previously stated, one of the disadvantages of group decision making, is that this process normally takes longer and is therefore costlier than individual decision making. It is easy to see during this consensus process that a significant amount of time and resources must be used in order to influence the experts' opinions in order to converge to a group opinion. There is clearly a cost related to the resources being used during this consensus process. An informal survey conducted in 1998 found that on average a moderator costs between $50 to $190 per hour based on his or her expertise. Moreover, Susskind et. al (1999) believe that the average costs for this consensus process is about $15,000.

In this context we address the issues of product design compromise, develop algorithms to solve these problems and show how these algorithms can be used in the product design process. This paper first discusses the cost associated with product design compromise. Next, it describes the benefits of applying group consensus to aid in the product design compromise process. Lastly, a method to minimize this cost is provided.

2 Background

2.1 Decision Theory

The realm of decision theory can normally be broken down into two areas of study. These paradigms are known as normative (or prescriptive) decision theory and descriptive decision theory (Wang et al., 2004). Normative decision theory focuses on making the best decision assuming the decision maker is fully informed and rational. This area of study normally uses mathematical or statistical methodology or tools to calculate these optimal decisions. The second area of study focuses on how individuals make decisions, since individuals rarely act optimally or even in a rational manner. Furthermore, the use of decision theory spreads across many disciplines including computer science, psychology, political science, management science, economics, sociology, and statistics.

Decision theory does not only focus on individual decisions but allows for the analysis of group decisions as well. This area of study is known as Multi-Person Decisions Making (MPDM). For MPDM problems the idea of consensus has become a major topic of research over the past few years. The bulk of this research has been in the area of normative decision theory, which aims at determining ways to calculate experts' opinion numerically, and then aggregate those opinion scores into a group consensus. This can be accomplished by using preference orderings, utility functions, fuzzy preference scores, and multiplicative preference relations (Kwok et al, 2002; Herrara-Viedma et al, 2002; Seidenfeld and Schervish, 1990; Van Dew Torre, 2001). This type of consensus calculation typically uses an average or weighted average of the expert's opinions. The primary assumption is that the experts do not actually change their opinion to converge to the group opinion.

Conversely, numerous researchers have studied the descriptive side of MPDM, which focuses on the processes in which experts must modify their opinion in order to reach group consensus. This consensus process definition is used in this paper and in many of the structured group decision making processes such as the Delphi Method (Hartman, and Baldwin, 1995; Wilson, 2005) or Nominal Group Techniques (Eng Shwe Sein Aye et al, 1999; Teltumbde, 2000).

Most of the research performed to date lacks the consideration of the cost of aggregating the opinions of the experts. This paper's contribution considers this information and finds a minimum cost consensus. Moreover, it can be shown that the costs of influencing each expert's opinion can differ based on each expert's propensity to change his/her opinion.

In this paper we do not consider the specific method used by the experts to express their opinions, but assume that an expert's opinion can be captured numerically. This paper further assumes that this propensity to change can be measured, and is known prior to the analysis and is symmetric and quadratic for each expert. Thus, the incremental cost increases quadratically as the expert moves farther from his/her original opinion. Also, it is assumed that there exists a continuous opinion measure that is inherent to the decision process.

This research describes three situations of single criterion group consensus and gives algorithms to solve all cases. The first case is an unconstrained consensus in which the decision maker is searching for the optimal group consensus with minimum cost. The second case is also an unconstrained consensus which allows for all the experts to be within a threshold of the consensus. The final case is a budget constrained consensus in which the decision maker is trying to maximize the number of experts within the consensus based on a predetermined limited budget.

2.2 Product Design Compromise

Here, product design is defined as the idea generation or concept development of a physical product or service. Furthermore, product design compromise is defined as an alteration to any aspect of the preliminary product design based on information gained through the product design process.

Product design compromise is an integral part to any business' success. One example of a successful product design case involves the Apple's digital music players, the iPod. Reppel, et al., (2006) explains that Apple secured 75 percent of the market segment for digital music players using an agile design environment. Additionally Apple has a very customer-centered product development process, which allows for easy external integration. Apple has had the ability to distinguish between technology-driven and market-led products focusing on decreasing the size of their product with the iPod Mini and iPod Nano.

Some products succeed, while other products, which perform the same function, fail. This success or failure reflects a company's ability to measure and evaluate the following concepts (among other factors):

- cost versus quality trade-offs,
- internal versus external integration, and
- technology driven versus market led products.

All of these concepts related to design decisions require compromises from various departments in the design process. That is, each department desires certain product properties in order to more easily fulfill its mission. These desires frequently conflict and compromises leading to a consensus must be made prior to a product be producible.

One of the main issues within product design compromise is the need for more information. From the definition stated previously, the final design of the product is based on information gained through the product design process. As previously discussed, MPDM or group decision making provides several advantages, one is the creation of more information or knowledge through group decision making. The more designers or "experts" that participate in the process, the more information becomes available, but this also increases the number of opportunities for conflicting interests.

One way in which a company can benefit from this added information is through a cost versus quality trade-off analysis. This follows the notion that a higher quality product comes at a higher cost. It is important to balance these conflicting needs so the company can provide a high quality product at a reasonable price to its customers. Jiao and Zhang (2005) report that most researchers try to maximize the surplus: the margin between the customer-perceived utility and the price of the product. This incorporates the quality perception of the customer versus the cost of the product.

Another common objective for more information within the product development process is the internal/external integration analysis. Koufteros, et al., (2005) states that an effective product design process requires firms to unify internal and external participants. This requires compromise between external suppliers, internal users and external consumers.

A third domain for a compromise is the decision whether to follow a technology driven or a product based market. Here the decision is to follow a strategy in which the product is being driven to the public by the development of new technology or the market is leading the push for a new product. Curtis (2000) explains that 80% of new products fail with the main reason being that engineers do not understand the customer's desires and the business people do not understand the technical aspects of the product, both being unable to compromise their points of view.

One tool that many businesses are using to integrate the customer "wants" with the engineering requirements is Quality Function Deployment (QFD). QFD was developed in Japan in the early 1970s by Mitsubishi and enhanced over the next decade by Toyota (ReVelle et al., 1998). QFD uses a House of Quality structure to display certain data to help designers visualize the process. Although the structure may vary depending on the design process or objectives, there is a set of standard components that include the following (Iranmanesh et al., 2005):

- customer attributes (product requirements) and their relative importance,
- design characteristics (product specifications),
- the relationship matrix between customer attributes and design characteristics,
- a correlation matrix among design characteristics,
- computed absolute/relative importance ratings of design characteristics, and
- marketing and technical benchmarking data from customer and technical competitive analysis.

ReVelle, et al., (1998), also established two objectives associated with QFD:

- to convert the users' needs for product benefits into quality characte-ristics at the design stage, and
- to deploy the quality characteristics identified at the design stage to the production activities, thereby establishing the necessary control points and check points prior to production start-up.

There have been numerous researchers that have analyzed QFD and its useful-ness in improving customer perception (Iranmanesh, 2005), QFD best practices (Miguel, 2005), and organizational creativity and productivity (Politis, 2005).

Another tool suggested to reduce conflicts during the design process is the use of Design Rationale (DR), which intends to increase communication within an or-ganization. Lee (1997) supports the importance of DR by explaining that it in-cludes not only the reasons behind the decisions, but also the justification for the decisions. Molavi, et al., (2003) also supports the DR approach since it necessi-tates the communication of why certain decisions are being made between de-partments. Therefore, it requires that the downstream operations understand the justifications made by the upstream operations. Hence, DR is another way to help integrate an organization's departments in order to reduce conflicts and support compromise.

2.3 Product Design Compromise Cost

An important aspect of product design is the compromise cost. As stated before, one of the important reasons for design compromise is finding a good balance of quality and costs. It is well accepted that "product development is a large and complex task, and early design decisions can have a crucial impact on the quality, cost, robustness, and reusability of the product." (Ulman and D'Ambrosio, 1995).

One measure that has been highly researched is Taguchi's quadratic loss func-tion. Moorhead and Wu (1998), define this equation as:

$$R_{t_0} = E[(y - t_0)^2]$$

with y being the observed value and t_0 being the targeted value. This equation supports our notion that individuals change their opinion according to a quadratic cost function. Some researchers have suggested improvements to this cost func-tion; Leon and Wu (1992) suggest that the function is not necessarily quadratic and symmetric. Wagner (1973) suggests some improvements to Taguchi's loss function for inventory control models. This method shows that there is a loss (op-portunity cost) associated with not hitting the target value.

In order to better understand the cost of design decision compromise it is neces-sary to understand the cost implications of the various design decisions. The cost difference between two design choices represents the cost of shifting from one de-cision to the other (as in compromising). Duverlie and Castelain (1999) discuss a parametric method of cost estimation that attempts to evaluate the cost of a product based on its parameters. This method assumes that the cost of the product is directly related to its parametric characteristics.

Cost estimation is one of the major barriers that has limited the integration of financial aspects into the design process. There have been numerous methods researched to help estimate costs including neural networks and regression models, etc. (Khoshnevis et al. 1994; Smith and Mason, 1997; Weirda, 1988; Shehab and Abdalla, 2002). The consensus models used in this paper do not address the cost estimation methods used, but assumes that the cost can be estimated and the cost estimation function meets the assumptions of the particular model, which are symmetric and quadratic.

In order to reduce the design costs, there is a need to establish a consensus quickly. If the costs are correctly modeled, then the best consensus is the consensus that minimizes the total cost required for all experts to agree with that "best" consensus.

3   Consensus at Minimum Cost

Formally, let $E = \{e_1,..., e_n\}$ be a set of n experts and $o_i \in \mathfrak{R}$ be the opinion of the ith expert. Furthermore, the expert's initial opinion shall be shown as o_i^0, whereas their current opinion shall be shown as o_i'. Without loss of generality we can assume that the initial opinions are ordered on the number line, thus, $o_1^0 \leq o_2^0 ... \leq o_n^0$. Define a consensus to occur if and only if all the experts have the same current opinion, whereas $o_1' = o_2' = ... = o_n' = o'$, in which o' is the current group opinion. For each expert, define f_i to be the cost function of moving expert i's opinion to some other opinion o'. Furthermore, f_i is assumed to be quadratic and can be formulated as $f_i = c_i(o' - o_i^0)^2$, where c_i is a cost constant assigned to each expert to represent their propensity to change their opinion. The minimum quadratic cost consensus problem seeks to obtain a consensus o^*, at minimum quadratic cost (CMQC). Throughout the remainder of the paper, we assume that $o_i^0 \neq o_j^0$ for some $i,j \in \{1,...,n\}$ or the problem is trivial.

3.1   Consensus at Minimum Quadratic Cost (CMQC) Methodology

The initialization of this process sets all the cost functions so that they are all relative to the same origin and then sums the functions so that there is one cost function for the line interval. The main step of the process finds the optimal point based on the first derivative. It can be noted that the second derivative of the cost function is positive, and thus the function is convex. Consequently, setting the first derivative equal to 0 and solving yields a global minimum.

3.2 CMQC Algorithm

Initialization: Set $f_c(o') = \sum_{i=1}^{n} f_i$

Main Step: Find $\dfrac{df_c(o)}{do'}$, set equal to 0, and solve for o.

Termination: Report o as o^*, the optimal consensus value and calculate $f_c(o)$, the overall cost.

3.3 CMQC Example

Consider an automobile company where the performance characteristic being examined is the miles per gallon of a given automobile. Based on subjective probabilities and information available to each department, the departments differ in the desired mpg performance. The design department would like the performance characteristic set as 25 mpg, the engineering department would like it set at 28, the management department would like it set as 31, and the marketing department would like it set at 35. It is further assumed that the cost to move any of these opinions is quadratic and with a scaling cost coefficient of 1, 2, 3 and 1 for the design, engineering, management and marketing departments, respectively. This cost coefficients can represent the conceived consequences if the design parameter varies from the recommended. Thus the marketing department estimates marketing loses of $1*(o'-35)^2$ due to customer dissatisfaction or other similar consequences. The initial opinions and costs can now be set to $o_1^0=25$ with $f_1=(o'-25)^2$, $o_2^0=28$ with $f_2=2(o'-28)^2$, $o_3^0=31$ with $f_3=3(o'-31)^2$, and $o_4^0=35$ with $f_4=(o'-35)^2$. This is graphically shown in Figure 1.

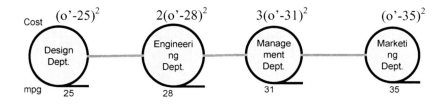

Fig. 1 Graphical Interpretation of Product Design Compromise Example

Thus the overall cost function is $f_c = (o'-25)^2 + 2(o'-28)^2 + 3(o'-31)^2 + (o'-35)^2$. After finding the derivative and setting it to 0, we have $\dfrac{df_c(o')}{do'} = 2(o'-25)+4(o'-28)+6(o'-31)+2(o'-35) = 0$. Solving for o'

we get $o^* = 29.857$. This gives us a minimum cost consensus value with an over-all cost of $f_c = 60.857$.

4 The ε Consensus at Minimum Quadratic Cost (εCMQC)

One extension to the CMQC is to allow the experts to be sufficiently close to the point of consensus. A consensus opinion $o' \in \mathfrak{R}$ is called an ε common consensus if $|o'\text{-}o_i'| \leq \varepsilon$ for all $i=1,..,n$. Then the ε consensus at minimum quadratic cost problem (εCMQC) seeks an o^* such that the cost to reach this ε consensus is minimized. For any o', it is clearly a waste of money to change the opinion of an expert that is within ε of o'. Furthermore, any expert with initial opinion further than ε from o' should only be moved until that individual is exactly ε away from o'. These facts lead to the ε consensus cost of any opinion, $f_c(o')$. The experts opinions can then be placed in one of three categories based on the following rules:

$$o_i' \in L \ \ \text{if} \ \ o_i' < o' \text{-} \varepsilon$$

$$o_i' \in N \ \ \text{if} \ \ o' \text{-} \varepsilon \leq o_i' \leq o' + \varepsilon$$

$$o_i' \in R \ \ \text{if} \ \ o_i' > o' + \varepsilon$$

where L and R are the set of experts whose opinion are further than ε to the left or right of o'. Given an opinion o', define $f_c(o') = \Sigma_{\{i \in L\}} \ c_i(o' \text{-} \varepsilon \text{-} o_i^0)^2 + \Sigma_{\{i \in R\}} \ c_i(o_i^0 \text{-} (o' + \varepsilon))^2$. Consequently, εCMQC seeks to obtain an opinion o^* such that $f_c(o^*) \leq f_c(o')$ for all $o' \in \mathfrak{R}$.

The problem with the εCMQC algorithm when using non-linear cost functions is that the overall cost function is not uniform over the entire interval. The individual cost functions can not simply be added together, due to the fact that the expert's opinions differ relative to a specific o' value. Thus, we break the overall interval into smaller intervals in which each expert's cost function does not change within the interval.

Furthermore, this εCMQC methodology assumes that the costs are again quadratic with the initial opinions again ordered. Trivially, if $o_n^0 \text{-} o_1^0 \leq 2\varepsilon$, then $o^* = (o_n^0 + o_1^0)/2$ is a trivial answer. So we will assume that $o_n^0 \text{-} o_1^0 > 2\varepsilon$.

4.1 εCMQC Methodology

The εCMQC methodology is very similar to the CMQC methodology except that in order to assure that each expert's opinion is only within one category (L, N, and R) at any given interval, the line must be segmented into smaller intervals. Furthermore, for each expert two critical points are established. Those points occur at the expert's original opinion plus and minus ε. This set of critical points will be defined by $C = \{o_i^0\text{-}\varepsilon, \ o_i^0+\varepsilon : \text{for all } i \in \{1,...,n\}\}$. Now let CP be a sorted set of ascending numbers that contain all of the elements of C. Thus

$CP=(cp_1, cp_2, ..., cp_{2n})$. By computing the cost of the overall function at each of the line's critical points (cp_j), the line segment at which $f_c(cp_j) > f_c(cp_{j-1})$ can be found. The line can then be analyzed within the line intervals $I_1 =[cp_{j-2}, cp_{j-1}]$ and $I_2 =[cp_{j-1}, cp_j]$. Since the overall cost function is the sum of convex functions it is also convex. Therefore, by finding the critical point at which this function increases assures that the minimum cost is located somewhere within the previous two critical points. The derivative can then be calculated over both intervals and the minimum cost consensus can be found.

This methodology produces two essential properties for the two intervals. The first is that the minimum cost is located somewhere within the two intervals. Secondly, it assures that the overall cost function does not change between any of the given critical points and therefore the categories L, N, and R are mutually exclusive in each of the intervals.

4.2 εCMQC Algorithm

Initialization: Form an ordered set of all Critical Points
$$CP = (cp_1, cp_2, ..., cp_{2n})$$
$$j = 1$$

Main Step 1: Do Until $f_c(cp_j) > f_c(cp_{j-1})$
Define   for all experts i

$$o_i' \in L \text{ if } o_i' < cp_j - \varepsilon$$
$$o_i' \in N \text{ if } cp_j - \varepsilon \le o_i' \le cp_j + \varepsilon$$
$$o_i' \in R \text{ if } o_i' > cp_j + \varepsilon$$

$$f_c(cp_j) \leftarrow \sum_{i \in L} c_i(cp_j - o_i^o - \varepsilon)^2 + \sum_{i \in R} c_i(o_i^o - cp_j - \varepsilon)^2$$

$$j \leftarrow j+1$$

Main Step 2:   Define interval $I_1 = [cp_{j-2}, cp_{j-1}]$ and $I_2 = [cp_{j-1}, cp_j]$

Find $\dfrac{df_c(o_{I_1}')}{do_{I_1}'} = 0$, and solve for o_{I_1}'. Where $o_{I_1}' \in I_1$, if

$o_{I_1}' \notin I_1$, set $f_c(o_{I_1}')= \infty$.

Find $\dfrac{df_c(o_{I_2}')}{do_{I_2}'} = 0$, and solve for o_{I_2}'. Where $o_{I_2}' \in I_2$, if

$o_{I_2}' \notin I_2$, set $f_c(o_{I2}')= \infty$.

Calculate $f_c(o_{I_1}')$ and $f_c(o_{I_2}')$,

If $f_c(o_{I_1}') < f_c(o_{I_2}')$ then $o^* = o_1'$

If $f_c(o_{I_2}') < f_c(o_{I_1}')$ then $o^* = o_2'$

Termination: Report o^* and calculate $f_c(o^*)$

4.3 εCMQC Example

Returning to data from the automobile example from Section 3.3 and setting $\varepsilon = .5$ gives the following critical points CP = (24.5, 25.5, 27.5, 28.5, 30.5, 31.5, 34.5, 35.5). This then gives us our first function as:

$$f_c(24.5) = 2(28 + .5 - 24.5)^2 + 3(31 + .5 - 24.5)^2 + (35 + .5 - 24.5)^2 = 226.$$

The total cost continues to decrease until $f_c(28.5) = 65.00$, $f_c(30.5) = 53.64$, and $f_c(31.5) = 56.44$. This satisfies the termination criteria that $f_c(cp_j) > f_c(cp_{j-1})$. The algorithm then moves to Main Step 2 in which the intervals are set as $I_1 = [28.5, 30.5]$ and $I_2 = [30.5, 31.5]$. With these intervals the following cost functions are:

$$f_c(o_{I_1}') = (o_{I_1}' - 25 - .5)^2 + 2(o_{I_1}' - 28 - .5)^2 + 3(31 - o_{I_1}' - .5)^2 + (35 - o_{I_1}' - .5)^2$$

$$f_c(o_{I_2}') = (o_{I_2}' - 25 - .5)^2 + 2(o_{I_2}' - 28 - .5)^2 + (35 - o_{I_2}' - .5)^2 .$$

The derivative of these cost functions are then found and set equal to 0. The resultant derivatives are:

$$\frac{df_c(o_{I_1}')}{do'} = 2(o_{I_1}' - 25.5) + 4(o_{I_1}' - 28.5) - 6(30.5 - o_{I_1}') - 2(34.5 - o_{I_1}') = 0$$

$$\frac{df_c(o_2')}{do'} = 2(o_2' - 25.5) + 4(o_2' - 28.5) - 2(34.5 - o_2') = 0 .$$

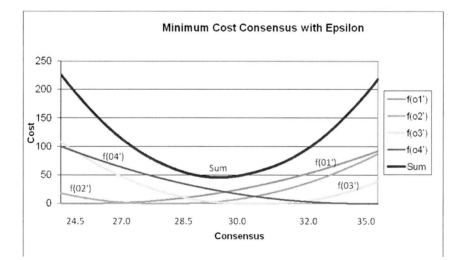

Fig. 2 Minimum Cost Consensus with $\varepsilon = .5$

The resulting optimal interval values and their resulting cost function values are as follows:

$$o_{I_1}' = 29.7857 \text{ with } f_c(o_{I_1}') = 45.4286$$

$$o_{I_2}' = 29.25 \text{ with } f_c(o_{I_2}') = \textit{Not Feasible} \text{ since } o_{I_2}' \text{ is not within the } I_2 \text{ in-}$$
terval.

Thus, an optimal consensus value is $o^* = 29.7857$ and $f_c(o^*) = 45.4286$.
A summary is shown in Figure 2.

5 Quadratic Maximum Expert Consensus (QMEC)

Another natural extension to the consensus at minimum cost problem is to find the maximum number of experts that can fit within the consensus given a specified budget, which is represented by B'. This problem seeks to find a consensus o^* such that the maximum number of experts share this opinion at minimum cost. The consensus group is represented by the set E', and its compliment is represented by $E = N/E'$.

5.1 QMEC Methodology

The algorithm we present maintains the cheapest list of experts (consensus group) that can have a given opinion o' and still be under the budget B'. Furthermore, the algorithm starts at an opinion $o' = o_I^0$ and increases o' until the first critical point is reached. The critical points, again denoted by cp_j, are the only locations where a potential change could occur to the solution (experts are added, subtracted or swapped from the consensus group). Therefore, it can be shown that the consensus group does not change between any two critical points. There are only three ways that the solution to the algorithm can change and each change corresponds to a critical point.

1. A solution could change if an expert, e_i, in the consensus swaps places with an expert, e_j, not in the consensus. This occurs when e_j becomes more beneficial than e_i. These points are called the *switching points of i and j* (sp_{ji} and sp_{ij}) and occur at the opinion where both i and j have identical costs to reach that opinion. Once all the switching points are found, for a given opinion o' a priority set $P_{o'}$ is created which is an ordering of the experts according to which experts require the least cost to change their opinion to o'. A closed expression for each switching point is

$$c_i(sp_{ij}o_i^0)^2 - c_j(sp_{ij} - o_j^0)^2 = 0$$

$$(c_i - c_j)sp_{ij}^2 - 2(c_io_i^0 - c_jo_j^0)sp_{ij} + (c_i(o_i^0)^2 - c_j(o_j^0)^2) = 0$$

$$sp_{ij} = \frac{2(c_io_i^0 - c_jo_j^0) - \sqrt{(2(c_io_i^0 - c_jo_j^0))^2 - 4(c_i - c_j)(c_i(o_i^0)^2 - c_j(o_j^0)^2)}}{2(c_i - c_j)}$$

$$sp_{ji} = \frac{2(c_io_i^0 - c_jo_j^0) + \sqrt{(2(c_io_i^0 - c_jo_j^0))^2 - 4(c_i - c_j)(c_io_i^{0^2} - c_j(o_j^0)^2)}}{2(c_i - c_j)}.$$

2. Since the algorithm always increases o', there may not be enough budget to keep all of the existing experts that agree with o'. Therefore, another critical point occurs when some expert must leave the consensus group. Clearly, this expert is the expert that costs the most money to move to o', which can easily be determined from $P_{o'}$. This critical point is called a *temporary critical point for the marginal cost* (x_M) and occurs at the opinion where the cost of all of the experts in the consensus group equals the budget, which reduces to

$$\sum_{i \in E'} c_i(x_M - o_i^0)^2 = B'$$

$$(\sum_{i \in E'} c_i)x_M^2 - (2\sum_{i \in E'} c_io_i^0)x_M + (\sum_{i \in E'} c_i(o_i^0)^2 - B') = 0$$

$$\alpha = (\sum_{i \in E'} c_i)$$

$$\beta = -(2\sum_{i \in E'} c_io_i^0)$$

$$\lambda = (\sum_{i \in E'} c_i(o_i^0)^2 - B')$$

$$x_M = \frac{-\beta \pm \sqrt{\beta^2 - 4\alpha\lambda}}{2\alpha}.$$

3. As o' moves, there could be enough budget to add another expert to the consensus group. As in 2, the expert should be added that is the cheapest, which is easy to determine from $P_{o'}$. These critical points are called *temporary critical point of the need for expert i* $(x_{N(i)})$ and can be calculated by temporarily adding expert i to the consensus group and determining the opinion where the cost to move this whole group to this opinion equals the budget B', which is calculated as follows.

$$\sum_{j \in E'} c_j (x_M - o_j^0)^2 + c_i (x_{N(i)} - o_i^0)^2 = B'$$

$$\alpha x_{N(i)}^2 + \beta x_{N(i)} + \lambda + c_i (x_{N(i)} - o_i^0)^2 = 0$$

$$(\alpha + c_i) x_{N(i)}^2 + (\beta - 2c_i o_i^0) x_{N(i)} + (\lambda + c_i (o_i^0)^2) = 0$$

$$x_{N(i)} = \frac{-(\beta - 2c_i o_i^0) \pm \sqrt{(\beta - 2c_i o_i^0)^2 - 4(\alpha + c_i)(\lambda + c_i (o_i^0)^2)}}{2(\alpha + c_i)}$$

where α, β, and λ are defined above.

An additional temporary critical point in the QMEC algorithm is one that finds the minimum cost for any consensus group which occurs by taking the derivative of their cost functions and setting it equal to 0. Again it can be stated that the sum of convex functions is a convex function and that the derivative of the function represents the global minimum on the function. This point is called the *temporary critical point of the derivative* (x_D). Realistically, this critical point does not need to be calculated, but it does provide the opinion where the minimum cost for the experts in the consensus group occurs. Thus, the algorithm can not only find an opinion that the maximum number of experts, but can also maximize the amount of unused budget. A temporary cirtical point of the derivative is calculated as follows.

$$\sum_{i \in E'} 2c_i (x_D - o_i^0) = 0$$

$$x_D = \frac{\sum_{i \in E'} c_i o_i^0}{\sum_{i \in E'} c_i}.$$

Since the algorithm moves along consecutive critical points and the consensus group does not change between any consecutive critical points this algorithm optimally solves the QMEC problem.

5.2 QMEC Algorithm

Initialization:

1. For all $i \neq j$ find the switching points sp_{ij}.
2. Set $o' \leftarrow o_1^o$.
3. Form a priority set $P_{o'}$ of experts by calculating their cost at o' and sorting according to the minimum cost.
4. Set $z^* \leftarrow \varnothing$, $s^* \leftarrow 0$ and $o^* \leftarrow o'$.

Main: While $o' \leq o_n^o$

1. Build the consensus group E' from $P_{o'}$ as long as the budget allows to obtain the sets E' and E and the slack s'.

2. If, $|E'| > z^*$ and $s' > s^*$, then $z^* = |E'|, s^* = s'$, and $o^* \leftarrow o'$.

3. Find all Temporary Critical Points $(x_M, x_{N(i)}, x_D)$ associated with o' and set $p = min(o_i^0, sp_{ij}, x_M, x_{N(i)}, x_D)$ such that $p > o'$.

4. Set $o' \leftarrow p$.

5. Update the priority set $P_{o'}$.

Termination:

1. Report o^*, z^*, and s^*.

5.3 QMEC Example

QMEC has many more details than the previous algorithms and so we will not continue with the miles per gallon example. Now suppose that a car maker has recently found an error in the design of a car and the cars need to be retrofitted in order for the car to perform safely. This problem effects only three departments, Legal, Management and Engineering. Each department has an opinion regarding the time that the customers should be notified of the problem and the customer can then take his/her car to a dealership to be fixed.

Naturally, the lawyers want the problem corrected immediately to avoid any law suits and so $o_1^0=0$. The management wants the problem corrected soon, $o_2^0=3$ (notify customers in 3 weeks). The engineering department wants time to build and test parts that will fix the cars and would like 10 weeks, $o_3^0=10$. Furthermore, assume that these three departments have the following cost functions $f_1 = 2x^2$, $f_2 = 15x^2$, and $f_3 = x^2$, where x represents the amount of change in the expert's opinion in order to reach the group consensus. Notice how much weight is given to Management. Finally the maximum budget used to create a consensus is 50.

Begin by calculating the switching points, which are $sp_{12} = 4.7255$, $sp_{21} = 2.1976$, $sp_{31} = 4.1421$, $sp_{23} = .5635$, $sp_{32} = 4.4365$ and the initial priority set of $P_0 = (1,3,2)$. Note that $sp_{13} = -24.1421$, which is less than o_1^0 and therefore will not affect the algorithm. Furthermore, o' is set to 0 and the initial expert group is $\{E_1\}$. Then after calculating the temporary critical points (x) from the formulas given we determine our p value to be .5635 and set that to o'. Since .5635 is the switching point between e_2 and e_3 (sp_{23}) the priority set changed to $P_{0.5635} = (1,2,3)$. The next iteration used .5635 as the o' value and since the consensus group did not change the temporary critical points will not change. So the p value is now 1.2304, which is the $x_{N(2)}$. This is the point at which expert 2 can enter the consensus group. The consensus group is therefore changed to $\{E_1, E_2\}$. The consecutive iterations perform similarly and a summary of the iterations of the algorithm can be found in Table 1.

Table 2 Summary of the QMEC

Step	O'	Consensus Group(E')	$P_{o'}$	Slack	x_M	x_D	x_{N1}	x_{N2}	x_{N3}
0	0	E_1	1, 3,2	50	5	∞	c	1.2304 4.0637	c
1	.5635	E_1	1, 2,3	49.365	5	0	c	1.2304 4.0637	c
2	1.2304	E_1, E_2	1, 2,3	0	4.0637	2.6471	c	∞	c
3	2.1976	E_2, E_1	2, 1,3	30.68	4.0637	2.6471	c	∞	c
4	2.6471	E_2, E_1	2, 1,3	34.118	4.0637	∞	c	∞	c
5	4.0637	E_2	2, 1,3	33.03	4.8257	∞	c	∞	c
6	4.1421	E_2	2, 3,1	30.41	4.8257	∞	c	∞	c
7	4.4365	E_3	3, 2,1	19.05	∞	10	c	∞	c
8	4.7255	E_3	3, 1,2	22.18	∞	10	c	∞	c
9	10	E_3	3, 1,2	50	∞	∞	c	∞	c

Therefore, the maximum number of departments that can share a consensus is 2 and the departments are legal and management (E_2 and E_1). Furthermore, the consensus should be 2.6471 weeks with a cost of 15.88, equivalently a slack of 34.12. Figure 3 outlines the path taken by this methodology.

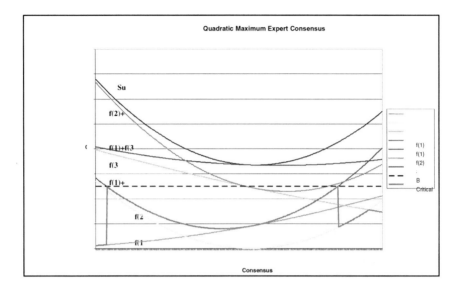

Fig. 3 QMEC Summary

6 Conclusion

Based primarily on its effectiveness in allowing businesses to analytically evaluate their decision-making processes, the concept of consensus has been a heavily researched topic in the realm of multi-person group decision making over the last several years.

This paper defined the consensus process as a dynamic and interactive group decision process, which is coordinated by a moderator, who helps the experts gradually move their opinions closer to that of the group opinion. This definition assumes that each expert within the decision group will regularly change his/her opinion such that a consensus is found within the group. This definition also implies that resources must be used in order to influence each expert's opinion. Therefore, there must be a cost involved in finding a group consensus.

Furthermore, this paper shows how and why consensus models should be used in the product design compromise process. This is based primarily on the integration of information throughout the organization.

Most of the research that has been conducted in the field of consensus theory has failed to incorporate any cost function. Therefore, it is the intent of this paper to find the minimum cost consensus.

This paper provides 3 separate cases of consensus and the methodology that can be used to solve each. The first is a Consensus at Minimum Quadratic Cost (CMQC) model. The second is a Consensus at Minimum Cost with Epsilon (εCMQC) model. The third and final model is the Quadratic Maximum Expert Consensus (QMEC) model, which considers a consensus under a budget constraint. Numerical examples demonstrate each case.

References

1. Curtis, T.: Technology Driven or Market Led. Engineering Management Journal 10(4), 197–204 (2000)
2. Duverlie, P., Castelain, J.M.: Cost Estimation During Design Step: Parametric Method Versus Case Based Reasoning Method. The International Journal of Advanced Manufacturing Technology 15, 895–906 (1999)
3. Eng Shwe Sein Aye, H., Young-Jou, L., Shing, I.C.: An Integrated Group Decision Making Approach to Quality Function Deployment. IIE Transactions 31, 553–567 (1999)
4. Griffin, R.W.: Management, 8th edn. Houghton Mifflin Company, Boston (2005)
5. Hartman, F.T., Baldwin, A.: Using Technology to Improve Delphi Method. Journal of Computing in Civil Engineering 9(4), 244–249 (1995)
6. Herrera-Viedma, E., Herrera, F., Chiclana, F.: A Consensus Model for Multiperson Decision Making Under Different Preference Structures. IEEE Transactions on Systems, Man and Cybernetics, Part A 32(3), 394–402 (2002)
7. Iranmanesh, S.H., Thomson, V., Salimi, M.H.: Design Parameter Estimation using a Modified QFD Method to Improve Customer Perception. Concurrent Engineering: Research and Applications 13(1), 57–67 (2005)
8. Jiao, J., Zhang, Y.: Product Portfolio Planning with Customer-Engineering Interaction. IIE Transactions 37, 801–814 (2005)
9. Khoshnevis, B., Park, J.Y., Sormaz, D.: A Cost Based System for Concurrent Part and Process Design. The Engineering Economist 40(1), 101–124 (1994)
10. Koufteros, X., Vondermbse, M., Jayaram, J.: Internal and External Integration for Product Development: The Contingency Effects of Uncertainty, Equivocality, and Platform Strategy. Decision Sciences 36(1), 97–133 (2005)
11. Kwok, C.-W.R., Ma, J., Zhou, D.: Improving Group Decision Making: A Fuzzy GSS Approach. IEEE Transactions on Systems, Man and Cybernetics, Part C 32(1), 54–63 (2002)
12. Lee, J.: Design Rationale Systems: Understanding The Issues. IEEE Intelligent Systems and Their Applications 12(3), 78–85 (1997)
13. Leon, R.V., Wu, C.F.J.: A Theory of Performance Measures in Parameter Design. Statistica Sinica 2, 335–358 (1992)
14. Miguel, P.A.C.: Evidence of QFD best practices for product development: a multiple case study. International Journal of Quality and Reliability Management 22(1), 72–82 (2005)
15. Molavi, J.M., McCall, R., Songer, A.: A New Approach to Effective Use of Design Rationale in Practice. Journal of Architectural Engineering 9(2), 62–69 (2003)
16. Moorhead, P.R., Wu, C.F.J.: Cost-Driven Parameter Design. Technometrics 40(2), 111–119 (1998)

17. Ness, J., Hoffman, C.: Putting sense into consensus: solving the puzzle of making team decisions. Vista Associates, Tacoma, Wash (1998)
18. Politis, J.D.: QFD, Organisational creativity and productivity. International Journal of Quality and Reliability Management 22(1), 59–71 (2005)
19. Reppel, A.E., Szmigan, I., Gruber, T.: The iPod phenomenon: identifying a market leader's secrets through qualitative marketing research. Journal of Product and Brand Management 15(4), 239–249 (2006)
20. ReVelle, J.B., Moran, J.W., Cox, C.A.: QFD Handbook. Wiley, New York (1998)
21. Seidenfeld, T., Schervish, M.J.: Two Perspectives on Consensus for (Baysian) Inference and Decisions. IEEE Transactions on Systems, Man and Cybernetics 20(1), 318–325 (1990)
22. Shanteau, J.: What does it mean when experts disagree? In: Klein, G., Salas, E. (eds.) Naturalistic Decision Making. Lawrence Erlbaum Associates, Hillsdale (2001)
23. Shanteau, J., Weiss, D., Thomas, R.P., Pounds, J.C.: Performance based assessment of expertise: How to decide if someone is an expert or not. European Journal of Operational Research 136, 253–263 (2002)
24. Shehab, E., Abdalla, H.: An Intellegent Knowledge based System for Product Cost Modelling. The International Journal of Advanced Manufacturing Technology 19, 49–65 (2002)
25. Smith, A.E., Mason, A.K.: Cost Estimation Predictive Modeling: Regression Versus Neural Network. The Engineering Economist 42(2), 137–161 (1997)
26. Susskind, L., McKearnan, S., Thomas-Larmer, J.(eds.): The Consensus Building Handbook. Sage Publications Inc., Thousand Oaks (1999)
27. Teltumbde, A.: A Framework for evaluating ERO Projects. International Journal of Production Research 38(17), 4507–4520 (2000)
28. Ullman, D.G., D'Ambrosio, B.: An Introduction to the consensus model of engineering design decision making. Interactive and Mixed-Initiative Decision Theoretic Systems, 131–139 (1995)
29. Van Der Torre, L.: Parameters for Utilitarian Desires in A Qualitative Decision Theory. Applied Intelligence 14, 285–301 (2001)
30. Wagner, H.M.: Principles of Operations Research. Prentice-Hall, Englewood Cliffs (1973)
31. Wang, Y., Liu, D., Ruhe, G.: Formal Description of the Cognitive Process of Decision Making. In: Proceedings of the Third International Conference on Cognitive Informatics, pp. 124–130 (2004)
32. Wierda, L.S.: Product Cost-Estimation by the Designer. Engineering Costs and Production Economics 13, 189–198 (1988)
33. Wilson Orndoff, C.J.: Promising New Tool for Stakeholder Interaction. Journal of Architectural Engineering 11(4), 139–146 (2005)

Index

Author Index